MOLECULAR BIOLOGY
INTELLIGENCE
UNIT

# Wnt Signaling in Development

Michael Kühl, Ph.D.
Abt. Biochemie
Universität Ulm
Ulm, Germany

LANDES BIOSCIENCE / EUREKAH.COM
GEORGETOWN, TEXAS
U.S.A.

KLUWER ACADEMIC / PLENUM PUBLISHERS
NEW YORK, NEW YORK
U.S.A.

# Wnt Signaling in Development

## Molecular Biology Intelligence Unit

Eurekah.com / Landes Bioscience
Kluwer Academic / Plenum Publishers



Printed in the U.S.A.

Kluwer Academic / Plenum Publishers, 233 Spring Street, New York, New York, U.S.A. 10013
http://www.wkap.nl/

Please address all inquiries to the Publishers:
Eurekah.com / Landes Bioscience, 810 South Church Street
Georgetown, Texas, U.S.A. 78626
Phone: 512/ 863 7762; FAX: 512/ 863 0081
www.Eurekah.com
www.landesbioscience.com

Wnt Signaling in Development edited by Michael Kühl, Landes / Kluwer dual imprint / Landes series: Molecular Biology Intelligence Unit

ISBN: 0-306-47838-2



## Library of Congress Cataloging-in-Publication Data

Kühl, Michael, 1965-
Wnt signaling in development / Michael Kühl.
p. ; cm.
Includes bibliographical references and index.
ISBN 0-306-47838-2
1. Wnt proteins. 2. Wnt genes. 3. Cellular signal transduction. 4.
Developmental cytology. 5. Developmental genetics.
[DNLM: 1. Glycoproteins. 2. Proto-Oncogene Proteins. 3.
Morphogenesis. 4. Signal Transduction. QU 55 K94w 2003] I. Title.
QP552.W58K84 2003
572'.68--dc21 2003012621

# CONTENTS

# EDITOR


**Michael Kühl, Ph.D.**
Abt. Biochemie
Universität Ulm
Ulm, Germany
michael.kuehl@medizin.uni-ulm.de
*Chapters 1, 11*


# CONTRIBUTORS


Denise Aragnol
Laboratoire de Génétique et Physiologie du Développement
IBDM
Campus de Luminy
Marseille, France
aragnol@ibdm.univ-mrs.fr
*Chapter 3*

Jürgen Behrens
University Erlangen-Nürnberg
Nikolaus-Fiebiger-Center
Erlangen, Germany
jbehrens@molmed.uni-erlangen.de
*Chapter 1*

André W. Brändli
Institute of Pharmaceutical Sciences
Department of Chemistry & Applied Biosciences
Swiss Federal Institute of Technology Zurich
Zurich, Switzerland
brandli@pharma.anbi.ethz.ch
*Chapter 8*

Bodo Christ
Institute of Anatomy II
University of Freiburg
Freiburg, Germany
christb@uni-freiburg.de
*Chapter 9*

Yacine Graba
Laboratoire de Genetique et Physiologie du Developpement
IBDM
Campus de Luminy
Marseille, France
graba@ibdm.univ-mrs.fr
*Chapter 3*

Christine Hartmann
I.M.P.
Institute for Molecular Pathology
Vienna, Austria
hartmann@nt.imp.univie.ac.at
*Chapter 10*

Xi He
Division of Neuroscience
Children's Hospital/Harvard Medical School
Department of Neurology
Boston, Massachusetts, U.S.A.
xi.he@tch.harvard.edu
*Chapter 2*

Michael A. Herman
Division of Biology
Kansas State University
Manhattan, Kansas, U.S.A.
mherman@ksu.edu
*Chapter 12*

Linda Z. Holland
Marine Biology Research Division
Scripps Institution of Oceanography
University of California San Diego
La Jolla, California, U.S.A.
lzholland@ucsd.edu
*Chapter 13*

Ruijin Huang
Institute of Anatomy II
University of Freiburg
Freiburg, Germany
*Chapter 9*

Clemens Kiecker
MRC Centre for Developmental Neurobiology
King's College, Guy's Hospital Campus
London, England, U.K.
clemens.kiecker@kcl.ac.uk
*Chapter 5*

Leung Kim
Laboratory of Cellular and Developmental Biology
NIDDK
National Institutes of Health
Bethesda, Maryland, U.S.A.
*Chapter 15*

Alan R. Kimmel
Laboratory of Cellular and Developmental Biology
NIDDK
National Institutes of Health
Bethesda, Maryland, U.S.A.
ark1@helix.nih.gov
*Chapter 15*

Andreas Kispert
Institut für Molekularbiologie
Medizinische Hochschule Hannover
Hannover, Germany
Kispert.Andreas@MH-Hannover.de
*Chapter 8*

Jeffrey R. Miller
Department of Genetics, Cell Biology and Genetics
University of Minnesota–Twin Cities
Minneapolis, Minnesota, U.S.A.
mille380@mail.med.umn.edu
*Chapter 4*

Marek Mlodzik
Mount Sinai School of Medicine
Dept. of Cell Biology and Anatomy
New York, New York, U.S.A.
Marek.Mlodzik@mssm.edu
*Chapter 6*

Christof Niehrs
Division of Molecular Embryology, DKFZ
Heidelberg, Germany
Niehrs@DKFZ-Heidelberg.de
*Chapter 5*

Petra Pandur
Abt. Biochemie
Universität Ulm
Ulm, Germany
petra.pandur@medizin.uni-ulm.de
*Chapter 11*

Jacques Pradel
Laboratoire de Génétique et Physiologie du Développement
IBDM
Campus de Luminy
Marseille, France
pradel@lgpd.univ-mrs.fr
*Chapter 3*

Ute Rothbächer
Laboratoire de Génétique et Physiologie du Développement
IBDM
Campus de Luminy
Marseille, France
rothbach@ibdm.univ-mrs.fr
*Chapter 3*

Jean-Pierre Saint-Jeannet
Department of Animal Biology
School of Veterinary Medicine
University of Pennsylvania
Philadelphia, Pennsylvania, U.S.A.
saintj@vet.upenn.edu
*Chapter 7*

Alexandra Schambony
University of Karlsruhe (TH)
Institute of Zoology II
Karlsruhe, Germany
alexandra.schambony@zi2.uni-karlsruhe.de
*Chapter 14*

Michael Schubert
Ecole Normale Supérieure de Lyon
Laboratoire de Biologie Moléculaire et Cellulaire
Lyon, France
Michael.Schubert@ens-lyon.fr
*Chapter 13*

Mikhail V. Semënov
Division of Neuroscience
Children's Hospital/Harvard Medical School
Department of Neurology
Boston, Massachusetts, U.S.A.
mikhail.semenov@TCH.Harvard.edu
*Chapter 2*

Doris Wedlich
University of Karlsruhe (TH)
Institute of Zoology II
Karlsruhe, Germany
doris.wedlich@zi2.uni-karlsruhe.de
*Chapter 14*

# PREFACE

It has been 20 years since Roel Nusse and Harold Varmus discovered the murine proto-oncogene int-1 which is upregulated by integration (hence the name int-1) of the mouse mammary tumor virus (MMTV). Later it turned out that int-1 is a homolog of the Drosophila segment polarity gene wingless. As often, these two genes are the founding members of a large family of secreted glycoproteins, and new functions of this gene family are published every other week. With the description of the first Wnt genes it turned out that Wnt genes do not have redundant properties but the molecular basis for this observation remained obscure. Despite the discovery of the Wnt gene family in the early 80s, it took 15 years to discover the corresponding transmembrane Frizzled receptor for Wnts in 1996. And again, we meanwhile have a whole family of Frizzled receptors. At the same time it turned out that Wnt proteins can activate different intracellular pathways, and it was tempting to speculate that these different biological activites of different Wnts are due to activation of different intracellular pathways. Indeed, there is mounting evidence for this hypothesis. Wnt genes have important functions during development of an organism and in the adult. Misregulation of Wnt function or signal transduction in the adult might lead to cancer. Thus understanding Wnts not only allows us to have a deeper look into biology but also helps us to develop new drugs to fight diseases.

During development, Wnt genes influence functions as different as determination and differentiation, proliferation, apoptosis, cell migration, and cell polarity in different tissues and organs. Within this book, established experts in the Wnt field give overviews on Wnt triggered signaling pathways and the role of Wnt genes in development. As many readers will be interested in the development of a selected organism or an organ, the book is divided into 15 chapters based on different experimental systems or topics. Model systems covered include *C. elegans*, Drosophila, Xenopus, chicken, mice and even Dictyostelium. We will focus on the development of neural tissue, kidney, limbs, heart, and sceletal muscle. Cellular aspects, like epithelial cell polarity or cell migration, are also covered.

We hope the reader will find these chapters of general interest.

*Michael Kühl*
*Ulm, August 2002*

# CHAPTER 1

# Wnt Signal Transduction Pathways: An Overview

**Jürgen Behrens and Michael Kühl**

## Abstract

Wnt proteins form a family of extracellular growth factors that are involved in aspects of development as diverse as cell determination, cell differentiation, cell migration, cell polarity and cell proliferation. Wnt proteins have been shown to activate different intracellular signaling cascades. We here present an outline of each of these pathways and summarize our current knowledge in this field. Misregulation of these pathways leads to developmental defects as well as tumor formation in the adult.

## Introduction

Members of the Wnt gene family of secreted glycoproteins are defined by their homology to the first identified members of this family, *wingless* in *Drosophila* and Wnt-1 in mouse (at that time named int-1).[1] Characteristically, Wnt proteins show a conserved pattern of cysteine residues which are thought to be linked through disulfide bridges. Nowadays we know of at least 19 members of this family based on the data of the human genome projects. A summary of identified Wnt genes and related references is given in Table 1. To initiate intracellular signaling, Wnt proteins interact with seven pass transmembrane receptors of the Frizzled family, of which 12 members can be identified in the human genome database.[75,76] Given the large number of ligands and receptors it is of no surprise that in the past years much evidence suggests the existence of divergent Wnt-triggered signaling pathways. For instance, in the chicken limb bud Wnt-7A activates Lmx-1 whereas in the same tissue Wnt-3A induces the expression of LEF-1.[32] In the forming muscle of early mouse embryos we find a similar situation. Whereas Wnt-7A activates MyoD, Wnt-1 stimulates the transcription of Myf-5.[78] Historically, Wnt proteins were subdivided into two classes based on their biological activities in different, specific assays. Members of the Wnt-1/Wg class are able to transform mammary C57MG epithelial cells and elicit formation of a secondary body axis when injected ventrally into Xenopus embryos.[8] Based on these assays, Wnt-1, Wnt-3A and Wnt-8 fall into this class. In contrast, members of the Wnt-5A class fail in both of these assays but instead influence cell adhesion and cell migration. Beside Wnt-5A, Wnt-4 and Wnt-11 belong to this class.[46] Considering these data and the fact that a rather large number of Wnt ligands and Frizzled receptors exist, it was of no surprise when it turned out that Wnts can activate different intracellular signaling pathways. Based on the major intracellular mediators, these pathways are depicted as the Wnt/β-catenin, Wnt/$Ca^{2+}$ and Wnt/JNK pathways. The Wnt/JNK pathway is also known as the

*Wnt Signaling in Development*, edited by Michael Kühl.

***Table 1. Known members of the Wnt gene family in vertebrates***

| Gene | Organisms | Mouse Knock-Out Phenotype | Xenopus 2° Axis | Chicken Effect of Overexpression | Reference |
|---|---|---|---|---|---|
| *Wnt-1* | H, M, X, C, Z | • loss of midbrain<br>• loss of cerebellum<br>• loss of neural crest in Wnt-1/Wnt-3A knock-out | yes | - | 2-11 |
| *Wnt-2* | H, M, X, - , Z | • placenta defects | yes | - | 12-16 |
| *Wnt-2b/13* | H, M, X, C, - | - | n.d. | - | 17-20 |
| *Wnt-3* | H, M, X, - , Z | • gastrulation defekts<br>• affects axis formation<br>• hair growth | n.d. | - | 21-26 |
| *Wnt-3A* | -, M, X, C, - | • disturbed somite formation<br>• loss of neural crest in Wnt-1/Wnt-3A knock-out<br>• loss of hippocampus | yes | •interferes with apical ectodermal ridge formation | 7, 27-32 |
| *Wnt-4* | H, M, X, C, Z | • kidney defects<br>• disturbed female development<br>• defect in breast duct formation | no | - | 33-41 |
| *Wnt-5A* | H, M, X, C, Z | • interferes with limb bud outgrowth<br>• shortened AP axis<br>• reduced cell proliferation | no | • chondrogenesis of limb buds | 38, 42-48 |
| *Wnt-5B* | - , M, - , - , - | - | n.d. | - | 49 |
| *Wnt-6* | H, - , X, - , - | - | n.d. | - | 19,50 |
| *Wnt-7A* | H, M, X, C, Z | • limb bud polarity defects<br>• female infertility<br>• delayed synapse maturation | n.d. | • interferes with limb bud polarity<br>• disturbed feather formation | 32, 51-57 |

*continued on next page*

***Table 1. Known members of the Wnt gene family in vertebrates (continued)***

| Gene | Organisms | Mouse Knock-Out Phenotype | Xenopus 2° Axis | Chicken Effect of Overexpression | Reference |
|---|---|---|---|---|---|
| *Wnt-7B* | H, M, X, C, - | - | no | - | 19, 33, 39, 58 |
| *Wnt-7C* | -, -, X, -, - | - | n.d. | - | 19 |
| *Wnt-8A* | -, M, X, -, Z | - | yes | - | 59-61 |
| *Wnt-8B* | H, M, X, C, Z | - | yes | - | 19, 62, 63 |
| *Wnt-8C* | -, -, -, C, - | - | n.d. | - | 64 |
| *Wnt-9* | Eptatretus stoutii<br>Alopius vulpinus | - | n.d. | - | 141 |
| *Wnt-10A* | -, M, X, -, Z | - | n.d. | - | 19,66 |
| *Wnt-10B/12* | H, M, -, -, Z | • inhibition of adipogenesis (LOF in cell culture) | n.d. | - | 67, 68 |
| *Wnt-11* | H, M, X, C, Z | • interferes with kidney function | no | • differentiation of QCE-6 cells (avian)<br>• development of derma | 38, 65, 69-72 |
| *Wnt-14* | H, -, -, C, - | - | n.d. | - | 73 |
| *Wnt-15* | H, M, -, -, - | - | n.d. | - | 73 |
| *Wnt-16* | H, M, -, -, - | - | n.d. | - | 74 |

Biological functions as predicted by knock-out mice and overexpression studies in chicken are also given. The *Xenopus* secondary axis induction assay has been used to group different members of the Wnt gene family into different classes (see text). For more detailed and routinously updated tables on vertebrate and invertebrate Wnts see: www.stanford.edu/~rnusse/Wntwindow.html. Abbreviations used: H: human, M: mouse, X: *Xenopus laevis*, C: chicken, Z: zebrafish. *Eptatretus stoutii* is the hagfish, Alopius vulpinus is the thresher shark.

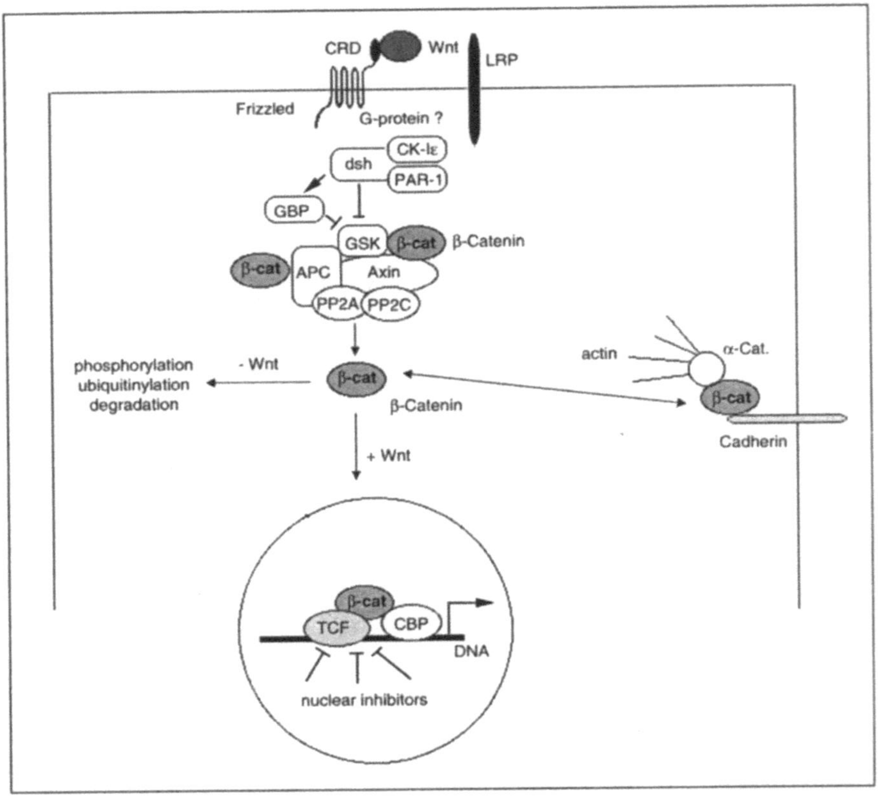


Figure 1. Outline of the Wnt/β-catenin pathway. For details see main text. Abbreviations used: CRD: cystein-rich domain, LRP: lipoprotein receptor related protein, dsh: dishevelled, CK-Iε: casein kinase Iε, GBP: GSK3β binding protein, GSK: glycogen synthase kinase 3β, APC: adenomatous polyposis coli, PP: protein phosphatase, TCF: T-cell factor.

planar cell polarity pathway in *Drosophila*. In the following, we will describe the characteristics of each of these intracellular cascades.

## Wnt/β-Catenin Signaling

β-Catenin is a cytoplasmic protein that fulfils two distinct functions: on one hand, β-catenin binds to the cadherin family of cell adhesion molecules and establishes an essential link to the actin cytoskeleton, on the other hand, β-catenin transmits Wnt signals to the nucleus. The involvement of β-catenin in Wnt signaling became evident from different experimental approaches. In *Xenopus* embryos, ectopic expression of Wnts induces the formation of a secondary body axis, resulting in double-headed tadpoles. This effect reflects the capacity of the Wnt pathway to induce dorsalization of the embryo. Expression of β-catenin also resulted in double axis formation indicating a functional connection to Wnts. Similarly, in *Drosophila*, mutants of the β-catenin homolog *armadillo* had the same phenotype as *wingless* mutants, and epistasis experiments indicated that *armadillo* is downstream of wingless.[78] Finally, treatment of various cell lines with Wnt conditioned medium results in accumulation of the cytoplasmic "free"

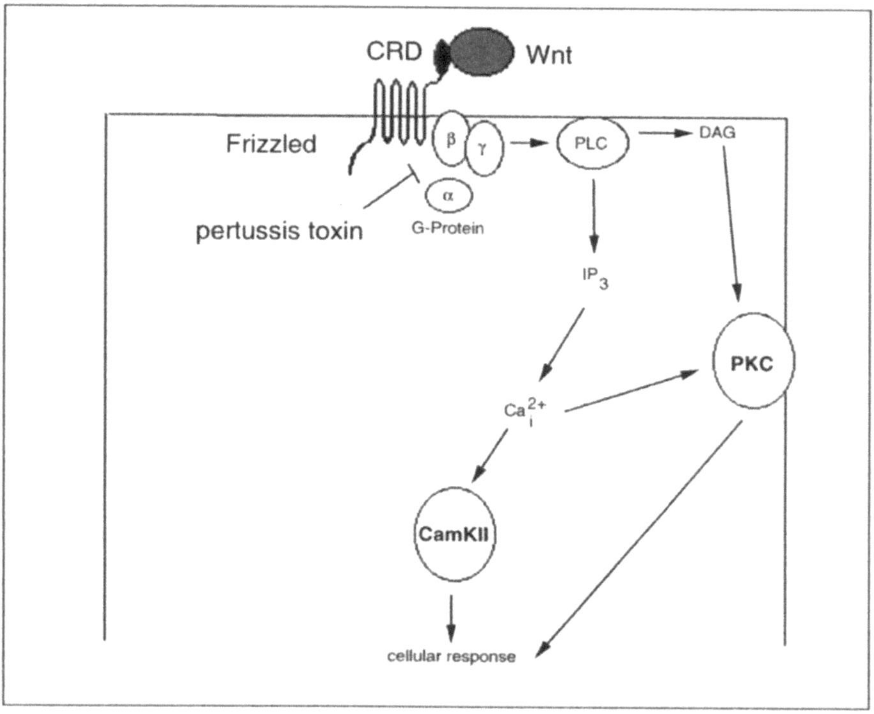


Figure 2. The Wnt/$Ca^{2+}$-pathway. For details see main text. Abbreviations used: α,β,γ: α,β, and γ subunits of heterotrimeric G-proteins, CRD: cystein rich domain, PLC: phospholipase C, DAG: diacylglycerol, $IP_3$: inositoltrisphosphate, PKC: protein kinase C, CamKII: calcium-calmodulin-dependent kinase II.

form of β-catenin, i.e., of β-catenin that is not bound to cadherins.[79] Further experiments indicated that Wnts induce stabilization of β-catenin by preventing its degradation in proteasomes.[80] Finally, the identification of the interaction of β-catenin with TCF transcription factors revealed its role in gene transcription.[81,82] A combination of biochemical and genetic experiments led to the identification of further components of the pathway and resulted in the complex picture that will be briefly illustrated here.

Wnt signaling blocks the function of a multi-protein β-catenin degradation complex, which consists of the scaffold proteins axin or conductin, the tumor suppressor protein APC (adenomatous polyposis coli), the serine/threonine kinase GSK3β (glycogen synthase kinase 3β) and β-catenin (Fig. 1).[83,84,85] APC, GSK3β, and β-catenin bind to axin/conductin at different domains.[84] GSK3β phoshorylates β-catenin, which is then ubiquitinated (a process involving the F-box protein slimb/βTrCP), and finally degraded in proteasomes (Fig. 1, "-Wnt").[86,87] Axin/conductin functions by bringing GSK3β and β-catenin into close proximity thus allowing efficient phosphorylation of β-catenin.[44] The role of APC in this complex is less clear. APC has multiple binding sites for β-catenin and may act as a "trap" that sequesters free β-catenin and delivers it to the degradation complex.[88,89] Further evidence indicates that APC also induces nuclear export of β-catenin.[90] Taken together the results indicate that in the absence of Wnt, accumulation of cytoplasmic and nuclear β-catenin is efficiently prevented by the axin/conductin-based degradation complexes. In the presence of Wnt, the cytoplasmic

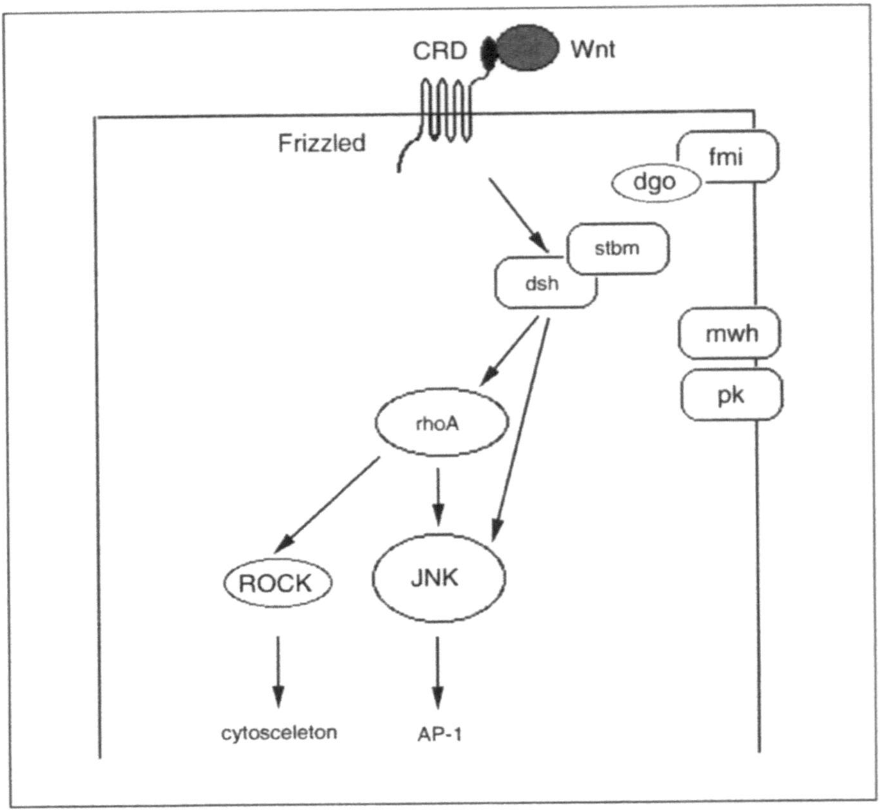


Figure 3. Synopsis of the planar cell polarity pathway, also named Wnt/JNK pathway. For details see main text. Abbreviations used: CRD: cystein rich domain, dsh: dishevelled, stbm: strabismus, dgo: diego, fmi: flamingo, mwh: multiple wing hair, pk: prickle, ROCK: rho kinase, JNK: jun-N-terminal kinase.

phosphoprotein dishevelled is activated to interfere with the β-catenin degradation complex. Dishevelled inhibits GSK3β activity and thereby blocks phosphorylation of β-catenin (Fig. 1, "+Wnt"). Hypophosphorylated β-catenin is no longer degraded, it accumulates in the cytoplasm and can enter the nucleus, where it teams up with TCF transcription factors to activate Wnt target genes (see below).

It is not known how frizzled receptors activate the pathway after stimulation by Wnts. Recently, chimeric receptors containing the ligand-binding and transmembrane portions of the β2-adrenergic receptor fused to the intracellular domains of rat Frizzled-1 were shown to activate TCF/β-catenin signaling upon stimulation with β-adrenergic agonist. This effect could be blocked by pertussis toxin, indicating that Frizzleds transmit signals through stimulation of heterotrimeric G-proteins.[91] The LDL receptor-like proteins 5 and 6 (LRP 5 and 6, *arrow* in *Drosophila*) were shown to play an essential role as co-receptors for Wnts, possibly by sequestering axin.[92-95] Interestingly, an antagonist of Wnt signaling , Dkk-1 binds to LRP6 and thereby inhibits its activity as a co-receptor for Wnts.[96,97]

The exact mechanism by which dishevelled interferes with the function of the axin/conductin complex is ill-defined. Dishevelled interacts directly with axin and the GSK3β binding protein GBP/Frat-1, which might lead to inhibition of GSK3β.[98] Dishevelled also interacts with kinases such as casein kinase1ε (CK1ε) and PAR-1. CK1ε stimulates Wnt signaling, possibly by promoting the interaction of dishevelled with GBP.[99] CK1ε also stimulates interaction of β-catenin with TCFs. PAR-1 also potentiates Wnt signaling, possibly through phosphorylation of dishevelled, but blocks JNK activation, indicating that PAR-1 acts as a switch between these dishevelled-dependent pathways.[100] Interestingly, dishevelled associates with naked cuticle (Nkd), an EF-hand protein, which becomes upregulated as a consequence of Wnt signaling.[101] Nkd inhibits activation of TCF/β-catenin signaling but promotes dishevelled-induced JNK signaling. Thus, similar to PAR-1, but acting in opposite direction, Nkd might function as a switch between different Wnt pathways.[101] The activity of the β-catenin destruction complex may also be regulated by dephosphorylation, as subunits of protein phosphatase 2A have been shown to bind both axin and APC and block Wnt signaling in *Xenopus* embryos.[102]

The entry of β-catenin into the nucleus appears to be independent of the usual importin-mediated transport system.[103] Once arrived in the nucleus, β-catenin associates with transcription factors of the LEF-1/TCF family that include LEF-1, TCF-1, -3 and -4.[81,82,104] TCFs are HMG box DNA binding proteins which lack a classical transactivation domain.[105] Conversely, β-catenin does not bind DNA directly but can act as a strong transcriptional activator.[106] Thus, the TCF/β-catenin complexes can be considered as bipartite transcription factors in which the DNA binding and transactivation functions reside on separate molecules. Several Wnt target genes that are regulated by TCF/β-catenin complexes have been identified in different species. These include transcription factors, growth factors, extracellular matrix proteins, and proteases, indicating that Wnts trigger an entire program of changes in gene expression (for an updated list of Wnt target genes, see www.stanford.edu/~rnusse/Wntwindow.html). The transcriptional activity of the TCF/β-catenin complexes can be modulated by different means (Fig. 1). Wnt target genes can be silenced by the interaction of TCFs with transcriptional co-repressors such as TLE/groucho family members and CtBP.[107] The binding of β-catenin to TCFs relieves this repression through an unknown mechanism. In *Drosophila* the histone acetyltransferase CBP/p300 was shown to acetylate TCF, which reduces the interaction with β-catenin,[108] while in *Xenopus* and in mammalian cell culture p300/CBP acts as a co-activator of TCF/β-catenin complexes, probably by altering chromatin structure.[109] Moreover, β-catenin interacts with Brg-1, a component of the SWI/SNF complex of chromatin regulators, which might also promote the activation of target genes through its effects on nucleosome structure.[110]

A variety of regulatory co-factors that modulate the basic Wnt signaling cascade have been identified (reviewed in Ref. 111). These modulators act on different levels of the pathway and probably have tissue and developmental specificities. Various components regulate the Wnt-frizzled interaction by either acting as a co-receptors such as *dally*, or by preventing binding of Wnts to Frizzleds (e.g., cerberus, WIF-1, FrzB and others). β-Catenin binds to a variety of nuclear factors, such as the *Drosophila* zinc finger protein teashirt, which represses transcription,[112] ICAT and duplin, which block binding of TCFs to β-catenin,[113,114] and a pair of antagonistic regulators of β-catenin mediated transcription, pontin52 and reptin52.[115] Likewise, a number of TCF interactors have been found, including the HMG box protein HBP1 that inhibits binding of TCFs to DNA,[116] and PIASy, a nuclear matrix associated SUMO E3 ligase, which sequesters TCF into nuclear bodies.[117] It will be interesting to determine whether these interactions serve general roles during gene transcription or have functions specific to Wnt signaling. There is also evidence for cross-regulation of the Wnt pathway by p53, which induces degradation of β-catenin.[118,119]

Finally, Wnt signaling might also be influenced by the cadherin-based cell-adhesion system. β-Catenin associates with the cytoplasmic domain of cadherins, and binds the vinculin-related protein α-catenin, which in turn makes contact with actin filaments.[120]

Overexpression of the transmembrane and cytoplasmic domains of cadherins can interfere with Wnt signaling in experimental systems, presumably by depleting the cytoplasmic (signaling) pool of β-catenin. However, loss of E-cadherin, as found in various tumor cells, does not activate the pathway per se[121] and β-catenin's functions in cell adhesion and signaling can be genetically separated in *Drosophila*, raising the question whether cadherins are physiological regulators of Wnt signaling.[111]

Wnt signaling has a prominent role in cancer, a fact that has strongly stimulated research in this area. Mutations of APC in colorectal tumors result in stabilization of β-catenin and constitutive formation of TCF/β-catenin complexes that activate oncogenic target genes such as *c-myc* and *cyclin*D1.[122] In colorectal tumors that lack mutations in APC, mutations of the serine and threonine phosphorylation sites of β-catenin required for degradation have been identified. While mutations of APC are mainly found in colorectal cancer, mutations of β-catenin have also been detected in a wide variety of other tumor types, indicating a potent role for β-catenin as an oncogene.[123]

## The Wnt/$Ca^{2+}$ Pathway in Vertebrates

Initial hints how Wnt-5A class members might exert their effects came with the observation that overexpressing the 5-HT1c serotonin receptor in *Xenopus* or zebrafish embryos phenocopies the effect of overexpressing Wnt-5A, i.e., interferes with convergent extension movements during gastrulation.[124] As the 5-HT-1c receptor activates intracellular $Ca^{2+}$ signaling[125] it was tempting to speculate that Wnt-5A also signals through $Ca^{2+}$. Indeed, Wnt-5A is able to elicit intracellular calcium release in early zebrafish embryos.[126] Subsequently this was characterized in more detail showing that this $Ca^{2+}$-release is mediated through βγ subunits of heterotrimeric G-proteins, phospholipase C, and inositoltrisphosphate ($IP_3$).[127] One caveat here is that up to now a direct interaction between the 7TM Frizzled receptor and G-protein subunits has not been proven. Taking these observations into account it was plausible to analyze whether these $Ca^{2+}$ transients are sufficient to activate cytoplasmic, $Ca^{2+}$-sensitive enzymes. Sheldahl et al reported on the activation of protein kinase C (PKC) by Wnt and Frizzled homologs in *Xenopus* animal cap cells.[128] In addition, the same Wnts and Frizzleds turned out to be sufficient to activate the ubiquitously expressed, calcium-calmodulin dependent kinase II (CamKII) as shown by enzymatic activity assays in *Xenopus*.[129] It can be expected that other $Ca^{2+}$-sensitive cytoplasmic enzymes might also function as mediators of the Wnt/$Ca^{2+}$-pathway, e.g., the $Ca^{2+}$-regulated protein phosphatase calcineurin. In vertebrates, the Wnt/$Ca^{2+}$-pathway seems to be important for ventral development in *Xenopus* embryogenesis[129] and for regulating cell movements and tissue separation during gastrulation.[130,131] A similar pathway has not been described in invertebrates so far although it can not be excluded at the moment that the Wnt/$Ca^{2+}$ pathway and the Wnt/JNK pathway show overlapping activities.

## Planar Cell Polarity Signaling, or the Wnt/JNK Pathway

The planar cell polarity (PCP) pathway in *Drosophila melanogaster* determines the polarity of epithelial cells perpendicular to the apical-basal axis. Although it has been known for a while that this pathway is mediated by Frizzled a Wnt ligand that triggers this pathway remained obscure for long. Recently, Wnt-4, Wnt-5A, Wnt-7A and Wnt-11 have been shown to activate jun N-terminal kinase (JNK) in cell culture cell lines.[132,133] Interestingly, the first known intracellular component of this pathway, dishevelled, is shared by both, the Wnt/β-catenin and the planar cell polarity pathway[134] and different domains in dsh were identified that can activate either pathway.[135] It will be of interest to investigate how dishevelled is regulated to distinguish between these signaling choices and initial hints point towards an important role of phosphorylation of dishevelled. Candidate protein kinases are given by casein kinase Iε, PAR-1, and PKC.[98-101] Downstream of dsh in PCP signaling are the small GTPases rhoA and cdc42[136] that result in an activation of JNK.[137] Activation of rhoA by dsh involves the linker protein Daam-1.[138] Recently, the rhoA stimulated kinase (ROCK) has also been implicated in PCP signaling and cytoskeletal rearrangements.[139] It is not evident so far how other genes involved in establishing planar cell polarity in *Drosophila* are involved in this pathway, e.g., prickle, multiple wing hair, inturned or flamingo. The planar cell polarity pathway in *Drosophila* is required for establishing the right orientation of bristles and wing hairs as well as building up the omatidial eye.[140]

## Outlook

Twenty years after the first identification of a member of the Wnt genes family, it is more and more evident that Wnt proteins not only activate one single intracellular pathway. With more experimental data available it seems that a complex signaling network is activated depending on the commitment of the responding cell. In the future it will be of great interest to decipher how Wnt proteins elicit different cellular responses.

## References


1. Wodarz A, Nusse R. Mechanisms of Wnt signaling in development. Annu Rev Cell Dev Biol 1998; 14:59-88.
2. van Ooyen A, Kwee V, Nusse R. The nucleotide sequence of the human int-1 mammary oncogene; evolutionary conservation of coding and non-coding sequences. EMBO J 1985; 4:2905-2909.
3. Arheden K, Mandahl N, Strombeck B et al. Chromosome localization of the human oncogene INT1 to 12q13 by in situ hybridization. Cytogenet Cell Genet 1988; 47:86-87.
4. Nusse R, Varmus HE. Many tumors induced by the mouse mammary tumor virus contain a provirus integrated in the same region of the host genome. Cell 1982; 31:99-109.
5. Nusse R, van Ooyen A, Cox D et al. Mode of proviral activation of a putative mammary oncogene (int-1) on mouse chromosome 15. Nature 1984; 307:131-136.
6. McMahon AP, Bradley A. The Wnt-1 (int-1) proto-oncogene is required for development of a large region of the mouse brain. Cell 1990; 62:1073-1085.
7. Ikeya M, Lee SM, Johnson JE et al. Wnt signalling required for expansion of neural crest and CNS progenitors. Nature 1997; 389:966-970.
8. McMahon AP, Moon RT. Ectopic expression of the proto-oncogene int-1 in Xenopus embryos leads to duplication of the embryonic axis. Cell 1989; 58:1075-1084.
9. Noordermeer J, Meijlink F, Verrijzer P et al. Isolation of the Xenopus homolog of int-1/wingless and expression during neurula stages of early development. Nucleic Acids Res 1989; 17:11-18.
10. Hollyday M, McMahon JA, McMahon AP. Wnt expression patterns in chick embryo nervous system. Mech Dev 1995; 52:9-25.
11. Molven A, Njolstad PR, Fjose A. Genomic structure and restricted neural expression of the zebrafish wnt-1 (int-1) gene. EMBO J 1991; 10:799-807.
12. Wainwright BJ, Scambler PJ, Stanier P et al. Isolation of a human gene with protein sequence similarity to human and murine int-1 and the Drosophila segment polarity mutant wingless. EMBO J 1988; 7:1743-1748.
13. McMahon JA, McMahon AP. Nucleotide sequence, chromosomal localization and developmental expression of the mouse int-1-related gene. Development 1989; 107:643-650.
14. Monkley SJ, Delaney SJ, Pennisi DJ et al. Targeted disruption of the Wnt2 gene results in placentation defects. Development 1996; 122:3343-3353.
15. Landesman Y, Sokol SY. Xwnt-2b is a novel axis-inducing Xenopus Wnt, which is expressed in embryonic brain. Mech Dev 1997; 63:199-209.
16. Blader P, Strahle U, Ingham PW. Three wnt genes expressed in a wide variety of tissues during development of the zebrafish, danio rerio: developmental and evolutionary perspectives. Dev Genes Evol 1996; 206:3-13.
17. Katoh M, Hirai M, Sugimura T, Terada M. Cloning, expression and chromosomal localization of Wnt-13, a novel member of the Wnt gene family. Oncogene 1996; 13:873-876.

18. Zakin LD, Mazan S, Maury M et al. Structure and expression of Wnt13, a novel mouse Wnt2 related gene. Mech Dev 1998; 73:107-116.
19. Wolda SL, Moon RT. Cloning and developmental expression in Xenopus laevis of seven additional members of the Wnt family. Oncogene 1992; 7:1941-1947.
20. Jasoni C, Hendrickson A, Roelink H. Analysis of chicken Wnt-13 expression demonstrates coincidence with cell division in the developing eye and is consistent with a role in induction. Dev Dynam 1999; 215:215-224.
21. Roelink H, Wagenaar E, Lopes da Silva S, Nusse R. Wnt-3, a gene activated by proviral insertion in mouse mammary tumors, is homologous to int-1/Wnt-1 and is normally expressed in mouse embryos and adult brain. Proc Natl Acad Sci USA 1990; 87:4519-4523.
22. Roelink H, Wang J, Black DM et al. Molecular cloning and chromosomal localization to 17q21 of the human WNT3 gene. Genomics 1993; 17:790-792.
23. Liu P, Wakamiya M, Shea MJ et al. Requirement for Wnt3 in vertebrate axis formation. Nat Genet 1999; 22:361-365.
24. Kishimoto J, Burgeson RE, Morgan BA. Wnt signaling maintains the hair-inducing activity of the dermal papilla. Genes Dev 2000; 14:1181-1185.
25. Christian JL, Gavin BJ, McMahon AP, Moon RT. Isolation of cDNAs partially encoding four Xenopus Wnt-1/int-1-related proteins and characterization of their transient expression during embryonic development. Dev Biol 1991; 143:230-234.
26. Krauss S, Korzh V, Fjose A, Johansen T. Expression of four zebrafish wnt-related genes during embryogenesis. Development 1992; 116:249-259.
27. Roelink H, Nusse R. Expression of two members of the Wnt family during mouse development—restricted temporal and spatial patterns in the developing neural tube. Genes Dev 1991; 5: 381-388.
28. Takada S, Stark KL, Shea MJ et al. Wnt-3a regulates somite and tailbud formation in the mouse embryo. Genes Dev 1994; 8:174-189.
29. Lee SM, Tole S, Grove E, McMahon AP. A local Wnt-3a signal is required for development of the mammalian hippocampus. Development 2000; 127:457-467.
30. Wolda SL, Moody CJ, Moon RT. Overlapping expression of Xwnt-3A and Xwnt-1 in neural tissue of Xenopus laevis embryos. Dev Biol 1993; 155:46-57.
31. Hollyday M, McMahon JA, McMahon AP. Wnt expression patterns in chick embryo nervous system. Mech Dev 1995; 52:9-25.
32. Kengaku M, Capdevila J, Rodriguez-Esteban C et al. Distinct WNT pathways regulating AER formation and dorsoventral polarity in the chick limb bud. Science 1998; 280:1274-1277.
33. Huguet EL, McMahon JA, McMahon AP et al. Differential expression of human Wnt genes 2, 3, 4, and 7B in human breast cell lines and normal and disease states of human breast tissue. Cancer Res 1994; 54:2615-2621.
34. Stark K, Vainio S, Vassileva G, McMahon AP. Epithelial transformation of metanephric mesenchyme in the developing kidney regulated by Wnt-4. Nature 1994; 372:679-683.
35. Vainio S, Heikkila M, Kispert A et al. Female development in mammals is regulated by Wnt-4 signalling. Nature 1999; 397:405-409.
36. Brisken C, Heineman A, Chavarria T et al. Essential function of Wnt-4 in mammary gland development downstream of progesterone signaling. Genes Dev 2000; 14:650-654.
37. Ungar AR, Kelly GM, Moon RT. Wnt4 affects morphogenesis when misexpressed in the zebrafish embryo. Mech Dev 1994; 52:153-164.
38. Du SJ, Purcell SM, Christian JL et al. Identification of distict classes and functional domains of Wnts through expression of wild-type and chimeric protein in Xenopus embryos. Mol Cell Biol 1995; 15:2625-2634.
39. Hollyday M, McMahon JA, McMahon AP. Wnt expression patterns in chick embryo nervous system. Mech Dev 1995; 52:9-25.
40. Tanda N, Kawakami Y, Saito T et al. Cloning and characterization of Wnt-4 and Wnt-11 cDNAs from chick embryo. DNA Seq 1995; 5:277-281.
41. Blader P, Strahle U, Ingham PW. Three wnt genes expressed in a wide variety of tissues during development of the zebrafish, danio rerio: developmental and evolutionary perspectives. Dev Genes Evol 1996; 206:3-13.

42. Clark CC, Cohen I, Eichstetter I et al. Molecular cloning of the human proto-oncogene Wnt-5A and mapping of the gene (WNT5A) to chromosome 3p14-p21. Genomics 1993; 18:249-260.
43. Dealy CN, Roth A, Ferrari D et al. Wnt-5a and Wnt-7a are expressed in the developing chick limb bud in a manner suggesting roles in pattern formation along the proximodistal and dorsoventral axes. Mech Dev 1993; 43:175-186.
44. Yamaguchi TP, Bradley A, McMahon AP, Jones SA. Wnt-5A pathway underlies outgrowth of multiple structures in the vertebrate embryo. Development 1999; 126:1211-1223.
45. Moon RT, Campbell RM, Christian JL et al. Xwnt-5A: a maternal Wnt that affects morphogenetic movements after overexpression in embryos of Xenopus laevis. Development 1993; 119:97-111.
46. Torres MA, Yang-Snyder JA, Purcell SM et al. Activities of the Wnt-1 class of secreted signaling factors are antagonized by the Wnt-5A class and by a dominant negative cadherin in early Xenopus development. J Cell Biol 1996; 133:1123-1137.
47. Kawakami Y, Wada N, Nishimatsu SI et al. Involvement of Wnt-5a in chondrogenic pattern formation in the chick limb bud. Dev Growth Differ 1999; 41:29-40.
48. Rauch GJ, Hammerschmidt M, Blader P et al. Wnt5 is required for tail formation in the zebrafish embryo. Cold Spring Harb Symp Quant Biol 1997; 62:227-234.
49. Gavin BJ, McMahon JA, McMahon AP. Expression of multiple novel Wnt-1/int-1-related genes during fetal and adult mouse development. Genes Dev 1990; 4:2319-2332.
50. Rankin J, Strachan T, Lako M, Lindsay S. Partial cloning and assignment of WNT6 to human chromosome band 2q35 by in situ hybridization. Cytogenet. Cell Genet 1999; 84:50-52.
51. Ikegawa S, Kumano Y, Okui K et al. Isolation, characterization and chromosomal assignment of the human WNT7A gene. Cytogenet. Cell Genet 1996; 74:149-152.
52. Adamson MC, Dennis C, Delaney S et al. Isolation and genetic mapping of two novel members of the murine Wnt gene family, Wnt11 and Wnt12, and the mapping of Wnt5a and Wnt7a. Genomics 1994; 24:9-13.
53. Parr BA, McMahon AP. Dorsalizing signal Wnt-7a required for normal polarity of D-V and A-P axes of mouse limb. Nature 1995; 374:350-353.
54. Parr BA, McMahon AP. Sexually dimorphic development of the mammalian reproductive tract requires Wnt-7a. Nature 1998; 395:707-710.
55. Miller C, Sassoon DA. Wnt-7a maintains appropriate uterine patterning during the development of the mouse female reproductive tract. Development 1998; 125:3201-3211.
56. Hall AC, Lucas FR, Salinas PC. Axonal remodeling and synaptic differentiation in the cerebellum is regulated by WNT-7a signaling. Cell 2000; 100:525-535.
57. Widelitz RB, Jiang TX, Chen CW et al. Wnt-7a in feather morphogenesis: involvement of anterior-posterior asymmetry and proximal-distal elongation demonstrated with an in vitro reconstitution model. Development 1999; 126:2577-2587.
58. Van Bokhoven H, Kissing J, Schepens M et al. Assignment of WNT7B to human chromosome band 22q13 by in situ hybridization. Cytogenet. Cell Genet 1997; 77:288-289.
59. Bouillet P, Oulad-Abdelghani M, Ward et al. A new mouse member of the Wnt gene family, mWnt-8, is expressed during early embryogenesis and is ectopically induced by retinoic acid. Mech Dev 1996; 58:141-152.
60. Christian JL, McMahon JA, McMahon AP, Moon RT. Xwnt-8, a Xenopus Wnt-1/int-1-related gene responsive to mesoderm-inducing growth factors, may play a role in ventral mesodermal patterning during embryogenesis. Development 1991; 111:1045-1055.
61. Kelly GM, Greenstein P, Erezyilmaz DF, Moon RT. Zebrafish wnt8 and wnt8b share a common activity but are involved in distinct developmental pathways. Development 1995; 121:1787-1799.
62. Lako M, Strachan T, Curtis AR, Lindsay S. Isolation and characterization of WNT8B, a novel human Wnt gene that maps to 10q24. Genomics 1996; 35:386-388.
63. Cui Y, Brown JD, Moon RT, Christian JL. Xwnt-8b: A maternally expressed Xenopus Wnt gene with a potential role in establishing the dorsoventral axis. Development 1995; 121:2177-2186.
64. Hume CR, Dodd J. Cwnt-8C: a novel Wnt gene with a potential role in primitive streak formation and hindbrain organization. Development 1993; 119:1147-1160.
65. Ku M, Melton DA. Xwnt-11: a maternally expressed Xenopus wnt gene. Development 1993; 119:1161-1173.

66. Wang J, Shackleford GM. Murine Wnt10a and Wnt10b: cloning and expression in developing limbs, face and skin of embryos and in adults. Oncogene 1996; 13:1537-1544.
67 Bui TD, Rankin J, Smith K et al. A novel human Wnt gene, WNT10B, maps to 12q13 and is expressed in human breast carcinomas. Oncogene 1997; 14:1249-1253.
68. Ross SE, Hemati N, Longo KA et al. Inhibition of adipogenesis by Wnt signaling. Science 2000; 289:950-953.
69. Lako M, Strachan T, Bullen P et al. Isolation, characterisation and embryonic expression of WNT11, a gene which maps to 11q13.5 and has possible roles in the development of skeleton, kidney and lung. Gene 1998; 219:101-110.
70. Tanda N, Kawakami Y, Saito T et al. Cloning and characterization of Wnt-4 and Wnt-11 cDNAs from chick embryo. DNA Seq 1995; 5:277-281.
71. Eisenberg CA, Gourdie RG, Eisenberg LM. Wnt-11 is expressed in early avian mesoderm and required for the differentiation of the quail mesoderm cell line QCE-6. Development 1997; 24:525-536.
72. Heisenberg CP, Tada M, Rauch GJ et al. Silberblick/Wnt-11 mediates convergent extension movements during zebrafish gastrulation. Nature 2000; 405:76-81.
73. Bergstein I, Eisenberg LM, Bhalerao J et al. Isolation of two novel WNT genes, WNT14 and WNT15, one of which (WNT15) is closely linked to WNT3 on human chromosome 17q21. Genomics 1997; 46:450-458.
74. McWhirter JR, Neuteboom ST, Wancewicz EV et al. Oncogenic homeodomain transcription factor E2A-Pbx1 activates a novel WNT gene in pre-B acute lymphoblastoid leukemia. Proc Natl Acad Sci U S A 1999; 96:11464-11469.
75. Bhanot P, Brink M, Samos CH et al. A new member of the frizzled family from Drosophila functions as a Wingless receptor. Nature 1996; 382:225-230.
76. Yang-Snyder J, Miller J, Brown JD et al. A frizzled homolog functions in a vertebrate Wnt signaling pathway. Curr Biol 1996; 6:1302-1306.
77. Tajbakhsh S, Borello U, Vivarelli E et al. Differential activation of Myf5 and MyoD by different Wnts in explants of mouse paraxial mesoderm and the later activation of myogenesis in the absence of Myf5. Development 1998; 125:4155-4162.
78. Cadigan KM, Nusse R. Wnt signaling: a common theme in animal development. Genes Dev 1997; 11:3286-305.
79. Giarre M, Semenov MV, Brown AM. Wnt signaling stabilizes the dual-function protein beta-catenin in diverse cell types. Ann N Y Acad Sci. 1998; 857:43-55.
80. Aberle H, Bauer A, Stappert J et al. beta-catenin is a target for the ubiquitin-proteasome pathway. EMBO J 1997; 16:3797-804.
81. Behrens J, von Kries JP, Kuhl M et al. Functional interaction of beta-catenin with the transcription factor LEF-1. Nature 1996; 382:638-642.
82. Molenaar M, van de Wetering M, Oosterwegel M et al. XTcf-3 transcription factor mediates beta-catenin-induced axis formation in Xenopus embryos. Cell 1996; 86:391-399.
83. Behrens J, Jerchow BA, Wurtele M et al. Functional interaction of an axin homolog, conductin, with beta-catenin, APC, and GSK3beta. Science 1998; 280:596-599.
84. Ikeda S, Kishida S, Yamamoto et al. Axin, a negative regulator of the Wnt signaling pathway, forms a complex with GSK-3beta and beta-catenin and promotes GSK-3beta-dependent phosphorylation of beta-catenin. EMBO J 1998; 17:1371-84.
85. Zeng L, Fagotto F, Zhang T et al. The mouse Fused locus encodes Axin, an inhibitor of the Wnt signaling pathway that regulates embryonic axis formation. Cell 1997; 90:181-192.
86. Jiang J, Struhl G. Regulation of the Hedgehog and Wingless signalling pathways by the F-box/WD40-repeat protein Slimb. Nature 1998; 391:493-496.
87. Kitagawa M, Hatakeyama S, Shirane M et al. An F-box protein, FWD1, mediates ubiquitin-dependent proteolysis of beta-catenin. EMBO J 1999; 18:2401-2410.
88. Munemitsu S, Albert I, Souza B et al. Regulation of intracellular beta-catenin levels by the adenomatous polyposis coli (APC) tumor-suppressor protein. Proc Natl Acad Sci USA 1995; 92:3046-3050.
89. Polakis P. The adenomatous polyposis coli (APC) tumor suppressor. Biochim Biophys Acta 1997; 1332:F127-147.

90. Rosin-Arbesfeld R, Townsley F, Bienz M. The APC tumour suppressor has a nuclear export function. Nature 2000; 406:1009-1012.
91. Liu T, DeCostanzo AJ, Liu X et al. G protein signaling from activated rat frizzled-1 to the beta-catenin-Lef-Tcf pathway. Science 2001; 292:1718-1722.
92. Mao J, Wang J, Liu B et al. Low-density lipoprotein receptor-related protein-5 binds to Axin and regulates the canonical Wnt signaling pathway. Mol Cell 2001; 7:801-809.
93. Pinson KI, Brennan J, Monkley S et al. An LDL-receptor-related protein mediates Wnt signalling in mice. Nature 2000; 407:535-538.
94. Tamai K, Semenov M, Kato Y et al. LDL-receptor-related proteins in Wnt signal transduction. Nature 2000; 407:530-535.
95. Wehrli M, Dougan ST, Caldwell K et al. arrow encodes an LDL-receptor-related protein essential for Wingless signalling. Nature 2000; 407:527-530.
96. Mao B, Wu W, Li Y et al. LDL-receptor-related protein 6 is a receptor for Dickkopf proteins. Nature 2001; 411:321-325.
97. Semenov MV, Tamai K, Brott BK et al. Head inducer Dickkopf-1 is a ligand for Wnt coreceptor LRP6. Curr Biol 2001; 11:951-961.
98. Li L, Yuan H, Weaver CD et al. Axin and Frat1 interact with dvl and GSK, bridging Dvl to GSK in Wnt-mediated regulation of LEF-1. EMBO J 1999; 18:4233-4240.
99. Lee E, Salic A, Kirschner MW. Physiological regulation of [beta]-catenin stability by Tcf3 and CK1epsilon. J Cell Biol. 2001; 154:983-993.
100. Sun TQ, Lu B, Feng JJ et al. PAR-1 is a Dishevelled-associated kinase and a positive regulator of Wnt signalling. Nat Cell Biol 2001; 3:628-636.
101. Rousset R, Mack JA, Wharton KA et al. Naked cuticle targets dishevelled to antagonize Wnt signal transduction. Genes Dev 2001; 15:658-671.
102. Li X, Yost HJ, Virshup DM, Seeling JM. Protein phosphatase 2A and its B56 regulatory subunit inhibit Wnt signaling in Xenopus. EMBO J 2001; 20:4122-4131.
103. Wiechens N, Fagotto F. CRM1- and Ran-independent nuclear export of beta-catenin. Curr Biol 2001; 11:18-27.
104. Huber O, Korn R, McLaughlin J et al. Nuclear localization of beta-catenin by interaction with transcription factor LEF-1. Mech Dev 1996; 59:3-10.
105. Giese K, Kingsley C, Kirshner JR, Grosschedl R. Assembly and function of a TCR alpha enhancer complex is dependent on LEF-1-induced DNA bending and multiple protein-protein interactions. Genes Dev 1995; 9:995-1008.
106. van de Wetering M, Cavallo R, Dooijes D et al. Armadillo coactivates transcription driven by the product of the Drosophila segment polarity gene dTCF. Cell 1997; 88:789-799.
107. Roose J, Molenaar M, Peterson J et al. The Xenopus Wnt effector XTcf-3 interacts with Groucho-related transcriptional repressors. Nature 1998; 395:608-612.
108. Waltzer L, Bienz M. Drosophila CBP represses the transcription factor TCF to antagonize Wingless signalling. Nature 1998; 395:521-525.
109. Hecht A, Vleminckx K, Stemmler MP et al. The p300/CBP acetyltransferases function as transcriptional coactivators of beta-catenin in vertebrates. EMBO J 2000; 19:1839-1850.
110. Barker N, Hurlstone A, Musisi H et al. The chromatin remodelling factor Brg-1 interacts with beta-catenin to promote target gene activation. EMBO J 2001; 20:4935-4943.
111. Seidensticker MJ, Behrens J. Biochemical interactions in the Wnt pathway. Biochim Biophys Acta 2000; 1495:168-182.
112. Waltzer L, Vandel L, Bienz M. Teashirt is required for transcriptional repression mediated by high Wingless levels. EMBO J 2001; 20:137-145.
113. Sakamoto I, Kishida S, Fukui A et al. A novel beta-catenin-binding protein inhibits beta-catenin-dependent Tcf activation and axis formation. J Biol Chem 2000; 275:32871-32878.
114. Tago K, Nakamura T, Nishita M et al. Inhibition of Wnt signaling by ICAT, a novel beta-catenin-interacting protein. Genes Dev 2000; 14:1741-1749.
115. Bauer A, Chauvet S, Huber O et al. Pontin52 and reptin52 function as antagonistic regulators of beta-catenin signalling activity. EMBO J 2000; 19:6121-6130.
116. Sampson EM, Haque ZK, Ku MC et al. Negative regulation of the Wnt-beta-catenin pathway by the transcriptional repressor HBP1. EMBO J 2001; 20:4500-4511.

117. Sachdev S, Bruhn L, Sieber H et al. PIASy, a nuclear matrix-associated SUMO E3 ligase, represses LEF1 activity by sequestration into nuclear bodies. Genes Dev 2001; 15:3088-3103.
118. Matsuzawa SI, Reed JC. Siah-1, SIP, and Ebi collaborate in a novel pathway for beta-catenin degradation linked to p53 responses. Mol Cell 2001; 7:915-926.
119. Sadot, E., B. Geiger, M. Oren, and A. Ben-Ze'ev. 2001. Down-regulation of beta-catenin by activated p53. Mol Cell Biol 21:6768-81.
120. Hulsken J, Birchmeier W, Behrens J. E-cadherin and APC compete for the interaction with beta-catenin and the cytoskeleton. J Cell Biol 1994; 127:2061-2069.
121. van de Wetering M, Barker N, Harkes IC et al. Mutant E-cadherin breast cancer cells do not display constitutive Wnt signaling. Cancer Res 2001; 61:278-284.
122. Polakis P. Wnt signaling and cancer. Genes Dev 2000; 14:1837-1851.
123. Behrens J. Cadherins and catenins: role in signal transduction and tumor progression. Cancer Metastasis Rev 1999; 18:15-30.
124. Slusarki DC, Yang-Snyder J, Busa WB, Moon RT. Modulation of embryonic intracellular $Ca^{2+}$ signaling by Wnt-5A. Dev Biol 1997; 182:114-120.
125. Ungar AR, Kelly GM, Moon RT. Wnt4 affects morphogenesis when misexpressed in the zebrafish embryo. Mech Dev 1994; 52:153-164.
126. Ault KT, Durmowicz G, Galione A et al. Modulation of Xenopus embryo mesoderm-specific gene expression and dorsoanterior patterning by receptors that activate the phosphatidylinositol cycle signal transduction pathway. Development 1996; 122:2033-2041.
127. Slusarski DC, Corces V, Moon RT. Interaction of Wnt and a frizzled homologue triggers G-protein-linked phosphatidylinositol signalling. Nature 1997; 390:410-413.
128. Sheldahl LC, Park M, Malbon CC, Moon RT. Protein kinase C is differentially stimulated by Wnt and Frizzled homologs in a G-protein-dependent manner. Curr Biol 1999; 9:695-698.
129. Kühl M, Sheldahl LC, Malbon CC, Moon RT. $Ca^{2+}$/calmodulin-dependent protein kinase II is stimulated by Wnt and Frizzled homologs and promotes ventral cell fates in Xenopus. J Biol Chem 2000; 275:12701-12711.
130. Wallingford JB, Rowning BA, Vogeli KM et al. Dishevelled controls cell polarity during Xenopus gastrulation. Nature 2000; 405:81-85.
131. Winklbauer R, Medina A, Swain RK, Steinbeisser H. Frizzled-7 signaling controls tissue separation during Xenopus gastrulation. Nature 2001; 413:856-860.
132. Yamanaka H, Moriguchi T, Masuyama N et al. JNK functions in the non-canonical Wnt pathway to regulate convergent extension movements in vertebrates. EMBO Rep 2002; 3:69-75.
133. Cai Y, Lechner MS, Nihalani D et al. Phosphorylation of Pax2 by the c-Jun N-terminal kinase and enhanced Pax2-dependent transcription activation. J Biol Chem 2002; 277:1217-1222.
134. Boutros M, Mlodzik M. Dishevelled: at the crossroads of divergent intracellular signaling pathways. Mech Dev 1999; 83:27-37.
135. Boutros M, Paricio N, Strutt DI, Mlodzik M. Dishevelled activates JNK and discriminates between JNK pathways in planar polarity and wingless signaling. Cell 1998; 94:109-118.
136. Djiane A, Riou JF, Umbhauer M et al. Role of frizzled7 in the regulation of convergent extension movements during gastrulation in Xenopus laevis. EMBO J 2000; 127:3091-3100.
137. Strutt DI, Weber U, Mlodzik M. The role of RhoA in tissue polarity and Frizzled signalling. Nature 1997; 387:292-295.
138. Habas R, Kato Y, He X. Wnt/Frizzled activation of rho regulates vertebrate gastrulation and requires a novel formin homolgy protein Daam1. Cell 2001; 107:843-854.
139. Winter CG, Wang B, Ballew A et al. Drosophila Rho-associated kinase (Drok) links Frizzled-mediated planar cell polarity signaling to the actin cytoskeleton. Cell 2001; 105:81-91.
140. Mlodzik M. Planar polarity in the Drosophila eye: a multifaceted view of signaling specificity and cross-talk. EMBO J 1999; 18:6873-6879.
141. Sidow A. Diversification of the Wnt gene family on the ancestral lineage of vertebrates. Proc Natl Acad Sci USA 1992; 89:5098-5102.

# CHAPTER 2

# Secreted Antagonists/Modulators of Wnt Signaling

**Mikhail V. Semënov and Xi He**

## Abstract

Wnt signaling is controlled by a plethora of extracellular modulators that bind either Wnt molecules or Wnt receptors. These modulators in most cases function to antagonize Wnt signaling and in concept define the range, amplitude, and duration of Wnt signaling. Four conserved but structurally distinct families of Wnt antagonists are currently known from lower vertebrates to human: sFRP (secreted frizzled related protein), WIF-1 (Wnt-inhibitory factor 1), Cerberus, and Dickkopf (Dkk). sFRP proteins, WIF-1 and Cerberus have been shown to bind Wnt molecules and may inhibit multiple signaling pathways activated by these Wnt molecules. Dkk proteins bind to the Wnt co-receptor LRP5/LRP6, and specifically inhibit (but in some cases stimulate) LRP5/LRP6-dependent Wnt/β-catenin signaling. In Drosophila, secreted Wingful/Notum antagonizes Wingless (Wnt) signaling by functioning as a modifying enzyme for Dally and Dally-like, proteoglycans that may facilitate the Wnt receptor interaction. Secreted Wnt antagonists play critical roles in embryogenesis and are implicated in variety of physiological and pathological processes.

## Introduction

Secreted Wnt proteins, via binding to the cell surface Frizzled (Fz) receptor or to Fz and its co-receptor LRP5/LRP6, activate several distinct intracellular signaling pathways (Chapter 1). While these signaling pathways are subjected to feedback regulations at many steps within the cell, an extensive network of secreted modulators for Wnt signaling exists in the extracellular space. These modulators function as Wnt antagonists in most cases, but in some instances they can activate Wnt signaling. Thus far four structurally distinct families of secreted factors are known in vertebrates: sFRP (secreted frizzled related protein), WIF-1 (Wnt-inhibitory factor 1), Cerberus, and Dickkopf (Dkk). Members of these families have been found from lower vertebrates to human, but not in any invertebrates, such as in *Drosophila* or nematodes. However, a new Wnt antagonist in *Drosophila* was recently identified as the wingful/notum gene product, which has orthologs in mice and human. Here we discuss the molecular nature of these diverse Wnt modulators and the mechanism of their action in regulation of Wnt signaling. We will only briefly touch upon the biological role of these secreted Wnt modulators because in-depth discussion of Wnt signaling in various developmental processes are presented in other chapters.

*Wnt Signaling in Development*, edited by Michael Kühl.

## Secreted Frizzled-Related Protein (sFRPs)

sFRPs [1] represent the largest family of and the more extensively studied Wnt modulators. The human genome has five sFRP genes (Fig. 1) whose orthologs have been identified in various vertebrate species. Sizzled[2] and Crescent[3] represent additional sFRP members that are found in amphibians and birds but not in mammals. FrzB was the first identified sFRP.[4] Many of the sFRP genes were given different names due to almost simultaneous publications by different groups. Human Genome Organization (HUGO) has approved sFRP as the primary name for sFRP1, 2, 4, and 5. FRZB is allowed to keep its original name and given sFRP3 as an alias.

## Domain Structure of sFRP

sFRPs are composed of approximately 300 amino acid residues, harboring an amino terminal signal peptide that is followed by two structural domains (Fig. 1). The amino half has the Fz domain, also known as CRD or FRI domain[5] (hence the name "secreted frizzled-related protein"). The sFRP Fz domain, like the Fz domain in Frizzled receptors,[6] is able to bind Wnt proteins and contains the Wnt modulating activity.[7] The carboxyl half of sFRP has the Netrin C-terminal domain (NTR), also known as Complement C3, C4, and C5 C-terminal domain (C345C), whose function is not defined. A spacer of 18 to 32 amino acid residues lies between the last cysteine of the Fz domain and the first cysteine of the NTR domain.

The Fz domain consists of approximately 120 amino acid residues, including ten cysteines distributed throughout with conserved spacing and arrangements of disulfide bonds.[8,9] The distribution and position of hydrophobic, basic and acidic amino acid residues are also conserved (Fig. 2). Interestingly, the Fz domain is found not only in sFRP and Frizzled proteins, but also in Smoothened (a co-receptor for the secreted signaling molecule Hedgehog),[10] transmembrane serine protease Corin,[11] membrane-type Frizzled-related protein (mFRP),[12] Carboxypeptidase Z,[13] receptor tyrosine kinase MuSK, ROR1, and ROR2,[14-17] and a splicing variant of collagen XVIIIa[18] (Fig. 2). All these proteins, which are either secreted or transmembrane proteins, have a single Fz domain with the exception of Corin, which has two Fz domains. In addition, Corin[11] and mFRP[12] have LDLa domains in their extracellular regions. These domains are major structural components of the extracellular portion of LRP5 and LRP6 proteins, which are Frizzled co-receptors for Wnt proteins.[19] Thus far there is no evidence that Fz domains from these non-sFRP and non-Frizzled proteins can interact with Wnt.

sFRPs are clearly divided into two groups on the basis of Fz domain similarity. FrzB and sFRP4 form one group whose Fz domains are more similar to those of Frizzled proteins (Fig. 3). sFRP-1, sFRP 2, sFRP 5, Sizzled and Crescent form the other group, whose Fz domains, surprisingly, do not display a higher similarity with those of Frizzled proteins but rather appear to be more related to Fz domains of the non-Frizzled proteins in overall sequence comparisons (Fig. 3). Given the segregation of Fz domains of sFRPs from those of Frizzleds (Fig. 3), we may speculate that sFRPs' spectrum of interaction with Wnt proteins is different from that of Frizzled proteins. This speculation may be resolved when the structure of Wnt proteins is defined and the molecular basis of specificity of Fz-Wnt interaction is established.

Crystal structures of Fz domains of FrzB (sFRP-3) and Fz-8 reveal that Fz domain represents a novel protein fold.[9] Mutagenesis studies together with the analysis of the crystal structures of these two Fz domains suggest a single continued surface that may participate in interaction with Wnt.[9] This surface contains two key regions, which are characterized by consensus amino acid sequences that are distinct between sFRP s/Frizzleds versus non- sFRP/non-Frizzled proteins (Fig. 2). Therefore, we may expect that Fz domains of Frizzleds and sFRPs share a similar interface for Wnt interaction, but other Fz domain containing proteins likely lack such an interface and may not efficiently interact with Wnt proteins.

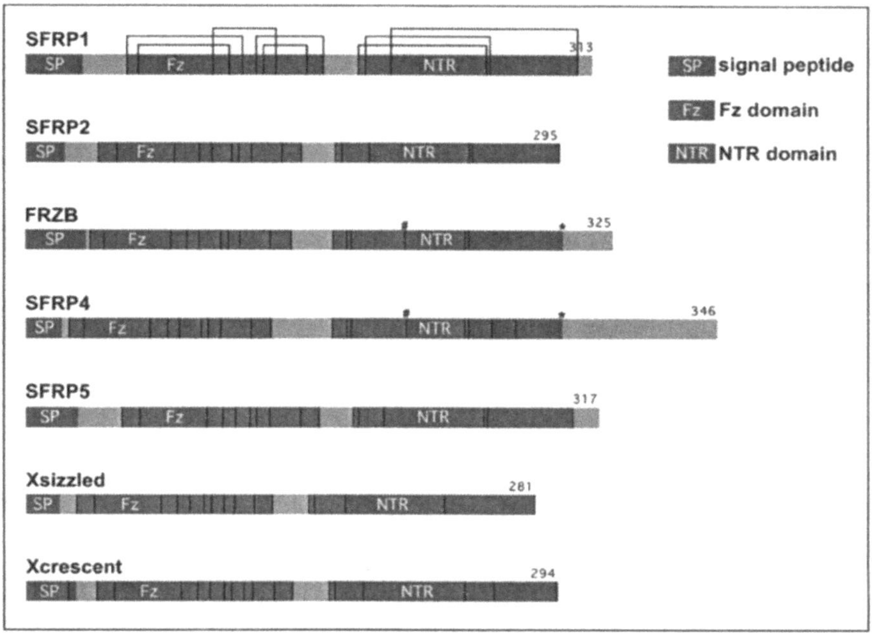


Figure 1. Domain structures of sFRPs. Black vertical bars indicate cysteine residues. Disulfide bonds in sFRP-1 are shown as black brackets. A cysteine residue absent from the C terminus of FRZB and sFRP4 NTR domains is indicated by (*). An extra cysteine residue present in the middle of NTR domains is indicated by (#).

The NTR domain consists of approximately 120 amino acid residues. A single NTR domain can be found at the carboxyl terminal region of sFRPs (but not Frizzleds), netrins, Complement C3, C4, and C5, and procollagen C-endopeptidase enhancers. This domain is characterized by six cysteine residues and a block of basic residues at the C terminus. Positions of cysteine residues are not as conserved as in Fz domains and in fact differ significantly between Complements and other NTR domain containing proteins[8] (Fig. 4). The three-dimensional structure of the NTR domain is a five-stranded beta-barrel[20-22] that is similar to the OB-fold found in the oligosaccharide/oligonucleotide binding proteins.[23] sFRPs can also be divided into two subgroups on the basis of NTR domain similarity: FrzB and sFRP-4 form one group whereas sFRP-1, sFRP-2, and sFRP-5 form the other (Fig. 5). This grouping is identical to that based on Fz domain similarity.

The function of the NTR domain in sFRPs is not defined. The NTR domain may interact with heparan-sulfate proteoglycans[24] and be responsible for sFRP association with the cell surface.[25] The NTR domain also bears some resemblance to the TIMP domain of tissue inhibitors of metalloproteinase.[26] TIMP domain forms complexes with and inactivates extracellular matrix metalloproteinases such as collagenases.[27] Thus NTR domain may possess some protease inhibitory activity.[28] Indeed, it was shown that sFRP decreases the activity of matrix metalloproteinase-2 (MMP-2) when over-expressed in glioma cells.[29]

sFRPs have been found in human, mice, chicken, *Xenopus* and other vertebrates. Chicken and Xenopus have two additional related proteins, Sizzled and Crescent.[2,3,30,31] Although these two proteins display significantly higher similarities in both Fz and NTR domains to sFRP-1, -2,

Figure 2. Sequence alignment of Fz domains of indicated human proteins and Xenopus Sizzled and Crescent. Conserved cysteine residues are indicated under the alignment. $C^A$ is an extra cysteine present only in Fz domain of *Xenopus* Sizzled and Crescent. Putative regions involved in Wnt interaction are underlined by the black bar. Amino acid residues are colored according to ClustalW: D, E (acidic) are in blue; R, H, K (basic) are in black; A, V, F, P, M, I, L, W (hydrophobic (including aromatic—Y)) are in red; S, T, Y, H, C, N, G, Q (Hydroxyl + Amine + others) are in green. In all figures with sequence alignment, "*" represents identical residues in all sequences; ":" represents conserved residues; and "." represents semi-conserved residues (http://www.ebi.ac.uk/clustalw/#). The full color version of this figure is available at the publisher's web site, www.eurekah.com. *(continued on next page)*

and -5 (Figs. 3 and 5), Sizzled and Crescent have one extra cysteine in the Fz domain but lack one cysteine in the NTR domain (Figs. 2 and 4). The extra cysteine in the Fz domain is located between cysteine 8 and 9, thus corresponds to the carboxyl end of the fourth alpha helix of the Fz domain and is on the opposite side to the putative Wnt binding surface.[9] Therefore, this extra cysteine in Fz domain should not affect its ability to interact with Wnt molecules. Whether this cysteine will form a disulfide bond with the unpaired cysteine from NTR domain is not known.

```
SFRP1                 -F-YWPE--MLK-C--DK-FP---EGD-----------VC 109
SFRP2                 -F-PWPD--MLE-C--DR-FPQ--DND-----------LC 113
FRZB                  -H-SWPE--NLA-C--EE-LP---VYD------R-G-VC 113
SFRP4                 -H-SWPE--SLA-C--DE-LP---VYD------R-G-VC 113
SFRP5                 -F-PWPE--MLH-C--HK-FPL--DND-----------LC 110
Xsizzled              -H-SWPE--SLD-C--DR-FPA--GED-----------MC 112
Xcrescent             -Y-PWPE--ILK-C--DK-FP----ED---------HGMC 110
FZD1                  -F-QWPD--TLK-C--EK-FP---VHG-------A-GELC 112
FZD2                  -F-QWPE--RLR-C--EH-FP---RHG-------A-EQIC 112
FZD3                  -V-PWPE--DME-C--SR-FP-----D------------C 106
FZD4                  -F-AWPE--SLN-C--SK-FPP--QND-------H-NHMC 114
FZD5                  -F-AWPE--RMS-C--DR-LP---VLGRD-----A-EVLC 115
FZD6                  -I-RWPE--ELE-C--DR-LQ-----Y------------C 106
FZD7                  -F-QWPE--RLR-C--EN-FP---VHG-------A-GEIC 112
FZD8                  -F-AWPD--RMR-C--DR-LP---EQG-N-----P-DTLC 114
FZD9                  -F-GWPD--SLD-C--AR-LPT--RND-------P-HALC 114
FZD10                 -F-KWPD--SLD-C--RK-LPN--KND-------P-NYLC 114
SMOH                  ----WPD--FLR-CTPDR-FPEGCTNEVQNIKFNSSGQC 124
MFRP                  -T-PWP----FN-C--NR-LP---EAD---------LEAC 110
Carboxypeptidase_Z    -M-AWPY--FLD-C--HRYFTR--EDE-----------GC 115
Corin_A               -Y-SWPD--FLR-C--SQ-FRN--QTESS-----NVSRIC 118
Corin_B               -L-QWPE--DTD-C--SQ-FP---E-ENS-----D-NQTC 116
MuSK                  -IGVYSAGLSLPDC--QR-LPS-IHHD---------PEAC 129
ROR1                  MI-LMRL--KLPNC--ED-LP---QPESP-----E-AANC 127
ROR2                  LI-LMRL--QLPKC--EA-LP---MPESP-----D-AANC 127
Collagen XVIIIa       GG-RLP---VA--C--AS-LPT--QEDG----------YC 110
                                   C9    :                    C10
```

Figure 2. Continued from previous page.

## Mechanism of Wnt Signaling Modulation

Initial findings that sFRPs antagonize Wnt activity were obtained mostly from experiments in *Xenopus* embryos. It was shown that FrzB and sFRP-1 each block ectopic axis formation induced by Xwnt8 when RNAs for FrzB or sFRP-1 and Xwnt-8 are co-injected, either in the same blastomere or in different neighboring blastomeres.[25,32,33] Similarly, transfected sFRP-1 can antagonize Wnt-1 expressed in the same cells or in co-cultured cells in mammalian tissue culture studies.[34] Further, purified sFRP-1 protein antagonizes Wnt signaling.[24]

The Fz domain of sFRPs is necessary and sufficient for interaction with Wnt proteins.[6,7,24] Different sFRPs appear to have distinct specificity towards Wnt molecules. For example, either FrzB or sFRP-1 inhibits signaling by Wnt-1, Wnt-2 or Wnt-8 but not Wnt-3A, Wnt-5A, and Wnt-11,[8,14,37-407,25,35-38] whereas sFRP-2 blocks the activity of Wnt-5A and Wnt-7A but not Wnt-4.[39,40] These data suggest that sFRPs may bind to specific Wnt proteins and prevent Wnt from interacting with cellular receptors, such as Frizzleds and LRP5/6. However, another mechanism of sFRP action can be envisioned based on the observation that Fz domains form dimers.[9,37] The heteroduplex formation between Fz domain of a Frizzled receptor and a sFRP molecule will most likely affect Wnt-Frizzled interaction and thus Wnt signaling.

When sFRPs are transfected or added as secreted proteins, they usually antagonize the activity of certain Wnt proteins. However, sFRPs in some instances were reported to synergize with Wnt or mimic Wnt activities. When added to *Drosophila* S2 cells expressing DFz-2, sFRP-1 at low concentrations can increase the accumulation of Armadillo (β-catenin) protein induced by Wg, but at higher doses antagonizes Wg activity.[24] In yet another example, the formation of tubules during kidney development depends on Wnt-4,[41] and is suppressed by added sFRP-1 protein in an in vitro culture model of metanephroi.[40] Somewhat surprisingly, such inhibition could be partially reversed by addition of sFRP-2 together with sFRP-1,[40] showing that sFRP-2 antagonizes sFRP-1 in this model system. One possibility is that sFRP-2, which does not antagonize Wnt-4,[40] forms a dimeric complex with sFRP-1, thereby preventing sFRP-1 from antagonizing Wnt-4 activity. Another possibility is that sFRP-2 may stabilize Wnt proteins in the extracellular space by preventing denaturation/degradation of free Wnt proteins or by preventing Wnt molecules from diffusion from the site of synthesis. In addition

***Table 1. References for expression patterns of sFRP genes***

| | Human | Mouse | Chicken | *Xenopus* |
|---|---|---|---|---|
| sFRP-1 | 25 | 1, 38, 94, 95 | 96 | 14 |
| sFRP-2 | 1, 97, 98 | 1, 42, 99, 100 | 101, 102 | 30 |
| FrzB | 1, 32 | 1, 32, 33, 94, 4, 103, 99, 104, 105, 106 | 101, 107, 108, 109 | 30 |
| sFRP-4 | 1 | 1 | | |
| sFRP-5 | 98 | | | |
| Crescent | | | 3. 65, 80 | 30, 31 |
| Sizzled | | | | 2, 30 |

to the aforementioned complex relationship between Wnt and sFRP molecules and the poorly understood Wnt-sFRP specificity, several other considerations need to be taken into account in the study of the biological roles of different sFRPs. These include complex and often overlapping expression of multiple sFRP genes together with multiple wnt and Frizzled genes and mutual or feedback regulation of their expression. For example, during kidney development Wnt-4 and sFRP-2 are expressed in condensed mesenchyme and simple epithelial bodies.[40,42] sFRP-2 expression depends on Wnt-4 as it does not occur in mesenchyme of Wnt-4 knockout mice and can be induced in cultured kidney mesenchyme by Wnt-4.[42]

## Expression and Function of sFRP

Like the Wnt gene family, members of the sFRP family exhibit dynamic and complex expression patterns during vertebrate development. The massive information of these expression data is beyond the scope of this review. Much of this information is covered in other chapters on specific developmental systems. We encourage readers who are interested in expression patterns of sFRP genes to refer to these chapters or go to original references, which are listed in Table 1. While overexpression studies clearly demonstrated sFRPs function in modulating Wnt signaling during development, no loss-of-function studies on any sFRP genes have been reported.

## Wnt-Inhibitory Factor-1 (WIF-1)

Wnt-inhibitory factor-1 (WIF-1) was originally identified as a cDNA expressed in human retina.[43] Human WIF-1 consists of 379 amino acid residues including an amino terminal signal peptide. The so-called WIF domain and five epidermal growth factor (EGF) domains occupy the amino and the carboxyl half of the protein, respectively (Fig. 6). The human genome appears to harbor a single WIF-1 without any other closely related family members. WIF-1 orthologs have been identified in mouse, rat, *Xenopus*, and zebrafish.[43]

The WIF domain has approximately 140 amino acids including two conserved cysteine residues (Fig. 6). The WIF domain binds to Wnt with a $K_d$ of 16 nM and antagonizes Wnt signaling via prevention of Wnt-Frizzled interaction.[43] Interestingly, the WIF domain can also be found in the extracellular portion of RYK (related to tyrosine kinase) tyrosine kinase receptor.[44-46] The identity between the WIF domain of human WIF-1 and RYK is 26% whereas that between WIF domain of human and zebrafish WIF-1 is 83% (Figs. 6 and 7). It is unknown whether WIF domain in RYK binds to any Wnt and whether such an interaction, if it exists, affects the tyrosine kinase activity of RYK.

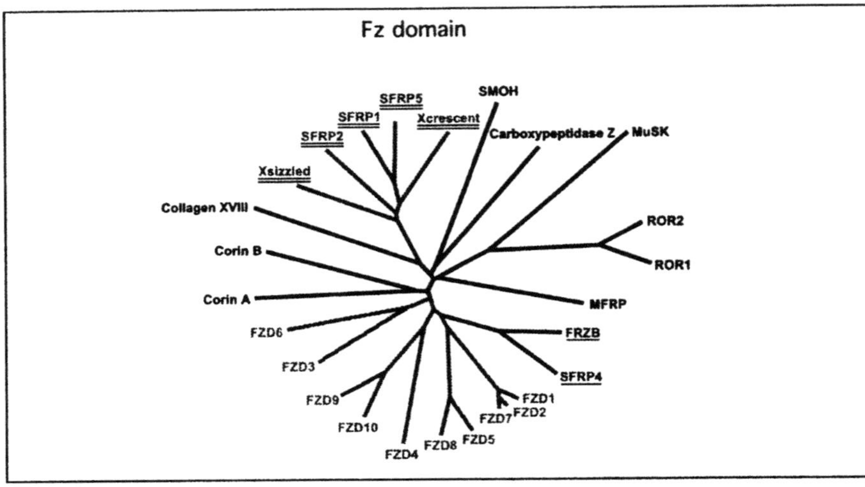


Figure 3. Homology tree of Fz domains of indicated human proteins and *Xenopus* Sizzled and Crescent. sFRPs in the sFRP4-FRZB group are underlined by a single line, and in the sFRP-1 group are underlined by a double line.

While the WIF domain is responsible for Wnt binding, five EGF domains of WIF-1 may have modulatory activities. Co-injection of Chordin (an antagonist of BMP) and WIF-1 RNAs into the early *Xenopus* embryo leads to ectopic head induction with two eyes, whereas co-injection of Chordin and the WIF domain alone leads to the formation of cyclopic head.[43] This significant phenotypic difference exists with various doses of injected RNAs,[43] thus likely reflecting qualitative rather than quantitative difference between the full length WIF-1 versus the WIF domain alone. As the EGF domains of WIF-1 are most closely related to those found in the extracellular matrix protein tenascin,[47] we speculate that EGF domains of WIF-1 may interact with extracellular matrix and affect distribution or accessibility of WIF-1 protein in the extracellular space.

WIF is expressed in *Xenopus* and zebrafish at the neurula stage. Expression first appears at the onset of somitogenesis in paraxial mesoderm and later in presomitic mesoderm and notochord. In *Xenopus* tadpoles, WIF-1 is expressed in visceral arches, otic vesicles, nasal placodes, and several regions of the anterior brain. In the adult mouse, WIF-1 mRNA level is high in heart and lung, and low in brain and eyes.[43] WIF-1 knock out mice have not been reported.

## Cerberus

*Xenopus* Cerberus (Xcer) was originally isolated as a gene highly expressed in Spemann's organizer.[48] Xcer encodes a secreted protein that is the founding member of the DAN/Cerberus family. Although a common feature for members of this family is their ability to antagonize signaling by specific members of the transforming growth factor-beta (TGFβ) family, such as Activin, Nodal, and BMPs (Bone Morphogenetic Proteins),[48-53] Xcer has also been demonstrated to be a potent antagonist for Wnt signaling.[50]

Dan/Cerberus proteins share a related carboxyl terminal cysteine knot (CT) domain (Fig. 8), which is about 90 amino acid residues and structurally resembles the TGFβ carboxyl terminal domain.[54] The CT domain has 9 cysteine residues that most likely form four intramolecular

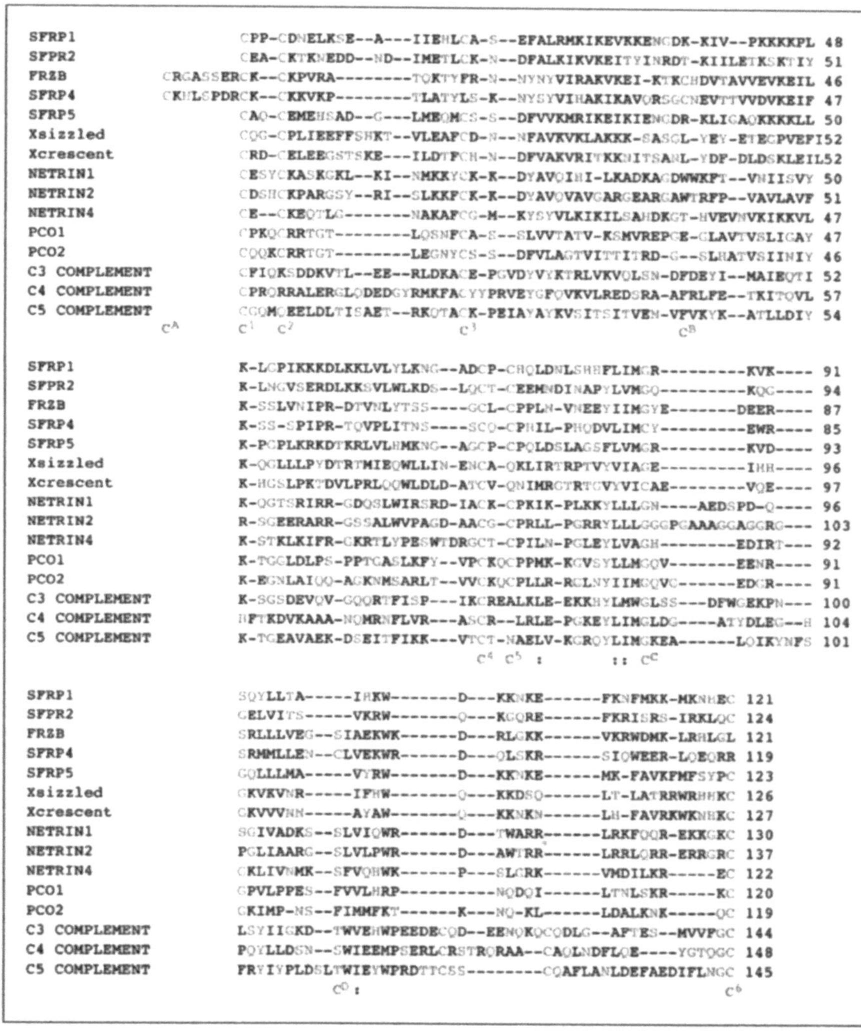


Figure 4. Sequence alignment of NTR domains of human proteins and Xenopus Sizzled and Crescent. $C^A$ and $C^B$ cysteines are specific for FRZB and sFRP4. Cysteines $C^C$ and $C^D$ are an extra pair of cysteines in sFRP4. The full color version of this figure is available at the publisher's web site, www.eurekah.com.

disulfide bonds,[52] with the remaining ninth cysteine potentially forming an inter molecular connection with another Dan/Cerberus molecule (Fig. 8). Such a Dan/Cerberus dimer perhaps is analogous to the TGFβ domain dimer.[55] Indeed, Xcer was demonstrated to exist as a dimer under non-reducing conditions.[52,53] CT domains of *Cerberus-like* proteins from different vertebrate species are more closely related to one another than to those of other members of the DAN/Cerberus family (Fig. 8).

The main difference between Xcer and other family members may be its ability to antagonize Wnt signaling.[50,56] All *Cerberus-like* proteins have a relatively long polypeptide chain at

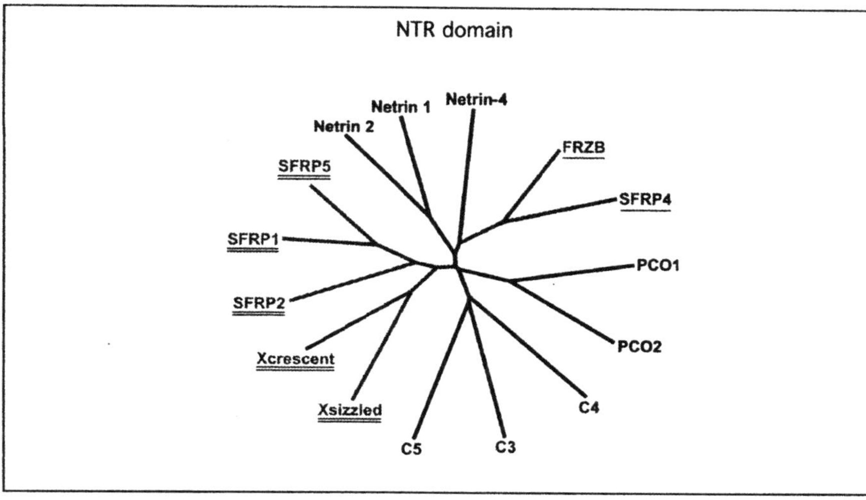


Figure 5. Homology tree of NTR domains of human proteins and Xenopus Sizzled and Crescent. sFRPs in the sFRP4-FRZB group are underlined by a single line, and in the sFRP-1 group are underlined by a double line.

the amino terminus before the CT domain (Fig. 8). This portion of Xcer is responsible for its Wnt-binding and Wnt-antagonizing activity.[50] However, this region shares very limited homology among Cerberus-like proteins (Fig. 9), leaving it an open question whether the Cerberus-like proteins from other species have anti-Wnt activity. In fact, mouse Cerberus-like is unable to inhibit Xwnt-8, which is bound and antagonized by Xcer.[57] The anti-TGFβ activity of Cerberus-like proteins is conserved, since the mouse Cerberus-like antagonizes BMP-4 and Nodal, as Xcer does.[57]

Xcer is expressed in Spermann's organizer in *Xenopus* embryos,[48] and the mouse Cerberus-like is expressed in the anterior visceral endoderm (AVE) at the pre-streak embryonic stage.[71-73,77,7851-53,58,59] These expression patterns together with observation that Xcer mRNA injection induces an ectopic head in *Xenopus* embryos suggest that Xcer and mCer-l might be head inducers. Mice lacking the *Cerberus-like* gene, however, develop normally.[76,79,8057,60,61] One possibility is that Cerberus may be functionally redundant with other organizer-specific antagonists.

During chick development Cerberus (also named Caronte) is symmetrically expressed at the anterior end of the mesendoderm layer at the onset of gastrulation. Its expression is subsequently confined to the left lateral plate mesoderm. Over-expression of Cerberus on the right side of embryo leads to reversal of heart orientation. That role of the chicken Cerberus in left-right asymmetry is probably mediated entirely via its anti-BMP activity, because expression of BMP antagonist Noggin on the right side of embryo produces the same asymmetry reversal.[62-64] While Wnt antagonists, including Xcer, are able to induce heart formation, chick Cerberus seems to lack this activity, again suggesting that Xcer may be unique in its ability to antagonize Wnt signaling.[65,66]

## Dickkopf

Dickkopf1 was identified in an expression screen for head inducers in *Xenopus*.[67] Injection of Xdkk-1 mRNA into *Xenopus* embryos promotes head development, hence came the name Dickkopf (German, big head).

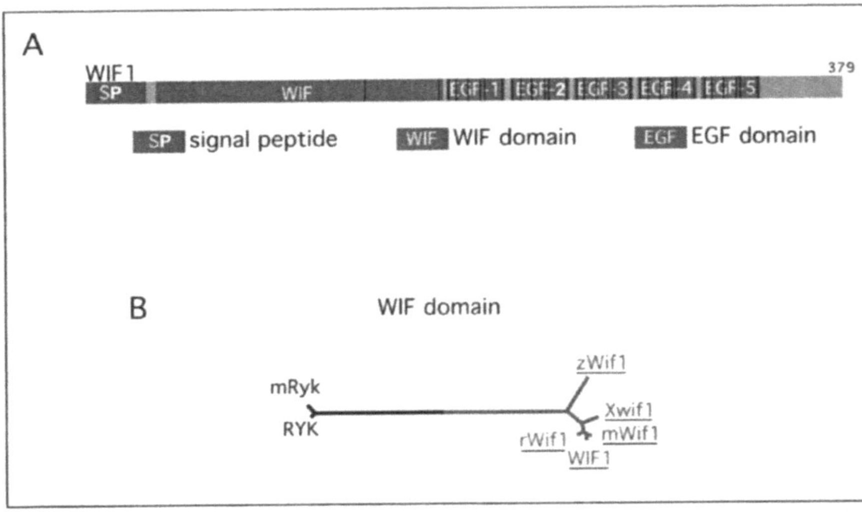


Figure 6. (A) Domain structure of WIF-1. Black vertical bars indicate cysteine residues. (B) Homology tree of WIF domains. WIF domains of WIF proteins are underlined. Z: zebra fish, x: Xenopus, m: mouse, r: rat.

The human Dkk family consists of four members that share two conserved cysteine-rich domains (Fig. 10).[67-69] The so called CYS1 domain, located closer to the amino terminus, is approximately 50 amino acids long and has 10 conserved cysteine residues, whereas the CYS2 domain, located closer to the carboxyl terminus, is approximately 75 amino acids long and also has 10 conserved cysteine residues. Thus five intra-domain disulfide bonds are possible in each domain. Dkk-3 is approximately 100 amino acids longer than other Dkk proteins and has two short (around 40 amino acids) coiled-coil domains at each terminus in addition to the two CYS domains (Fig. 10).

The CYS1 domains of Dkk-1, Dkk-2, and Dkk-4 have, on average, 45% identical amino acid residues. In contrast, the CYS1 domain of Dkk-3 is rather divergent from those of the other Dkks, sharing only about 35% identity. Identity among CYS2 domains of Dkk-1, Dkk-2, and Dkk-4 is 60% but only 27% among these Dkks and Dkk-3. Overall, Dkk-1, Dkk-2, and Dkk-4 clearly form one subgroup whereas Dkk-3 forms another subgroup (Fig. 10). Orthologs from other vertebrate species can be grouped with corresponding human proteins on the homology tree (Fig. 10). Interestingly, CYS2 domains of Dkks have the same cysteine spacing pattern as human colipase[69] and prokineticin and a poisonous snake toxin MIT1 (black mamba intestinal toxin 1). All Dkks (including Dkk-3), prokineticin and MIT1 share 29% amino acid identity, whereas Dkks and colipase share about 21% identical amino acids. The significance of this similarity between the CYS2 domain of Dkks and these other proteins is currently unknown.

Dkk-1 and Dkk-4 inhibit Wnt/β-catenin signaling,[67-70] but Dkk-3 does not.[69,70] Dkk-2 shows ambiguous properties. In *Xenopus* assays, mDkk-2 and Xdkk-2 inhibit neither gene induction nor axis duplication by Xwnt-8, but rather activate Wnt/β-catenin-inducible genes [69,70] and synergize with Xfz-8 in the axis induction assay. Moreover, this activity of mDkk-2 is antagonized by mDkk-1.[70] However, mDkk-2, like mDkk-1, inhibits Wnt/β-catenin signaling in the mammalian cell culture.[70,71]

```
WIF1      SLYLWIDAHQARVLIGFEEDILIVSEGKMAPPFTHDFRKAQQRMPAIPVNIHSMNFTWQAA 60
mWIF1     SLYLWIDAHQARVLIGFEEDILIVSEGKMAPPFTHDFRKAQQRMPAIPVNIHSMNFTWQAA 60
rWIF1     SLYLWIDAHQARVLIGFEEDILIVSEGKMAPPFTHDFRKAQQRMPAIPVNIHSMNFTWQAS 60
xWIF1     SLYMWIDAHQARVLIGFEEDILIVAEGKMAPPFTHDFRKAQQRMPAIPVNIHAMNFTWQAT 60
zWIF1     TMYMWIDANQARILIGFEEDILIVSEGKMAPPFTHDFRKAQQRMPAIPVNIHHVNFTWQAT 60
RYK       SVSLYLSEDEVRRLIGLDAELYYVRNDLISHYALSFN-----L-LVPSETNFLHFTWHAK 54
mRYK      SVSLYLSEDEVRRLLGLDAELYYVRNDLISHYALSFN-----L-LVPSETNFLHFTWHAK 54
          :: :::. .:.* *:*:: ::  * :. :: :: .*..:.. :. :* : : ::***:*

WIF1      GQAEYFYEFLSLRSLDKGI-MADPTVNVPLLGTVPHKASVVQVGFPCLGKQDGVAAFEVD 119
mWIF1     GQAEYFYEFLSLRSLDKGI-MADPTVNVPLLGTVPHKASVVQVGFPCLGKQDGVAAFEVN 119
rWIF1     GQAEYFYEFLSLRSLDKGI-MADPTVNVPRLGTVPHKASVVQVGFPCLGKQDGVAAFEVN 119
xWIF1     GQAEYFYEFLSLRSLDKGI-MADPTVNMPLLGTVPHKATVIQVGFPCLGNQDGVAAFEVN 119
zWIF1     DQAEYFYEFQTLRSLDKDI-MDDPTVNVPLLGSVPHKASVVQVGFPCRGDQDGVAAFEVT 119
RYK       SKVEY--K-LGFQ-VDNVLAMDMPQVNISVQGEVPRTLSVFRVELSCTGKVD--S--EV- 105
mRYK      SKVEY--K-LGFQ-VDNFVAMGMPQVNISAQGEVPRTLSVFRVELSCTGKVD--S--EV- 105
          .:.**  :   ::::*: : *  * **:.  * **:. :*.:* :.C *. *. :: **

WIF1      VIVMNSEGNTILKTPQN-AIF-FKT---CQQ 145
mWIF1     VIVMNSEGNTILRTPQN-AIF-FKT---CQQ 145
rWIF1     VIVMNSEGNPILRTPQN-AIF-FKT---CQQ 145
xWIF1     VIVMNSEGNVILQTPQN-AIF-FKT---CQQ 145
zWIF1     ILVMDAGGNIILRTPHN-AIF-FKT---CQR 145
RYK       MILMQL--NLTVNSSKNFTVLNFKRRKMCYK 134
mRYK      MILMQL--NLTVNSSKNFTVLNFKRRKMCYK 134
          :::*:  .*  :.:.:* ::: **    C :
```

Figure 7. Sequence alignment of WIF domains. The full color version of this figure is available at the publisher's web site, www. eurekah.com.

Distinct from other Wnt antagonists, such as sFRPs, WIF-1, and Xcer, Dkk-1 binds neither Wnt nor Frizzled, nor does it affect Wnt binding to Frizzled.[72,73] Instead, Dkk-1 binds, with high affinity, to LRP5 and LRP6,[72-74] which are Frizzled co-receptors.[19,75,76] Dkk-2 is also able to bind to LRP6, whereas Dkk-3, which is a more distant member of the Dkk family and does not have any Wnt antagonizing activities, cannot bind to LRP6.[72] A single amino acid substitution in the CYS2 domain abolishes the ability of Dkk-1 to interact with LRP5/6 and to antagonize Wnt signaling.[73] Conversely, the CYS2 domain of Dkk-1 alone is able to antagonize Wnt signaling.[71] Thus, the CYS2 domain is functionally critical for binding to LRP5/6 and inhibition of Wnt signaling. Residues in the CYS2 domain that are shared among Dkk-1, Dkk-2 and Dkk-4, but not Dkk-3 (Fig. 11), may contribute to Dkk interaction with LRP5/6. The function of the conserved CYS1 domain remains unknown.

How does Dkk-1 binding to LRP5/6 antagonize Wnt signaling? The simplest model is that Dkk-1 binding to LRP5/6 prevents the formation of Wnt-induced LRP5/6-Frizzled complex that is required for Wnt signaling.[73] Some in vitro binding data are consistent with this view.[73] On the other hand, a different, but not mutually exclusive, model was proposed based upon the recent identification of another class of distinct Dkk receptors. Kremen1 and Kremen2 are two related single-pass transmembrane receptors and bind with high affinity to Dkk-1 and Dkk-2, but not Dkk-3.[77] Dkk-1 induces the formation of a LRP6-Kremen complex that leads to the internalization of LRP6. Thus it is proposed that Dkk-1 promotes the removal of LRP6 from the cell surface via the Kremen protein, thereby preventing LRP5/6-Frizzled complex formation and initiation of Wnt signaling.[77] Further experiments will be required to test these two models.

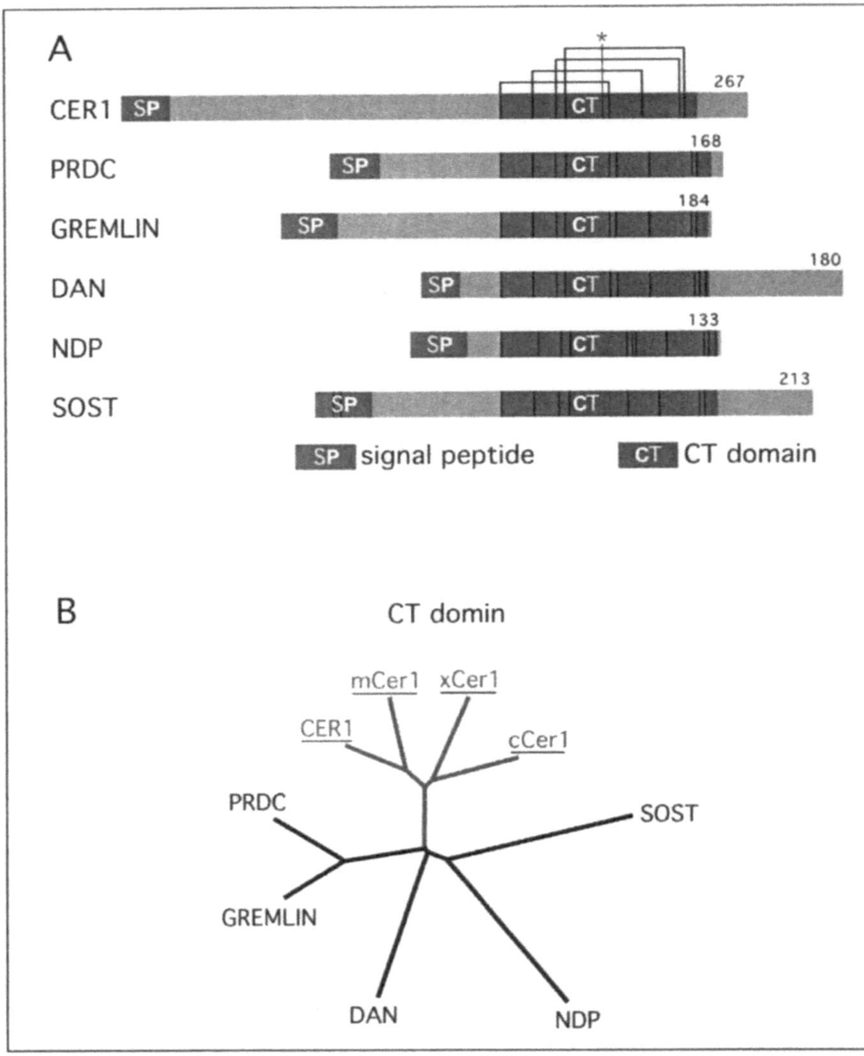


Figure 8. (A) Domain structure of proteins of Human Cerberus/DAN family. Black vertical bars indicate cysteine residues. Disulfide bonds in CER1 are shown as black brackets. A cysteine that is not involved in disulfide bond formation is marked by (*). (B) Homology tree of CT domains. CT domains of Cerberus-like proteins are underlined. M: mouse, x: Xenopus: c: chicken.

One particularly interesting aspect of Dkk-1 is its specificity in antagonizing Wnt/β-catenin signaling but not other Wnt signaling pathways.[73] This is because LRP5/6 functions as a co-receptor specifically for Wnt/β-catenin signaling,[73,75] but not for other Wnt/Fz pathways, such as the Wnt/Fz/Rho pathway.[78] This is in sharp contrast to sFRPs, WIF-1, and Xcer, which bind specific Wnts and inhibit, in principle, all signaling pathways triggered by these Wnt molecules.

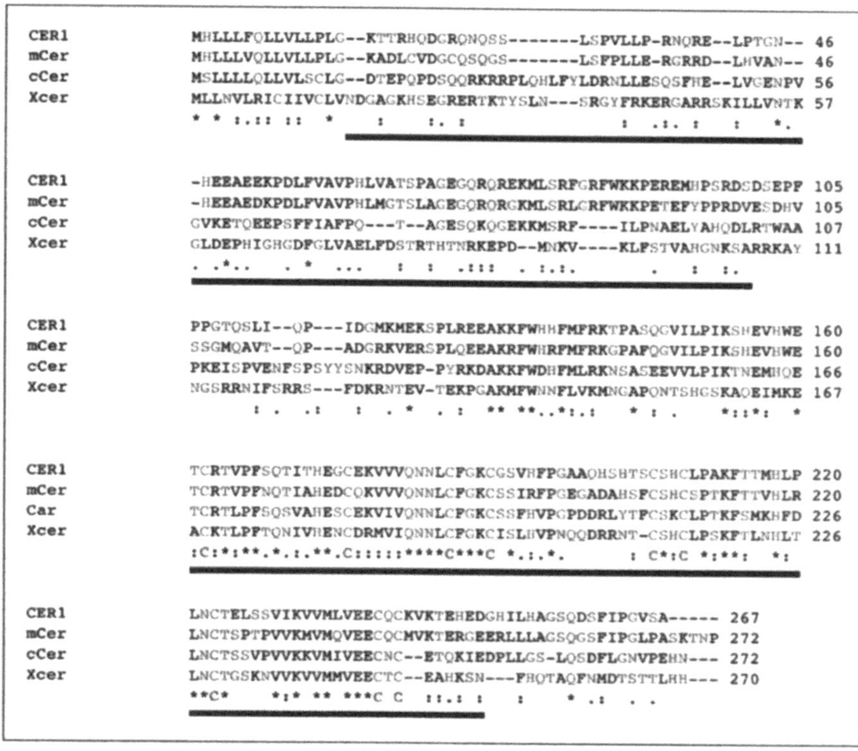


Figure 9. Sequence alignment of Cerberus proteins. The Wnt-binding region in Xcer is underlined by a black bar. CT domain is underlined by a green bar. The full color version of this figure is available at the publisher's web site, www.eurekah.com.

XDkk-1 is expressed in Spemann's organizer in the early gastrula in *Xenopus* embryos and later in the prechordal plate adjacent to prospective forebain and eyes, corresponding fully to regions with head organizing activities. In mouse embryo mDkk-1 is expressed in head mesoderm at E7 and E8, and later in posterior primitive-streak mesoderm and in developing somites.[67,79] Dkk-1 is also expressed in early stage of chicken and zebrafish embryogenesis in a dynamic pattern.[80-82] Dkk-1 plays a key role in head induction. Inactivation of XDkk1 by anti-XDkk1 antibodies or deletion of the mouse *Dkk-1* gene both result in a lack of head structures anterior to the midbrain.[67,83,84] In addition, Dkk-1 knockout mice also exhibit severe abnormality in limb development,[83] consistent with Dkk-1 expression in the developing limb bud.[79]

XDkk-2 is not expressed in early stages of *Xenopus* development. During organogenesis, XDkk-2 is expressed in head mesenchyme, lens, and somites.[70,79] Mouse Dkk-2 expression was first detected at E9 in the ventral diencephalon and the otic vesicle. Later mDkk-2 is expressed in branchial arches, heart, limb buds and other forming organs.[79] Mouse Dkk-3 expression appears at E9 in diencephalon.[79] At E14, mDkk-3 is expressed in developing eye, bones, and cartilage. In adult mice mDkk-3 is expressed in neurons of the cortex and hippocampus, in the retina's integrating bipolar neurons and ganglion cells, and in the atrioventricular valves and the atrial myocytes in heart.[69,79] The biological function of these Dkk genes remains to be established.

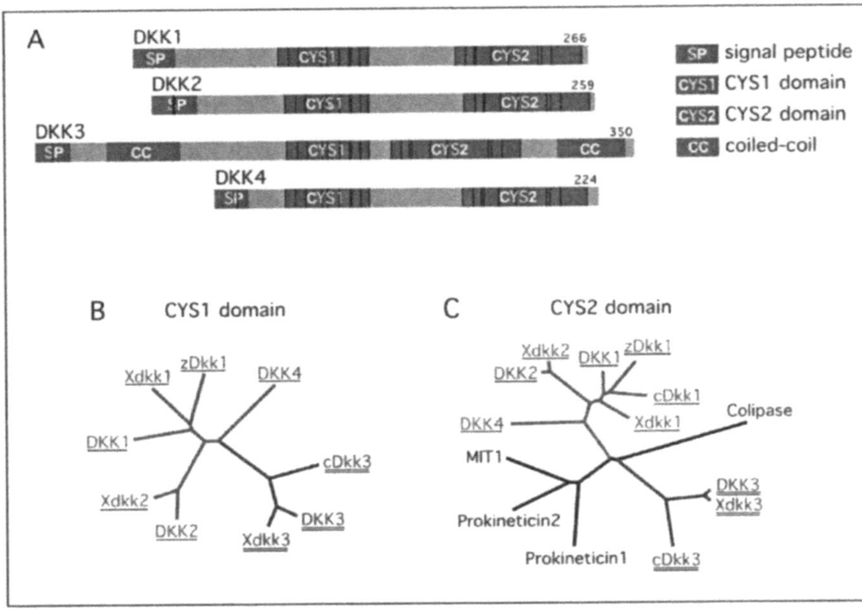


Figure 10. (A) Domain structure of human Dkks. Black vertical bars indicate cysteine residues. (B) Homology tree of CYS1 domains. Dkk-1, 2, and 4 are underlined by a single line, and Dkk3s are underlined by a double line. (C) Homology tree of CYS2 domains. Dkk-1, 2, and 4 proteins are underlined by a single line, Dkk3s are underlined by a double line.

## Wingful/Notum in *Drosophila*

Although homologues of sFRPs, WIF, Cerberus or Dkks have not been found in any invertebrates, genetic studies in Drosophila have revealed a novel secreted antagonist for Wingless (Wg) signaling, referred to as Wingful (Wf)/Notum.[85,86] Wf/Notum encodes a protein of 671 amino acid residues, which is related to plant pectin acetylesterases of the $\alpha/\beta$-hydrolase superfamily.[85,86] Overexpression of the *wf/notum* gene causes phenotypes similar to those from loss-of-function mutation of the wg gene, whereas mutations in *wf/notum* are associated with too much Wg signaling.[85,86] However, unlike vertebrate Wnt antagonists that bind either Wnt (or Fz) or LRP5/6, Wf/Notum binds neither Wg nor Fz nor Arrow,[75] the LRP5/6 ortholog.[86] The main target of Wf/Notum action appears to be cell surface proteoglycans Dally and/or Dally-like (Dlp).[85]

Wg signaling requires Dally and Dlp, but overexpression of Wg can bypass this requirement.[87] Dally and Dlp play important roles in the distribution of extracellular Wg and thus the formation of Wg morphogen gradient, most likely via binding to Wg and modulating Wg availability to Fz and Arrow receptors.[87-92] Indeed, co-expression of Wf/Notum together with Dlp blocks the ability of Dlp to stabilize Wg in the extracellular space[85] This is accompanied by profound effects of Wf/Notum expression on Dally and Dlp proteins. Dally protein level is significantly reduced by co-expression of Wf/Notum, whereas Dlp is converted into a distinct form as is judged by gel electrophoresis.[85] A single amino acid substitution in the putative enzymatically active motif (based on its homology to a/b-hydrolases) abolishes the effect of Wf/Notum on Dally and Dlp proteins.[85] Therefore, these data suggest that Wf/Notum

```
CYS1 domain alignment

DKK1           CAEDEECGTDEYCASPTRGGDAGVQICLACRKRRKRCMRHAMCCPGNYCKNGIC 54
DKK2           CSSDKECEVGRYCHSPHQGSSA----CMVCRRKKKRCHRDGMCCPSTRCNNGIC 50
DKK3           CIIDEDCGPSMYCQFAS-FQYT----CQPCRGQRMLCTRDSECCGDQLCVWGHC 49
DKK4           CLSDTDCNTRKFCLQPRDEKPF----CATCRGLRRRCQRDAMCCPGTLCVNDVC 50
               C   * :C    :C  .        . . C  C*  :  C *.. CC .  C  . C

CYS2 domain alignment

DKK1           CLRSSDCASGLCCARHFWS---KICKPVLKEGQVCTKHRRKG------SHG-LEIFQRCY 50
DKK2           CLRSSDCIEGFCCARHFWT---KICKPVLHQGEVCTKQRKKG------SHG-LEIFQRCD 50
DKK3           CDNQRDCQPGLCCAFQRGLL-FPVCTPLPVEGELCHDPASRLLDLITWELEPDGALDRCP 59
DKK4           CLRTFDCGPGLCCARHFWT---KICKPVLLEGQVCSRRGHKDT-----AQA-PEIFQRCD 51
Prokineticin1  CERDVQCGAGTCCAISLWLRGLRMCTPLGREGEECHPGSHKVP-----FFR-KRKHHTCP 54
Prokineticin2  CDKDSQCGGGMCCAVSIWVKSIRICTPMGKLGDSCHPLTRKVP-----FFG-RRMHHTCP 54
Colipase1      CMNSAQCKSN-CCQHSSALG-LARCTSMASENSECSVKT---------LYG---IYYKCP 46
MIT1           CERDLQCGKGTCCAVSLWIKSVRVCTPVGTSGEDCHPASHKIP-----FSGQRKMHHTCP 55
               C .   :C   . CC             C..:     .. C                 C
                L         L    RHF       K  K  L     V                   E FQ

DKK1           CGEGLSCRIQKDHHQASNSSRLHTC 75
DKK2           CAKGLSCKVWKDATYSS-KARLHVC 74
DKK3           CASGLLCQPHSHS-------LVYVC 77
DKK4           CGPGLLCRSQLTSNRQH--ARLRVC 74
Prokineticin1  CLPNLLCSRFPD---------GRYRC 71
Prokineticin2  CLPGLACLRTSF---------NRFIC 71
Colipase1      CERGLTCEGDKTIVGSITNTNFGIC 71
MIT1           CAPNLACVQTSP---------KKFKC 72
               C   .* C           ..:    C
                                   L
```

Figure 11. (A) Sequence alignment of CYS1 domains of human Dkk proteins. (B) Sequence alignment of CYS2 domains of indicated human proteins and black mamba intestinal toxin 1 (MIT1). Residues in CYS2 domain that are sheared among Dkk-1, 2, and 4, but not Dkk3 and other proteins are shown under alignment. The full color version of this figure is available at the publisher's web site, www.eurekah.com.

antagonizes Wg signaling through Dally and Dlp, probably via modification of their glycoaminoglycan (GAG) side chains.

Two interesting aspects of Wf/Notum function await further discussion. Firstly, Wf/Notum antagonizes Wg signaling in the wing imaginal disc, where Wg controls pattern formation via long range signaling through a morphogen gradient. However, Wf/Notum plays a minimal, if any, role in modulating Wg signaling in embryonic segment polarity determination, where Wg patterns each segment via short range signaling.[85,86] Indeed, Wf/Notum has no effect on signaling via an artificially membrane-tethered Wg.[86] This is consistent with the notion that Wf/Notum governs the extracellular Wg morphogen distribution via Dally and Dlp. Secondly, like some vertebrate Wnt antagonists, *wf/notum* expression is activated by Wg signaling, revealing the existence of a feedback control mechanism in which Wg modifies its own gradient distribution.[85,86]

In vertebrate, proteoglycans related to Dally and Dlp likely play important roles in Wnt signaling, as suggested by the study of Zebrafish *knypek* gene, which encodes a glypican that is required for Wnt-11 signaling.[93] A single ortholog of *Wf/Notum* exists in the mouse and the human genome,[85,86] implying that a similar regulatory mechanism may exist in vertebrates.

## Conclusion

We have reviewed several families of modulator/antagonist for Wnt signaling. In vertebrates, sFRPs, WIF-1, Xcer, and Dkks form a network of Wnt modulators with distinct specificity and mechanisms. sFRPs, WIF-1 and Xcer belong to "Wnt-specific antagonists", which bind specific Wnt molecules and inhibit multiple Wnt signaling pathways initiated by these Wnt proteins. By contrast, Dkks appear to represent "pathway-specific antagonists", which bind to LRP5/6 and specifically inhibit LRP5/6-dependent Wnt/β-catenin signaling. In Drosophila, secreted Notum/Wingful protein functions as a modifying enzyme for Dally and Dally-like, two cell surface proteoglycans that are required for Wingless/Wnt signaling. These diverse Wnt modulators ensure the precise fine-tuning of different Wnt signaling pathways that are essential for animal development. It is quite likely that additional Wnt modulators and novel regulatory mechanisms are yet to discovered.

### *Acknowledgments*

Research in X. He's laboratory is supported by grants from US National Institutes of Health and Department of Defense. X.H. is a Pew Scholar in Biomedical Sciences, Klingenstein Fellow in Neuroscience, and W. M. Keck Foundation Distinguished Young Scholar in Medical Research.

## References

1. Rattner A, Hsieh JC, Smallwood PM et al. A family of secreted proteins contains homology to the cysteine-rich ligand-binding domain of frizzled receptors. Proc Natl Acad Sci USA 1997; 94:2859-2863.
2. Salic AN, Kroll KL, Evans LM, Kirschner MW. Sizzled: a secreted Xwnt8 antagonist expressed in the ventral marginal zone of Xenopus embryos. Development 1997; 124:4739-4748.
3. Pfeffer PL, De Robertis EM, Izpisua-Belmonte JC. Crescent, a novel chick gene encoding a Frizzled-like cysteine-rich domain, is expressed in anterior regions during early embryogenesis. Int J Dev Biol 1997; 41:449-458.
4. Hoang B, Moos M, Jr., Vukicevic S, Luyten FP. Primary structure and tissue distribution of FRZB, a novel protein related to Drosophila frizzled, suggest a role in skeletal morphogenesis. J Biol Chem 1996; 271:26131-26137.
5. Wang Y, Macke JP, Abella BS et al. A large family of putative transmembrane receptors homologous to the product of the Drosophila tissue polarity gene frizzled. J Biol Chem 1996; 271:4468-4476.
6. Bhanot P, Brink M, Samos CH et al. A new member of the frizzled family from Drosophila functions as a Wingless receptor. Nature 1996; 382:225-230.
7. Lin K, Wang S, Julius MA et al. The cysteine-rich frizzled domain of FrzB-1 is required and sufficient for modulation of Wnt signaling. Proc Natl Acad Sci USA 1997; 94:11196-11200.
8. Chong JM, Uren A, Rubin JS, Speicher DW. Disulfide bond assignments of secreted frizzled related protein-1 provide insights about frizzled homology and netrin modules. J Biol Chem 2001; 277:5134-5144.
9. Dann CE, Hsieh JC, Rattner A et al. Insights into Wnt binding and signalling from the structures of two Frizzled cysteine-rich domains. Nature 2001; 412:86-90.
10. Alcedo J, Ayzenzon M, Von Ohlen T et al. The Drosophila smoothened gene encodes a seven-pass membrane protein, a putative receptor for the hedgehog signal. Cell 1996; 86:221-232.
11. Yan W, Sheng N, Seto M et al. Corin, a mosaic transmembrane serine protease encoded by a novel cDNA from human heart. J Biol Chem 1999; 274:14926-14935.
12. Katoh M. Molecular cloning and characterization of MFRP, a novel gene encoding a membrane-type Frizzled-related protein. Biochem Biophys Res Commun 2001; 282:116-123.

13. Song L, Fricker LD. Cloning and expression of human carboxypeptidase Z, a novel metallocarboxypeptidase. J Biol Chem 1997; 272:10543-10550.
14. Xu YK, Nusse R. The Frizzled CRD domain is conserved in diverse proteins including several receptor tyrosine kinases. Curr Biol 1998; 8:R405-406.
15. Masiakowski P, Yancopoulos GD. The Wnt receptor CRD domain is also found in MuSK and related orphan receptor tyrosine kinases. Curr Biol 1998; 8:R407.
16. Saldanha J, Singh J, Mahadevan D. Identification of a Frizzled-like cysteine rich domain in the extracellular region of developmental receptor tyrosine kinases. Protein Sci 1998; 7:1632-1635.
17. Rehn M, Pihlajaniemi T, Hofmann KBucher P. The frizzled motif: in how many different protein families does it occur? Trends Biochem Sci 1998; 23:415-417.
18. Rehn M, Pihlajaniemi T. Identification of three N-terminal ends of type XVIII collagen chains and tissue-specific differences in the expression of the corresponding transcripts. The longest form contains a novel motif homologous to rat and Drosophila frizzled proteins. J Biol Chem 1995; 270:4705-4711.
19. Tamai K, Semenov M, Kato Y et al. LDL-receptor-related proteins in Wnt signal transduction. Nature 2000; 407:530-535.
20. Muskett FW, Frenkiel TA, Feeney J et al. High resolution structure of the N-terminal domain of tissue inhibitor of metalloproteinases-2 and characterization of its interaction site with matrix metalloproteinase-3. J Biol Chem 1998; 273:21736-21743.
21. Tuuttila A, Morgunova E, Bergmann U et al. Three-dimensional structure of human tissue inhibitor of metalloproteinases-2 at 2.1 A resolution. J Mol Biol 1998; 284:1133-1140.
22. Fernandez-Catalan C, Bode W, Huber R et al. Crystal structure of the complex formed by the membrane type 1-matrix metalloproteinase with the tissue inhibitor of metalloproteinases-2, the soluble progelatinase A receptor. Embo J 1998; 17:5238-5248.
23. Murzin AG. OB(oligonucleotide/oligosaccharide binding)-fold: common structural and functional solution for non-homologous sequences. Embo J 1993; 12:861-867.
24. Uren A, Reichsman F, Anest V et al. Secreted frizzled-related protein-1 binds directly to Wingless and is a biphasic modulator of Wnt signaling. J Biol Chem 2000; 275:4374-4382.
25. Finch PW, He X, Kelley MJ et al. Purification and molecular cloning of a secreted, Frizzled-related antagonist of Wnt action. Proc Natl Acad Sci USA 1997; 94:6770-6775.
26. Jiang Y, Goldberg ID, Shi YE. Complex roles of tissue inhibitors of metalloproteinases in cancer. Oncogene 2002; 28:2245-52
27. Banyai L, Patthy L. The NTR module: domains of netrins, secreted frizzled related proteins, and type I procollagen C-proteinase enhancer protein are homologous with tissue inhibitors of metalloproteases. Protein Sci 1999; 8:1636-1642.
28. Docherty AJ, Lyons A, Smith BJ et al. Sequence of human tissue inhibitor of metalloproteinases and its identity to erythroid-potentiating activity. Nature 1985; 318:66-69.
29. Roth W, Wild-Bode C, Platten M et al. Secreted Frizzled-related proteins inhibit motility and promote growth of human malignant glioma cells. Oncogene 2000; 19:4210-4220.
30. Pera EM, De Robertis EM. A direct screen for secreted proteins in Xenopus embryos identifies distinct activities for the Wnt antagonists Crescent and FrzB-1. Mech Dev 2000; 96:183-195.
31. Shibata M, Ono H, Hikasa H et al. Xenopus crescent encoding a Frizzled-like domain is expressed in the Spemann organizer and pronephros. Mech Dev 2000; 96:243-246.
32. Leyns L, Bouwmeester T, Kim SH et al. FrzB-1 is a secreted antagonist of Wnt signaling expressed in the Spemann organizer. Cell 1997; 88:747-756.
33. Wang S, Krinks M, Lin K et al. FrzB, a secreted protein expressed in the Spemann organizer, binds and inhibits Wnt-8. Cell 1997; 88:757-766.
34. Dennis S, Aikawa M, Szeto W et al. A secreted frizzled related protein, FrzA, selectively associates with Wnt-1 protein and regulates wnt-1 signaling. J Cell Sci 1999; 112:3815-3820.
35. Xu Q, D'Amore PA, Sokol SY. Functional and biochemical interactions of Wnts with FrzA, a secreted Wnt antagonist. Development 1998; 125:4767-4776.
36. Wang S, Krinks M, Moos M, Jr. FrzB-1, an antagonist of Wnt-1 and Wnt-8, does not block signaling by Wnts -3A, -5A, or -11. Biochem Biophys Res Commun 1997; 236:502-504.

37. Bafico A, Gazit A, Pramila T et al. Interaction of frizzled related protein (FRP) with Wnt ligands and the frizzled receptor suggests alternative mechanisms for FRP inhibition of Wnt signaling. J Biol Chem 1999; 274:16180-16187.
38. Jaspard B, Couffinhal T, Dufourcq P et al. Expression pattern of mouse sFRP-1 and mWnt-8 gene during heart morphogenesis. Mech Dev 2000; 90:263-267.
39. Bergwitz C, Wendlandt T, Kispert A, Brabant G. Wnts differentially regulate colony growth and differentiation of chondrogenic rat calvaria cells. Biochim Biophys Acta 2001; 1538:129-140.
40. Yoshino K, Rubin JS, Higinbotham KG et al. Secreted Frizzled-related proteins can regulate metanephric development. Mech Dev 2001; 102:45-55.
41. Stark K, Vainio S, Vassileva G, McMahon AP. Epithelial transformation of metanephric mesenchyme in the developing kidney regulated by Wnt-4. Nature 1994; 372:679-683.
42. Lescher B, Haenig B, Kispert A. sFRP-2 is a target of the Wnt-4 signaling pathway in the developing metanephric kidney. Dev Dyn 1998; 213:440-451.
43. Hsieh JC, Kodjabachian L, Rebbert ML et al. A new secreted protein that binds to Wnt proteins and inhibits their activities. Nature 1999; 398:431-436.
44. Katso RM, Russell RB, Ganesan TS. Functional analysis of H-Ryk, an atypical member of the receptor tyrosine kinase family. Mol Cell Biol 1999; 19:6427-6440.
45. Wang XC, Katso R, Butler R et al. H-RYK, an unusual receptor kinase: isolation and analysis of expression in ovarian cancer. Mol Med 1996; 2:189-203.
46. Hovens CM, Stacker SA, Andres AC et al. RYK, a receptor tyrosine kinase-related molecule with unusual kinase domain motifs. Proc Natl Acad Sci USA 1992; 89:11818-11822.
47. Jones FS, Jones PL. The tenascin family of ECM glycoproteins: structure, function, and regulation during embryonic development and tissue remodeling. Dev Dyn 2000; 218:235-59.
48. Bouwmeester T, Kim S, Sasai Y et al. Cerberus is a head-inducing secreted factor expressed in the anterior endoderm of Spemann's organizer. Nature 1996; 382:595-601.
49. Hsu DR, Economides AN, Wang X et al. The Xenopus dorsalizing factor Gremlin identifies a novel family of secreted proteins that antagonize BMP activities. Mol Cell 1998; 1:673-683.
50. Piccolo S, Agius E, Leyns L et al. The head inducer Cerberus is a multifunctional antagonist of Nodal, BMP and Wnt signals. Nature 1999; 397:707-710.
51. Belo JA, Bouwmeester T, Leyns L et al. Cerberus-like is a secreted factor with neutralizing activity expressed in the anterior primitive endoderm of the mouse gastrula. Mech Dev 1997; 68:45-57.
52. Biben C, Stanley E, Fabri L et al. Murine cerberus homologue mCer-1: a candidate anterior patterning molecule. Developmental Biology 1998; 194:135-151.
53. Pearce JJ, Penny G, Rossant J. A mouse cerberus/Dan-related gene family. Developmental Biology 1999; 209:98-110.
54. Meitinger T, Meindl A, Bork P et al. Molecular modelling of the Norrie disease protein predicts a cystine knot growth factor tertiary structure. Nat Genet 1993; 5:376-380.
55. McDonald NQ, Hendrickson WA. A structural superfamily of growth factors containing a cystine knot motif. Cell 1993; 73:421-424.
56. Glinka A, Wu W, Onichtchouk D et al. Head induction by simultaneous repression of Bmp and Wnt signalling in Xenopus. Nature 1997; 389:517-519.
57. Belo JA, Bachiller D, Agius E et al. Cerberus-like is a secreted BMP and nodal antagonist not essential for mouse development. Genesis 2000; 26:265-270.
58. Shawlot W, Deng JM, Behringer RR. Expression of the mouse cerberus-related gene, Cerr1, suggests a role in anterior neural induction and somitogenesis. Proc. Natl. Acad. Sci. USA 1998; 95:6198-6203.
59. Stanley EG, Biben C, Allison J et al. Targeted insertion of a lacZ reporter gene into the mouse Cerl locus reveals complex and dynamic expression during embryogenesis. Genesis: the Journal of Genetics & Development 2000; 26:259-264.
60. Simpson EH, Johnson DK, Hunsicker P et al. The mouse Cerl (Cerberus related or homologue) gene is not required for anterior pattern formation. Developmental Biology 1999; 213:202-206.
61. Shawlot W, Min Deng J, Wakamiya M, Behringer RR. The cerberus-related gene, Cerrl, is not essential for mouse head formation. Genesis: the Journal of Genetics & Development 2000; 26:253-258.

62. Zhu L, Marvin MJ, Gardiner A et al. Cerberus regulates left-right asymmetry of the embryonic head and heart. Current Biology 1999; 9:931-938.
63. Yokouchi Y, Vogan KJ, Pearse RV 2nd, Tabin CJ. Antagonistic signaling by Caronte, a novel Cerberus-related gene, establishes left-right asymmetric gene expression. Cell 1999; 98:573-583.
64. Rodriguez Esteban C, Capdevila J, Economides AN et al. The novel Cer-like protein Caronte mediates the establishment of embryonic left-right asymmetry. Nature 1999; 401:243-251.
65. Marvin MJ, Di Rocco G, Gardiner A et al. Inhibition of Wnt activity induces heart formation from posterior mesoderm. Genes & Development 2001; 15:316-327.
66. Schneider VA, Mercola M. Wnt antagonism initiates cardiogenesis in Xenopus laevis. Genes Dev 2001; 15:304-315.
67. Glinka A, Wu W, Delius H et al. Dickkopf-1 is a member of a new family of secreted proteins and functions in head induction. Nature 1998; 391:357-362.
68. Fedi P, Bafico A, Nieto Soria A et al. Isolation and biochemical characterization of the human Dkk-1 novel inhibitor of mammalian Wnt signaling. J Biol Chem 1999; 274:19465-19472.
69. Krupnik VE, Sharp JD, Jiang C et al. Functional and structural diversity of the human Dickkopf gene family. Gene 1999; 238:301-313.
70. Wu W, Glinka A, Delius H, Niehrs C. Mutual antagonism between dickkopf1 and dickkopf2 regulates Wnt/beta-catenin signalling. Current Biology 2000; 10:1611-1614.
71. Li L, Mao J, Sun L et al. Second cysteine-rich domain of Dickkopf-2 activates canonical Wnt signaling pathway via LRP-6 independently of dishevelled. Journal of Biological Chemistry 2002; 277:5977-5981.
72. Mao B, Wu W, Li Y et al. LDL-receptor-related protein 6 is a receptor for Dickkopf proteins. Nature 2001; 411:321-325.
73. Semenov MV, Tamai K, Brott BK et al. Head inducer Dickkopf-1 is a ligand for Wnt coreceptor LRP6. Curr Biol 2001; 11:951-961.
74. Bafico A, Liu G, Yaniv A et al. Novel mechanism of Wnt signalling inhibition mediated by Dickkopf-1 interaction with LRP6/Arrow. Nature Cell Biology 2001; 3:683-686.
75. Wehrli M, Dougan ST, Caldwell K et al. arrow encodes an Ldl-receptor-related protein essential for Wingless signalling. Nature 2000; 407:527-530.
76. Pinson KI, Brennan J, Monkley S et al. An LDL-receptor-related protein mediates Wnt signalling in mice. Nature 2000; 407:535-538.
77. Mao B, Wu W, Davidson G et al. Kremen proteins are Dickkopf receptors that regulate Wnt/ -catenin signalling. Nature 2002; 417:664-667.
78. Habas R, Kato Y, He X. Wnt/Frizzled activation of Rho regulates vertebrate gastrulation and requires a novel Formin homology protein Daam1. Cell 2001; 107:843-854.
79. Monaghan AP, Kioschis P, Wu W et al. Dickkopf genes are co-ordinately expressed in mesodermal lineages. Mechanisms of Development 1999; 87:45-56.
80. Chapman SC, Schubert FR, Schoenwolf GC, Lumsden A. Analysis of spatial and temporal gene expression patterns in blastula and gastrula stage chick embryos. Dev Biol 2002; 245:187-199.
81. Hashimoto H, Yabe T, Hirata T et al. Expression of the zinc finger gene fez-like in zebrafish forebrain. Mech Dev 2000; 97:191-195.
82. Shinya M, Eschbach C, Clark M et al. Zebrafish Dkk-1, induced by the pre-MBT Wnt signaling, is secreted from the prechordal plate and patterns the anterior neural plate. Mechanisms of Development 2000; 98:3-17.
83. Mukhopadhyay M, Shtrom S, Rodriguez-Esteban C et al. Dickkopf1 is required for embryonic head induction and limb morphogenesis in the mouse. Dev Cell 2001; 1:423-434.
84. Kazanskaya O, Glinka A, Niehrs C. The role of Xenopus dickkopf1 in prechordal plate specification and neural patterning. Development 2000; 127:4981-4992.
85. Giraldez AJ, Copley RR, Cohen SM. HSPG Modification by the Secreted Enzyme Notum Shapes the Wingless Morphogen Gradient. Dev Cell 2002; 2:667-676.
86. Gerlitz O, Basler K. Wingful, an extracellular feedback inhibitor of Wingless. Genes Dev 2002; 16:1055-1059.
87. Baeg GH, Lin X, Khare N et al. Heparan sulfate proteoglycans are critical for the organization of the extracellular distribution of Wingless. Development 2001; 128:87-94.

88. Lin X, Perrimon N. Dally cooperates with Drosophila Frizzled 2 to transduce Wingless signalling. Nature 1999; 400:281-284.
89. Tsuda M, Kamimura K, Nakato H et al. The cell-surface proteoglycan Dally regulates Wingless signalling in Drosophila. Nature 1999; 400:276-280.
90. Strigini M, Cohen SM. Wingless gradient formation in the Drosophila wing. Curr Biol 2000; 10:293-300.
91. Selleck SB. Genetic dissection of proteoglycan function in Drosophila and C. elegans. Semin Cell Dev Biol 2001; 12:127-134.
92. Perrimon N, Bernfield M. Specificities of heparan sulphate proteoglycans in developmental processes. Nature 2000; 404:725-728.
93. Topczewski J, Sepich DS, Myers DC et al. The zebrafish glypican knypek controls cell polarity during gastrulation movements of convergent extension. Dev Cell 2001; 1:251-264.
94. Kim AS, Lowenstein DH, Pleasure SJ. Wnt receptors and Wnt inhibitors are expressed in gradients in the developing telencephalon. Mech Dev 2001; 103:167-172.
95. Augustine C, Gunnersen J, Spirkoska V, Tan SS. Place- and time-dependent expression of mouse sFRP-1 during development of the cerebral neocortex. Mech Dev 2001; 109:395-397.
96. Esteve P, Morcillo J, Bovolenta P. Early and dynamic expression of csFRP-1 during chick embryo development. Mech Dev 2000; 97:217-221.
97. Hu E, Zhu Y, Fredrickson T et al. Tissue restricted expression of two human FrzBs in preadipocytes and pancreas. Biochem Biophys Res Commun 1998; 247:287-293.
98. Chang JT, Esumi N, Moore K et al. Cloning and characterization of a secreted frizzled-related protein that is expressed by the retinal pigment epithelium. Hum Mol Genet 1999; 8:575-583.
99. Lee CS, Buttitta LA, May NR et al. SHH-N upregulates sFRP-2 to mediate its competitive interaction with WNT1 and WNT4 in the somitic mesoderm. Development 2000; 127:109-118.
100. Kim AS, Anderson SA, Rubenstein JL et al. Pax-6 regulates expression of sFRP-2 and Wnt-7b in the developing CNS. J Neurosci 2001; 21:RC132.
101. Ladher RK, Church VL, Allen S et al. Cloning and expression of the Wnt antagonists sFRP-2 and FrzB during chick development. Dev Biol 2000; 218:183-198.
102. Terry K, Magan H, Baranski M, Burrus LW. sFRP-1 and sFRP-2 are expressed in overlapping and distinct domains during chick development. Mech Dev 2000; 97:177-182.
103. Hoang BH, Thomas JT, Abdul-Karim FW et al. Expression pattern of two Frizzled-related genes, FrzB-1 and sFRP-1, during mouse embryogenesis suggests a role for modulating action of Wnt family members. Dev Dyn 1998; 212:364-372.
104. Borello U, Coletta M, Tajbakhsh S et al. Transplacental delivery of the Wnt antagonist FrzB1 inhibits development of caudal paraxial mesoderm and skeletal myogenesis in mouse embryos. Development 1999; 126:4247-4255.
105. Sarkar L, Sharpe PT. Inhibition of Wnt signaling by exogenous MFrzB1 protein affects molar tooth size. J Dent Res 2000; 79:920-925.
106. Mayr T, Deutsch U, Kuhl M et al. Fritz: a secreted frizzled-related protein that inhibits Wnt activity. Mech Dev 1997; 63:109-125.
107. Baranski M, Berdougo E, Sandler JS et al. The dynamic expression pattern of FrzB-1 suggests multiple roles in chick development. Dev Biol 2000; 217:25-41.
108. Duprez D, Leyns L, Bonnin MA et al. Expression of FrzB-1 during chick development. Mech Dev 1999; 89:179-183.
109. Wada N, Kawakami Y, Ladher R et al. Involvement of FrzB-1 in mesenchymal condensation and cartilage differentiation in the chick limb bud. Int J Dev Biol 1999; 43:495-500.

# Chapter 3

# Wnt/Wingless Signaling in *Drosophila*

**Yacine Graba, Denise Aragnol, Ute Rothbächer and Jacques Pradel**

## Abstract

Wnt proteins form a group of secreted ligands that regulate a number of events in development and disease. In Drosophila, many of these processes are mediated through a canonical pathway, which is initiated by Wingless (Wg), the prototypic Wnt signal. Recent genetic and molecular studies have uncovered new aspects on how Wg is produced, received, transduced and converted into target gene control, and on how the Wg canonical pathway gains specificity to elicit distinct cellular responses. Another Wnt-activated pathway, initiated by DWnt-4, facilitates cell movement during ovarian morphogenesis. The cell motility pathway is distinct from both the canonical Wnt pathway and from the Planar Cell Polarity (PCP) pathway.

## Introduction

Wnt genes encode evolutionary conserved secreted glycoproteins that act as signaling molecules essential for a number of fundamental processes in development and disease. The *wg* gene, the *Drosophila* counterpart of vertebrate *Wnt-1*, was first identified as a hypomorphic mutation leading to wing deficient flies. Further characterization demonstrated that *wg* is involved in many developmental processes such as embryonic segment polarity and patterning, midgut morphogenesis, malpighian tubule organogenesis or limb development. The segment polarity phenotype was later instrumental for the identification of genes modifying the cellular or developmental response to Wg, thus of molecules required for transducing the signal, which resulted in the definition of a pathway specifically activated by Wg/Wnt-1 molecules, the so-called "canonical Wnt pathway" (see Chapter 1). The pathway includes reception by Frizzled transmembrane proteins, transduction through a cascade of cytoplasmic components, nuclear translocation of Armadillo (Arm), and control of Wg-responsive genes by a transcriptional effector Arm-dTCF complex (for a review see ref.1).

Unlike *wg*, the six additional *Wnt* genes existing in *Drosophila* were identified from approaches other than genetics (see Table 1). *DWnt-2* and *DWnt-3/5*, sequences homologous to vertebrate *Wnt-7* and to *Wnt-5*, were isolated through a PCR-based molecular screen. *DWnt-4*, which is very divergent from all other *Wnt* sequences, was discovered as a target gene of Hox proteins in the developing embryo. Finally, the last three *Wnt* genes, respectively related to vertebrates *Wnt-6*, *Wnt-8* and *Wnt-10*, were identified by genome wide sequencing. Out of the six genes, mutations are available for *DWnt-2* and *DWnt-4* only and genetics has provided evidence for a function of DWnt-2 in the determination of male-specific pigment cells in the developing gonad and for a function of *DWnt-4* in the control of cell movement during oogenesis.

*Wnt Signaling in Development*, edited by Michael Kühl.

*Table 1.* **Drosophila *Wnt genes have structural homologs in vertebrates***

| *Drosophila* genes | *wingless* | *DWnt-2* | *DWnt3/5* | *DWnt-4* | *DWnt-6* | *DWnt-8* | *DWnt-10* |
|---|---|---|---|---|---|---|---|
| Related vertebrate homologs | *Wnt-1* | *Wnt-7* | *Wnt-5* | *Wnt-9** <br> *Wnt-14* <br> *Wnt-15* | *Wnt-6* | *Wnt-8* | *Wnt-10* |

Vertebrate structural orthologs have been identified for all *Drosophila Wnt* genes, but for *DWnt-4. DWnt-4, Wnt-9, Wnt-14* and *Wnt-15* share unusual cysteine spacing pattern, suggesting these related genes arose from an ancestral sequence that diverged from the remainder of the Wnt family prior the divergence of deuterosomes and arthropods.[57,58] *Only isolated from hagfisch and shark.

Most of the review focuses on Wg signaling. We shall first briefly present the molecules that constitute the core of the canonical Wg pathway and emphasize recent findings that highlight the mechanisms operating at some steps. We will next discuss how the canonical pathway gains specificity to elicit distinct cellular responses and then will revisit the question of Wg functioning as an instructive morphogen. Finally, we will comment on the signaling cascade activated by DWnt-4, as it constitutes at present the only pathway currently proposed in *Drosophila* which is distinct from both the canonical Wnt pathway and from the PCP pathway.

## The Wg Pathway

The prototypic Wnt gene *wg* is required in a variety of developmental events. Most of the genes coding for intermediates in the induced signaling pathway were then screened on the criterion that their loss of function phenotypes shared traits with the *wg* phenotype. Combination of genetic and molecular studies attributed these molecules with particular functions in the production, reception, transduction or transcriptional translation of the Wg signal. Here we shall focus on newly discovered components or mechanisms operating in the cascade and shall discuss the implications for the current view of the canonical Wg pathway.

Correct localization of extracellular signaling molecules is often decisive to bring about normal development of animals. Several levels of intracellular and extracellular control accomplish the targeted production and release of Wg. Some of the molecular players are mentioned here and more mechanistic aspects, especially how to generate a Wg gradient, are discussed in a separate section.

It is well established that *wg* transcripts are asymmetrically localized at the apical side of embryonic cells.[2] Only recently, however, it was shown that this localization is important for the function. Minimal sequences that direct this localization could be identified in the *wg* 3'UTR and their manipulation in living embryos demonstrated that apical localization of *wg* mRNA is essential for signaling. Thus, the segment polarity phenotype of *wg* mutant embryos, loss of naked cuticle in each segment, can be rescued with apically but not basally localizing *wg* transcripts.[3] This localization requires an active transport of *wg* transcripts along cytoskeleton elements towards the apical end of cells. Intracytoplasmic transport proceeds along a microtubule network, using the motor protein ATPase dynein as a carrier, instead of myosin motors and actin filaments.[4] These localization events therefore result in the synthesis of Wg protein in close proximity to where it will be secreted.

Processing and export of Wg from the producing cells is a rather complex process, not fully understood yet. Known pieces in this puzzle are that Wg processing engages Porcupine, a multipass transmembrane protein of the endoplasmic reticulum,[5] and that domains in Wg critical for secretion are identified.[6] Once secreted, Wg establishes strong molecular interactions with heparan sulphate proteoglycans (HSPGs) at the cell surface and extracellular matrix. This is important for both signaling and diffusion of Wg away from secreting cells. Interestingly, some of the HSPG mutations, like *sugarless* and *sulfateless* affect both Wg and FGF signaling, while others, like *tout-velu* affect Hedgehog (Hh) signaling but not Wg and FGF, and *dally* affects both Wg and Decapentaplegic (Dpp) signaling but not in all tissues. This indicates that HSPGs bind extracellular ligands of different pathways in a developmentally regulated fashion (for a review see ref. 7). New studies of membrane dynamics in *Drosophila* imaginal discs have resolved the apparent paradox that molecules such as Wg and Hh bind tightly to membranes and travel long distances. Greco et al[8] described a cell biology mechanism that disperses membranes vesicles, which they named argosomes, over a large distance through the disc epithelium and showed that Wg colocalizes with argosomes derived from Wg-producing cells. This led the authors to suggest that a primary function of the interaction with HSPGs is to allow the incorporation of Wg into argosomes, which would then serve as vehicles for the spread of Wg. Another important advance regards the routing modulation of Wg signaling. Dubois et al.[9] showed that directional clues for Wg to signal originate from a lysosomal protein degradation system operating posterior to Wg-producing cells in each parasegment of the embryo and from spatially regulated receptor binding and endocytosis on the other side.

The first Wg receptors identified were dFz-2 and Fz, two members of the Frizzled family, seven pass transmembrane molecules with an extracellular cysteine-rich domain shown to bind Wnts. Mutant phenotypes indicate that Fz acts in PCP (see chapter 6), through the Jun kinase (JNK) pathway and as a receptor for Wg using the canonical pathway, whereas dFz-2 is restricted to Wg signaling.[10] While redundant roles for the two receptors were first discussed upon genetic experiments, dFz-2 was finally proposed to be a higher affinity receptor for Wg than Fz.[11,12] The respective roles of Fz and dFz-2 in Wg reception still is, however, a matter of debate, especially whether they could participate in different receptor complexes to mediate the distinct cellular responses to Wg. Particularly interesting in this context is the discovery of Arrow, a co-receptor used by dFz-2 to activate the pathway.[13] *arrow* mutants develop a phenotype identical to *wg* loss of function but do not show any PCP phenotype, which clearly links its function to the canonical signaling. Arrow is a single pass transmembrane protein which harbors in its extracellular domain four EGF-like repeats and three LDL-like receptor repeats, features shared with its mammalian LRP homologs. Consistent with a function of Arrow in Wg reception, epistasis experiments showed that it acts downstream of Wg and upstream of Dishevelled (Dsh), the first intracellular component of the transduction pathway. Direct evidence for a LRP/Fz co-receptor complex, comes from vertebrates where the extracellular domains of LRP6 and Fz interact in a Wnt-dependent fashion (for a review see ref. 14).

Dsh is the most immediate intracellular mediator of the canonical pathway. Its role is to antagonize a multiprotein complex that leads to Arm degradation in the absence of signaling. Although it is well known that Dsh transduces the Wg signal to Axin and the serine threonine kinase Shaggy/Zeste white3 (Zw3), there are still molecular and mechanistic gaps to be filled. Especially the signal relay to Dsh is enigmatic, since Dsh does not directly bind Frizzleds.

Dsh was first identified and named for its loss of function phenotype that revealed a role in PCP. A wealth of information is currently being generated about this aspect of Dsh function (see Chapter 6), and also for this pathway Dsh is most proximal to the receptor so far. It is moreover required for signaling by DWnt-4 (see below). Dsh thus appears to function as a dispatching molecule that directs Fz-mediated signals towards distinct transduction routes. Three evolutionary conserved domains are characteristic of Dsh, the DIX, PDZ and DEP domains. Most interestingly, a differential domain usage guides the function in Wg signaling versus the PCP pathway, whereby DIX signals to the canonical pathway, DEP to PCP, and

PDZ possibly to both.[15-18] More molecules are to be discovered around Dsh to understand in mechanistic terms how the selection of the appropriate pathway is achieved.

Two recently identified genes, *arrow* and *PAR-1*, likely play a role in this process. The presence of Arrow in the receptor complex (see above) commits Dsh to specify the Wg pathway, although whether Arrow directly contacts Dsh or acts through the recruitment of unknown cytoplasmic proteins remains to be established. *PAR-1* is a Dsh-associated kinase that in cell culture assays potentiates signaling by Wg and blocks the JNK pathway.[19] Gain of function and gene inactivation by RNA interference in fly consistently involved *PAR-1* in the naked/denticled cell fate choice mediated by Wg in the embryonic ectoderm. These results suggest that Dsh phosphorylation by *PAR-1* should be the molecular switch committing the Wg pathway. Further studies are needed, however, to fix the point since Dsh phosphorylation also correlates with the Dsh PCP function in vertebrates[17,20] and *Drosophila*.[18] *PAR-1* mutant isolation and epistasis analysis of functional interactions with mutations in *dsh* and *arrow* would thus unravel how a Fz activated receptor communicates with Dsh to ultimately activate either the Wg or the PCP pathway.

Naked (Nkd) is another recently identified factor interacting with Dsh through the PDZ domain.[21] Nkd acts as a regulatory component in a feedback loop that restrains signaling in cells distant from the Wg source by affecting the transduction step between Dsh and Zw3 in a cell-autonomous manner. Although both Zw3 and Nkd kinases antagonize the pathway, their roles are clearly distinct. Zw3, which phosphorylates and causes degradation of Arm in the absence of Wg, becomes inactivated upon Wg reception. Nkd is induced in cells receiving Wg signal and functions directly through Dsh to limit its activity by setting the levels of Zw3 activity and therefore of Arm stabilization and downstream gene activation. Whether Nkd could play in PCP also was suggested, but not proved, by overexpression studies in fly and in vertebrate cell lines where the JNK pathway seems to be activated.[22] Nkd might thus serve as a molecular switch converse to PAR-1, to limit Wg signaling and possibly activate PCP.

Armadillo, β-catenin in vertebrates, is a central player of the canonical Wnt pathway. This multifunctional protein has a pivotal role since, apart from its role as Wg transcriptional effector, it is involved in cadherin-mediated cell adhesion as a component of adherens junctions. In the absence of Wg signaling, the cytoplasmic pool of Arm is maintained at a low level following a process that depends on the formation of a multimeric complex composed of Zw3 and of the scaffolding proteins Axin and adenomatous polyposis coli (APC). This so-called "destruction complex" acts in two steps. Zw3 first phosphorylates critical sites in the N-terminal part of Arm which is then targeted for degradation by the ubiquitin-dependent proteasome. Wnt-induced inactivation of Zw3 prevents phosphorylation and subsequent proteosomal degradation, which stabilizes Arm in the cytoplasm in a cadherin free form. As a consequence of increased accumulation in the cytoplasm, Arm becomes translocated into the nucleus where it associates with the sequence-specific DNA binding protein dTCF/Pangolin to regulate Wg-responsive gene activity.[23] The sequence of cytoplasmic events in the canonical Wg/Wnt-1 pathway is solidly pictured and mechanistically well understood. Nuclear events, especially how nuclear import of Arm is controlled and how it operates in transcriptional regulation remain, however, far less well elucidated.

The ability of Arm to move back and forth between the nucleus and the cytoplasm is known for a long time.[24] Arm/β-catenin does not contain a nuclear localization sequence but shares homology with the importin/karyopherin family of transport receptors and was shown to bind directly to the nuclear pore machinery, consistent with an intrinsic importin-like activity required for nuclear import. Moreover, in vitro assays using permeabilized cells showed that the cytosol contains inhibitory activities anchoring β-catenin outside the nucleus, which suggests that its import may be regulated by upstream events in the signaling pathway.[25,26] This proposal has received strong support from recent findings in *Drosophila*: Axin titration to the cell membrane, or removal of *axin* in germ line clones, forces Arm into the nucleus, clearly demonstrating that Axin acts as a cytoplasmic anchor for Arm; dTCF was proposed to fulfill the converse function of anchoring Arm in the nucleus, based on the observation that the overexpression of a dominant negative form dTCF appears to block the nuclear accumulation of Arm which is observed when cytoplasmic Axin is depleted.[27] Although this observation clearly indicates that a nuclear anchoring system does exist to prevent export, it remains debatable whether dTCF may serve itself as an anchor, since Arm variants defective in dTCF binding still accumulate in the nucleus.[28] Arm nuclear retention could rather depend on interaction with other factor(s) involved in the formation of chromatin-bound Arm-dTCF transregulatory complexes (see below). As Arm/β-catenin can dynamically cross nuclear membrane pores in either directions,[29] a third piece in the mechanism was missing concerning the export process. Two recent reports showed that the APC nucleo-cytoplasmic protein contains a nuclear export signal, which is required for efficient export from the nucleus.[30,31] The emerging view therefore is that Arm/β-catenin subcellular localization is controlled by a combination of several mechanisms, Zw3-induced proteosomal degradation, cytoplasmic and nuclear anchoring as well as APC-mediated active export.

Another interesting emerging theme concerns the role of chromatin structural changes in the modulation of Wg/Wnt-1 responsive gene activation. In the absence of Wg signaling, dTCF recruits the corepressor Groucho (Gro) to maintain an inactive state of target genes.[32] Gro has been shown to repress transcription by directly interacting with histones and the histone deacetylase (HDAC) Rpd3.[33] As reversible acetylation of histone tails plays a major role in chromatin remodeling, with the general thought that HDACs act as transcriptional corepressors and histone acetyltransferases (HAT) as coactivators,[34] this connection indicates that the regulation of chromatin structure is important for dTCF target gene control. A recent study moreover demonstrated a role for the Osa/Brahma chromatin remodeling complex (the SWI/SNF complex) in this mechanism.[35] Loss of function of *osa*, as well as that of *brahma* and *moira*, induces ectopic expression of Wg targets; conversely, *osa* overexpression represses endogenous transcription of the same genes. Osa, Brahma and Moira belong to the Trithorax group of proteins that act together with Polycomb group proteins in epigenetic mechanisms of transcriptional control of a variety of genes during development.[36] The same study also revealed that *osa* genetically interacts with *gro* and *rpd3*, leading to the proposal that the Osa/Brahma complex functions in a larger complex containing dTCF, Gro and Rpd3 to silence Wg target genes. In this silencing process, an intriguing role has been attributed to the CBP HAT that would maintain dTCF in an acetylated state, preventing interaction with Arm.[37]

How does the nuclear accumulation of Arm, following cell stimulation by Wg, then relieve silencing? Arm/β-catenin was recently found to interact physically with both the SWI/SNF component Brg1 and the CBP HAT and evidence that these interactions are functional in vivo was provided from cell culture assays[38] and genetic interaction studies in fly.[39] Arm would thus displace Gro and Rpd3 from the silencing complex, redirect SWI/SNF and dTCF towards the formation of an activating complex able, by recruiting the HAT activity of CBP, to remodel the chromatin structure of target gene promoters into an open conformation accessible to the basal transcription machinery. Other chromatin proteins likely participate in this dynamic,[40] among which the RuvB-like DNA helicases Pontin52 and Reptin52 have been shown to exert an additional control level in the canonical pathway.[41] These two proteins, found in several chromatin modifying and remodeling complexes from yeast to mammals,[42,43] physically interact with Arm/β-catenin and with the TATA box binding protein TBP. Genetic experiments revealed opposite functions for the encoding genes, *pontin* producing a coactivator and *reptin* a corepressor of Arm. It has thus been proposed that the relative levels between the two proteins give rise to complexes that either promote or repress Wg target gene transcription.[41] The current view emerging from these studies therefore is that nuclear Arm recruits in multisubunit complex(es) a number of chromatin proteins that are required for the various aspects of chromatin dynamics related to transcriptional regulation, including histone tail modification, local relaxation, DNA unwinding, connection to transcriptional machinery and transcription initiation.

## Specificity in Wg Signaling

*Wingless* is involved in a number of different developmental processes. It is obvious that a number of interfering regulatory inputs cooperate with Wg signaling in the control of these processes. Relatively few mechanisms have been described, however, as to how stage or tissue specificity in the cellular response to Wg is achieved. We have selected below a few examples illustrating that distinct mechanisms can confer specificity.

### *Intrinsic Signaling Potentials of Wg and Distinct Receptor Machineries*

Wg specifies the naked cell fate in the posterior half of the embryonic segment and distinct denticled cell types in the anterior half. The mutant alleles $wg^{PE2}$ and $wg^{PE4}$ affect both phenotypes but in a different manner: $wg^{PE2}$, a missense mutation in the C-terminus of the protein, impairs the specification of naked cuticle but still generates the different denticled cells; conversely, $wg^{PE4}$, which produces a truncated protein missing the C-terminal half, specifies the naked fate but no denticle diversity. Thus, the two Wg functions are genetically separable and are attributable to distinct signaling activities of the protein variants,[44] indicating that intrinsic determinants in the Wg protein provide information for the different cell fate decisions. How is this information read by receiving cells? As mentioned above, Wg acts through the receptors Fz and dFz-2. A recent report showed that overexpression of Fz-2, but not Fz, is sufficient to rescue the inability of $Wg^{PE2}$ to produce naked cuticle, consistent with the idea that distinct ligand receptor complexes might be responsible for distinct Wg functions.[11] Further studies are required, however, to understand how different receptor complexes eventually use the same pathway to perform distinct cellular outputs.

### *Temporal Specification of Wg Signaling by the Nuclear Proteins Lines (Lin) and Teashirt (Tsh)*

During embryogenesis, early Wg signaling is essential for the definition and maintenance of parasegmental (PS) boundaries. Later on, by six hours of development, Wg is no longer required for this function but rather for cell fate specification in the epidermis, like alternation of naked and denticled ventral cuticle and stereotyped dorsal hair pattern. *lin* deficient embryos specifically lack hairs on the dorsal cuticle, a phenotype that results from a selective loss of late Wg signaling. Furthermore, the protein is nuclear and participates in the control of Wg target genes, *wg* itself and *rhomboid* (rho). Lin was therefore proposed to be a stage- and region-specific nuclear modulator of Wg signaling acting on late dorsal cell fate specification.[45] A somewhat similar function has been attributed to the zinc finger protein Tsh, but in the ventral epidermis and in the trunk only. As Lin, *tsh* is required for efficient Wg signaling late but not early in embryogenesis, since its loss of function results in an excess of denticled cells and its overexpression in naked cuticle, in a process dependent on Arm stabilization.[46]

Interesting from another point of view is that Lin and Tsh do not enter the nucleus when early signaling by Wg is lost. This indicates that the early Wg function is important not only for the definition of PS boundaries, but also for the induction of nuclear events occurring later during epidermal cell differentiation. The way Wg signaling controls Lin and Tsh nuclear translocation, and how these two nuclear proteins participate in the control of Arm responsive genes remains to be established. In the case of Tsh, there is some evidence supporting a direct involvement in the Wg pathway since the protein first becomes phosphorylated in response to Wg signaling and second physically interacts with Arm to enter the nucleus.[47]

### *Cooperation of Wg and Other Signaling Pathways in Target Gene Control*

Many developmental processes require simultaneous inputs from distinct signaling pathways. In mechanistic terms, one possibility is that transcriptional effectors from various pathways could cooperate in the transcriptional control of common target genes. Recent literature has provided clear evidence that Wg and other signaling pathways interact in such a manner and that these interactions generate reciprocal signaling specificities.

Wg signaling directly controls *Ubx* expression in PS7 of the embryonic visceral mesoderm (VM), via a specific enhancer that responds to the nuclear effectors Arm and dTCF.[48] Wg however does not efficiently induce this enhancer in the absence of Dpp signaling, and its activity is not only lower in PS7 but also occurs in a broader domain. The functional dissection of the enhancer led to the conclusion that it is directly controlled by the binding of dTCF and of the Dpp nuclear effectors Mad and Creb at critical sites. Therefore, the level and the spatio-temporal specificity of *Ubx* enhancer activation by Wg need the cooperation of the spatially restricted Dpp signaling pathway.

At least two other molecular studies showed that Wg signaling specificity results from the convergence with other signaling inputs at the level of target gene regulation. The first one concerns the restriction *stripe* (sr) gene in the embryonic ectoderm to determine correct muscle attachment sites.[49] The accurate expression of *sr* depends of Wg and Hh signaling, both acting directly through their transcriptional effectors on a *sr* enhancer. The Hh pathway activates this enhancer on both sides of the segment boundary while signaling by Wg imposes repression in anterior cells, which results in a restriction of *sr* transcription to cells lying in the posterior compartment. The second study concerns the restriction of *even skipped* (*eve*) expression to muscle and heart precursor cells, that depends on transcriptional effectors of Wg, Ras/MAPK and Dpp signaling pathways acting in concert with the tissue specific transcription factors Twist and Tinman on a cis-regulatory element of *eve*.[50]

### *Activation versus Repression of Target Gene Transcription*

A general question around Wg signaling specificity comes from the observation that it can result in gene activation and repression as well. The current model, whereby Wg signal transduction switches dTCF from a repressor into an activator of transcription, accounts for transcriptional stimulation only, which is indeed observed for most of the targets identified so far. However, three genes do not obey the rule, *rho, shaven baby* (*svb*) and *sr* that are repressed instead of activated by Wg. As there is no evidence for a direct control of *svb* and *rho* by Wg nuclear effectors, it is possible that the down-regulation of these two downstream genes results from the activation of a repressor. Such a mechanism has been proposed for the repression of the *Ubx* VM enhancer when high levels of Wg signaling are artificially provided.[48] In this case, the repression, which proceeds independent of dTCF, is thought to occur indirectly via the induction of Tsh that would in turn directly silence the *Ubx* enhancer. In this view, repression by Wg signaling might therefore be a secondary effect of direct target gene activation. This twist however does not apply to the regulation of *sr*, since its promoter contains two functional dTCF binding sites that mediate Wg-instructed transcriptional silencing.[49] Further characterization of elements acting in trans and in cis is needed to understand the mechanism by which *sr* is repressed by dTCF in Wg receiving cells.

## Does Wg Always Act as a Morphogen?

The concept of morphogen has long held attraction for developmental biologists because a single event, emission of a diffusible molecule from a source, provides parsimony for complex pattern formation. As a minimal definition, an extracellular morphogen is a molecule that can spread out from a localized source and directly instruct cell identities in a concentration-dependent manner. For secreted signaling proteins to fulfill these criteria implies the formation of a concentration gradient, a direct action at distance instead of an action by relay of a secondary signal and the commitment of responsive cells into distinct developmental fates.

It is now well established that Wg acts as a morphogen in the wing imaginal disc.[51,52] It is secreted from a narrow stripe of cells at the dorsoventral compartment boundary, forms a basally located gradient and acts in a long range manner to induce a graded expression of target genes such as *achaete*, *Distalless* and *vestigial*. Clonal analyses of loss and gain of function for Wg signaling moreover showed that target gene activation proceeds in direct response to Wg. New results emerge on the formation and maintenance of the gradient. Two mechanisms have been recently proposed. First, once secreted Wg spreads by extracellular diffusion along cell membranes through stabilizing interactions with dFz-2, which seems to adopt a graded distribution opposite to the Wg gradient and with components of the extracellular matrix such as the HSPG Dally-like.[53,54] The function of HSPGs, and possibly of dFz-2, would thus be to shape the gradient by limiting the diffusion of extracellular Wg. The second mechanism proposes a vesicle-mediated transport based on repeated cycles of endocytosis and secretion. This is supported by the recent demonstration that Wg is distributed in a gradient both intracellularly and extracellularly and is associated to endosomes/argosomes in cells distant from the Wg producing source.[8] One function for HSPGs would thus be to allow Wg incorporation into argosomes and release at the cell membrane. The two mechanisms are not mutually exclusive, however, and the role of HSPGs might be twofold, first in shaping the gradient, allowing a transcytosis-like transport throughout the disc epithelium and second in a progressively decreasing release of extracellular Wg and stabilization at the membrane of receiving cells.

Whether Wg acts as a morphogen in other developmental contexts is far from clear. In the embryonic ectoderm, where its function has been extensively studied, Wg is produced from posterior most cells in each parasegment, moves on both sides into adjacent cells and then adopts a graded distribution in anterior cells only. Significant progress was made to understand how the gradient becomes asymmetrical by mid-embryogenesis through a combination of differential degradation and regulated endoctosis,[9] which correlates with the asymmetrical instructive role played by Wg signaling in cell fate determination. Although it fulfils the criteria of a localized source, gradient formation and action at a distance, Wg however disobeys other major rules expected of a morphogen. First, there is no evidence for concentration-dependent signaling activity. Instead, a flat field of Wg protein provided by a thermo-inducible *HS-wg* transgene can considerably rescue the epidermal phenotype of *wg* deficient embryos.[55] Second, experiments using a membrane-tethered form of Wg showed that ligand transport mediated by cell movement is sufficient to account for the normal range of Wg, indicating that restricted diffusion is dispensable for embryonic ectoderm patterning.[56]

## DWnt-4 Controls Cell Movement through a Unique Signaling Pathway

Seven *Wnt* genes exist in *Drosophila*. Mouse structural orthologs have been identified for all of them, but for *DWnt-4* (Table 1). Functional data are available for only three, *wg*, *DWnt-2* and *DWnt-4*. Nothing but some expression data[59] is currently known about the three *Wnt* genes identified from *Drosophila* genome sequencing, *DWnt-6*, *DWnt-8* and *DWnt-10*. DWnt-3 is a secreted protein that accumulates in the embryonic nerve cord and whose ectopic expression leads to specific disruption of the commissural axon tracts of the central nervous system.[60] *DWnt-2* encodes a signal required for the determination of male-specific pigment cells in the developing gonad and should be responsible for this sexually dimorphic trait.[61] The signal transduction mechanism used by DWnt-2 has not been characterized.

*DWnt-4* has received more attention. *DWnt-4* and *wg* map to the same locus and show closely related developmental expression patterns, suggesting, despite of strong divergence in terms of coding sequence and of intron/exon structure, they have been maintained in a close physical linkage during evolution because of shared cis-regulatory elements.[61] When provided ectopically, DWnt-4 can have antagonistic, similar or distinct functions than Wg, depending on the developmental context: it antagonizes late embryonic Wg signaling in the *Drosophila* epidermis and blocks Wg-induced body axis duplication in *Xenopus*;[62] *wg* and *DWnt-4* induce different phenotypes in the dorsal embryonic epidermis;[63] the two genes elicit similar cellular responses during imaginal development.[64] These apparently conflicting results could be explained by invoking functional redundancy or competition of the Wnt ligands, or interfering interaction between overproduced DWnt-4 and the Wg reception machinery. It has been recently demonstrated that DWnt-4 indeed signals through the Fz-2 receptor.[65]

The message of the Cohen study[65] is actually of far broader interest in the Wnt field than only providing an explanation to the various phenotypes induced by ectopic DWnt-4. The message is that DWnt-4 facilitates cell movement through a unique signaling mechanism that is distinct from the canonical or planar polarity pathways. During vertebrate gastrulation Wnt signaling regulates lamellipodia behavior through Dsh[20] and controls convergent extension and cell movements.[66,67] The underlying cellular and molecular mechanisms are poorly defined, however. Cohen and colleagues have generated mutations in *DWnt-4* and shown that it facilitates cell movement during ovarian morphogenesis through a signaling mechanism that results in Focal Adhesion Kinase accumulation. This kinase is required for the formation and disassembly of focal adhesion complexes, which is critical in the control of cell movement.[68] Fz-2 is the primary receptor for DWnt-4-mediated cell movement. However, the downstream effector of canonical signaling, dTCF, is not required. A mutation in Dsh that specifically disrupts planar polarity but transduces canonical signals[69] does disrupt cell movement, but the planar polarity mutants *frizzled, prickled, inturned,* and fuzzy do not.[70] The signaling pathway used by DWnt-4 therefore requires Fz-2 and Dsh but is neither the canonical Wnt pathway nor the planar polarity pathway.

## References


1. Wodarz A, Nusse R. Mechanisms of Wnt signaling in development. Ann Rev Cell Dev Biol 1998; 14:59-88.
2. Baker NE. Molecular cloning of sequences from wingless, a segment polarity gene in Drosophila: The spacial distribution of a transcript in embryos. EMBO J 1987; 6:1765-1773.
3. Simmonds AJ, dos Santos G, Livne-Bar I et al. Apical localization of wingless transcripts is required for wingless signaling. Cell 2001; 105:197-207.
4. Wilkie GS, Davis I. Drosophila wingless and pair-rule transcripts localize apically by dynein-mediated transport of RNA particles. Cell 2001; 105:209-219.
5. Kadowaki T, Wilder E, Klingensmith J et al. The segment polarity gene porcupine encodes a putative multitransmembrane protein involved in Wingless processing, Genes & Dev 1996; 10:3116-3128.
6. Dierick HA, Bejsovec A. Functional analysis of Wingless reveals a link between intercellular ligand transport and dorsal-cell-specific signaling. Development 1998; 125:4729-4738.
7. Baeg GH, Perrimon N. Functional binding of secreted molecules to heparan sulfate proteoglycans in Drosophila. Curr Opin Cell Biol 2000; 12:575-580.

8. Greco V, Hannus M, Eaton S. Argosomes: a potential vehicle for the spread of morphogens through epithelia. Cell 2001; 106:633-645.
9. Dubois L, Lecourtois M, Alexandre C et al. Regulated endocytic routing modulates wingless signaling in Drosophila embryos. Cell 2001; 105:613-624.
10. Chen CM, Struhl G. Wingless transduction by the Frizzled and Frizzled2 proteins of Drosophila. Development 1999; 126:5441-5452.
11. Moline MM, Dierick HA, Southern C et al. Non-equivalent roles of Drosophila Frizzled and Dfrizzled2 in embryonic wingless signal transduction. Curr Biol 2000; 10:1127-1130.
12. Rulifson EJ, Wu CH, Nusse R. Pathway specificity by the bifunctional receptor frizzled is determined by affinity for wingless. Mol Cell 2000; 6:117-126.
13. Wehrli M, Dougan ST, Caldwell K et al. arrow encodes an LDL-receptor-related protein essential for Wingless signaling. Nature 2000; 407:527-530.
14. Pandur P, Kühl M. An arrow for Wingless to take-off. BioEssays 2001; 23:207-210.
15. Axelrod JD, JR Miller. Differential recruitment of Dishevelled provides signaling specificity in the planar cell polarity and Wingless signaling pathways. Genes & Dev 1998; 12:2610-2622.
16. Boutros M, Paricio N, Strutt DI et al. Dishevelled activates JNK and discriminates between JNK pathways in planar polarity and wingless signaling. Cell 1998; 94:109-118.
17. Rothbacher U, Laurent MN, Deardorff MA et al. Dishevelled phosphorylation, subcellular localization and multimerization regulate its role in early embryogenesis. EMBO J 2000; 19:1010-1022.
18. Axelrod JD. Unipolar membrane association of Dishevelled mediates Frizzled planar cell polarity signaling. Genes & Dev 2001; 15:1182-1187.
19. Sun TQ, Lu B, Feng JJ et al. PAR-1 is a Dishevelled-associated kinase and a positive regulator of Wnt signaling. Nat Cell Biol 2001; 3:628-636.
20. Wallingford JB, Rowning BA, Vogeli KM et al. Dishevelled controls cell polarity during Xenopus gastrulation. Nature 2000; 405:81-85.
21. Rousset R, Mack JA, Wharton KA et al. Naked cuticle targets dishevelled to antagonize Wnt signal transduction. Genes & Dev 2001; 15:658-671.
22. Yan D, Wallingford J, Sun TQ et al. Cell autonomous regulation of multiple Dishevelled-dependent pathways by mammalian Nkd, Proc Natl Acad Sci 2001; 98:3802-3807.
23. Van de Wetering M, Cavallo R, Dooijes D et al. Armadillo coactivates transcription driven by the product of the Drosophila segment polarity gene dTCF. Cell 1997; 88:789-799.
24. Peifer M, Sweeton D, Casey M et al. Wingless signal and Zeste-white 3 kinase trigger opposing changes in the intracellular distribution of Armadillo. Development 1994; 120:369-380.
25. Fagotto F, Glück U, Gumbiner B. Nuclear localization signal-independent and importin/karyopherin-independent nuclear import of β-catenin. Curr Biol 1998; 8:181-190.
26. Salic A, Lee E, Mayer L et al. Control of β-catenin stability: reconstitution of the cytoplasmic steps of the wnt pathway in Xenopus egg extracts. Mol Cell 2000; 5:523-532.
27. Tolwinski NS, Wieschauss E. Armadillo nuclear import is regulated by cytoplasmic anchor Axin and nuclear anchor dTCF/Pan. Development 2001; 128:2107-2117.
28. Orsulic S, Peifer M. An in vivo structure-function study of armadillo, the _-catenin homologue, reveals both separate and overlapping regions of the protein required for cell adhesion and for wingless signaling. J Cell Biol 1996; 134:1283-1300.
29. Yokoya F, Imamoto N, Tachibana T et al. β-catenin can be transported into the nucleus in a Ran-unassisted manner. Mol Biol Cell 1999; 10:1119-1131.
30. Henderson BR. Nuclear-cytoplasmic shuttling of APC regulates β-catenin subcellular localization and turnover. Nat Cell Biol 2000; 2:653-660.
31. Rosin-Arbesfeld R, Townsley F, Bienz M. The APC tumour suppressor has a nuclear export function. Nature 2000; 406:1009-1012.
32. Cavallo RA, Cox RT, Moline MM et al. Drosophila Tcf and Groucho interact to repress Wingless signal-ling activity. Nature 1998; 395:604–608.
33. Chen G, Fernandez J, Mische S et al. A functional interaction between the histone deacetylase Rpd3 and the corepressor Groucho in Drosophila development. Genes & Dev. 1999; 13:2218-2230.
34. Turner BM. Histone acetylation and an epigenetic code. BioEssays 2000; 2:836-845.
35. Collins RT, Treisman JE. Osa-containing Brahma chromatin remodeling complexes are required for the repression of wingless target genes. Genes & Dev 2000; 14:3140-3152.

36. Pirrotta V. Polycombing the Genome: PcG, trxG, and Chromatin Silencing. Cell 1998; 93:333-336.
37. Waltzer L, Bienz M. Drosophila CBP represses the transcription factor TCF to antagonize Wingless signaling. Nature 1998; 395:521-525.
38. Hecht A, Vleminckx K, Stemmler MP et al. The p300/CBP acetyltransferases function as transcriptional coactivators of β-catenin in vertebrates. EMBO J 2000; 19:1839–1850.
39. Barker N, Hurlstone A, Musisi H et al. The chromatin remodelling factor Brg-1 interacts with _-catenin to promote target gene activation. EMBO J 2001; 20:4935-4943.
40. Sharpe C, Lawrence N, Martinez Arias A. Wnt signaling: a theme with nuclear variations. BioEssays 2001; 23:311-318.
41. Bauer A, Chauvet S, Huber O et al. Pontin52 and Reptin52 function as antagonistic regulators of b-catenin signaling activity. EMBO J 2000; 19:6121-6130.
42. Shen X, Mizuguchi G, Hamiche A et al. A chromatin remodelling complex involved in transcription and DNA processing. Nature 2000; 406:541-544.
43. Ikura T, Ogryzko VV, Grigoriev M et al. Involvement of the TIP60 histone acetylase complex in DNA repair and apoptosis. Cell 2000; 102:463-474.
44. Bejsovec A, Wieschauss E. Signaling activities of the Drosophila wingless gene are separately mutable and appear to be transduced at the cell surface. Genetics 1995; 139:309-320.
45. Hatini V, Bokor P, Goto-Mandeville R et al. Tissue- and stage-specific modulation of Wingless signaling by the segment polarity gene lines. Genes & Dev 2000; 14:1364-1376.
46. Gallet A, Erkner A, Charroux B et al. Trunk-specific modulation of wingless signaling in Drosophila by teashirt binding to armadillo. Curr Biol 1998; 8:893-902.
47. Gallet A, Angelats C, Erkner A et al. The C-terminal domain of armadillo binds to hypophosphorylated teashirt to modulate wingless signaling in Drosophila. EMBO J 1999; 18:2208-2217.
48. Waltzer L, Vandel L, Bienz M. Teashirt is required for transcriptional repression mediated by high Wingless levels. EMBO J 2001; 20:137-145.
49. Piepenburg O, Vorbruggen G, Jackle H. Drosophila segment borders result from unilateral repression of hedgehog activity by wingless signaling. Mol Cell 2000; 6:203-209.
50. Halfon MS, Carmena A, Gisselbrecht S et al. Ras pathway specificity is determined by the integration of multiple signal-activated and tissue-restricted transcription factors. Cell 2000; 103:63-74.
51. Zecca M, Basler K, Struhl G. Direct and long-range action of a wingless morphogen gradient. Cell 1996; 87:833-844.
52. Neumann CJ, Cohen SM. Long-range action of Wingless organizes the dorsal-ventral axis of the Drosophila wing. Development 1997; 124:871-880.
53. Baeg GH, Lin X, Khare N et al. Heparan sulfate proteoglycans are critical for the organization of the extracellular distribution of Wingless. Development 2001; 128:87-94.
54. Strigini M, Cohen SM. A hedgehog activity gradient contributes to AP axial patterning of the Drosophila wing. Development 1997; 124:4697-4705.
55. Sampedro J, Johnston P, Lawrence PA. A role for wingless in the segmental gradient of Drosophila? Development 1993; 117:677-687.
56. Pfeiffer S, Alexandre C, Calleja M et al. The progeny of wingless-expressing cells deliver the signal at a distance in Drosophila embryos. Curr Biol 2000; 10:321-324.
57. Graba Y, Gieseler K, Laurenti P et al. DWnt-4, a new Drosophila Wnt gene involved in embryonic segmentation is regulated by HOM-C genes in the visceral mesoderm. Development 1995; 121:209-218.
58. Bergstein I, Eisenberg LM, Bhalerao J et al. Isolation of two novel WNT genes, WNT14 and WNT15, one of which (WNT15) is closely linked to WNT3 on human chromosome 17q21. Genomics 1997; 46:450-458.
59. Janson K, Cohen ED, Wilder EL. Expression of DWnt6, DWnt10, and DFz4 during Drosophila development. Mech Dev 2001; 103:117-20.
60. Fradkin LG, Noordermeer JN, Nusse R. The Drosophila Wnt protein DWnt-3 is a secreted glycoprotein localized on the axon tracts of the embryonic CNS. Dev Biol 1995; 168:202-213
61. Kozopas KM, Samos CH, Nusse R. DWnt-2, a Drosophila Wnt gene required for the development of the male reproductive tract, specifies a sexually dimorphic cell fate. Genes & Dev 1998; 12:1155-1165.

62. Gieseler K, Graba Y, Mariol MC et al. Antagonist activity of DWnt-4 and wingless in the Drosophila embryonic ventral ectoderm and in heterologous Xenopus assays. Mech Dev 1999; 85:123-131.
63. Buratovich MA, Anderson S, Gieseler K et al. DWnt-4 and Wingless Have Distinct Activities in the Drosophila Dorsal Epidermis. Dev Genes Evol 2000; 210:111-119.
64. Gieseler K, Wilder E, Mariol MC et al. DWnt-4 and wingless elicit similar cellular responses during imaginal development. Dev Biol 2001; 232:339-350.
65. Cohen ED, Mariol MC, Wallace RMH et al. DWnt-4 facilitates cell movement during Drosophila ovarian morphogenesis through Focal Adhesion Kinase. Dev Cell 2002; 2:437-448.
66. Heisenberg CP, Tada M, Rauch GJ et al. Silberblick/Wnt11 mediates convergent extension movements during zebrafish gastrulation. Nature 2000; 405:76-81.
67. Tada M, Smith JC. Xwnt11 is a target of Xenopus Brachyury: regulation of gastrulation movements via Dishevelled, but not through the canonical Wnt pathway. Development 2000; 127:2227-2238.
68. Parsons JT, Martin KH, Slack JK et al. Focal adhesion kinase: a regulator of focal adhesion dynamics and cell movement. Oncogene 2000; 19:5606-5613.
69. Boutros M,d Mlodzik M. Dishevelled: at the crossroads of divergent intracellular signaling pathways. Mech Dev. 1999; 83:27-37.
70. Mlodzik M. Spiny legs and prickled bodies: new insights and complexities in planar polarity establishment. BioEssays 2000; 22:311-315.

# CHAPTER 4

# Wnt Signaling and the Establishment of the Dorsal-Ventral Axis in *Xenopus*

**Jeffrey R. Miller**

## Abstract

Establishment of the dorsal-ventral axis in *Xenopus* is dependent on cytoplasmic rearrangements that result in the spatially restricted activation of the Wnt pathway along the dorsal-ventral axis. Prior to fertilization, a transplantable dorsal determinant resides in the cortical region at the vegetal pole and following fertilization this determinant is transported along microtubule tracks to the future dorsal side. The dorsal determinant functions as an activator of the Wnt pathway and stimulates the stabilization and nuclear accumulation of β-catenin. In the nucleus, β-catenin acts in concert with XTCF-3 to activate expression of dorsal-specific regulatory genes that induce formation of the Spemann Organizer in the dorsal equatorial region. On the ventral side, the destruction complex facilitates the rapid degradation of β-catenin thereby eliminating the signaling pool of β-catenin. In the absence of β-catenin, XTCF-3 acts as a transcriptional repressor of β-catenin-target genes. Ventral cell fates also appear to be dependent on the Wnt/$Ca^{2+}$ pathway, which might function to antagonize the canonical Wnt/β-catenin pathway on the ventral side of the embryo. This review will summarize our current understanding of the role the Wnt pathway plays in establishing the dorsal-ventral axis in *Xenopus*.

## Introduction

A central question in developmental biology is how the primary body axes are established. Typically, the unfertilized egg possesses few initial asymmetries and one or more of the primary body axes are established after fertilization by external cues. In *Xenopus*, specification of the animal-vegetal axis (which corresponds roughly to the future anterior-posterior axis of the embryo) occurs during oogenesis and can be clearly visualized by asymmetries in the distribution of yolk and pigment. At the molecular level, this asymmetry is manifest by the localization of specific mRNAs (e.g., Vg1, Xwnt-11, and VegT)[1-6] and a dorsal determinant[7,8] at the vegetal pole region.

The radial symmetry of the egg is broken shortly after fertilization by cortical rotation, a dramatic cytoplasmic rearrangement characterized by the 30° rotation of the cortex relative to the inner cytoplasmic mass.[9,10] Cortical rotation is dependent on the formation of a parallel array of microtubules present at the shear zone between the outer cortex and inner core. The microtubule array functions not only as tracks for the attachment of molecular motors that drive rotation but is also required for the transport of the dorsal determinant from the vegetal pole to the prospective dorsal side.[11] The result of transport of the dorsal determinant is the

*Wnt Signaling in Development*, edited by Michael Kühl.

establishment of a new axis of polarity in the egg. At the molecular level, it is now well established that the dorsal determinant functions as an activator of the canonical Wnt/β-catenin pathway and leads to the stabilization and nuclear accumulation of β-catenin in dorsal blastomeres (Fig. 1) (see for example refs. 12-14). β-catenin is both sufficient and required for dorsal development[12,15,16] and its accumulation on the dorsal side of the embryo is stimulated by the activity of XFz-7, Dsh, Casein Kinase Iε (CKIε), GBP, and XTCF-3 (see below). On the ventral side of the embryo, β-catenin levels are kept low due to the activity of the Axin/APC/GSK3 destruction complex. In the absence of β-catenin, XTCF-3 represses dorsal-specific target genes and development of ventral cell fates ensues. The goal of this review is to summarize the current model for the role of Wnt signaling in the specification of the dorsal-ventral axis in *Xenopus*. Particular emphasis is placed on integrating data from embryological and molecular analyses, highlighting recent and novel findings, and underscoring questions that remain to be answered. For additional information regarding pattern formation in *Xenopus*, readers are referred to the following reviews (see refs. 17-21).

## Cortical Rotation and Establishment of Asymmetry Along the Dorsal-Ventral Axis

The unfertilized *Xenopus* egg is polarized along the animal-vegetal axis but shows no discernable asymmetry along the dorsal-ventral axis. A variety of experiments have shown that the egg possesses the capacity to develop dorsal structures at any location (see for example ref. 22). Fertilization occurs anywhere in the animal hemisphere and the sperm entry point usually, but not always, identifies a meridian that predicts the future ventral midline of the embryo. Following fertilization, the egg undergoes a dramatic cytoplasmic rearrangement, called cortical rotation, which leads to the establishment of dorsal cell fates at a position approximately opposite the sperm entry point.[9,23,11] Cortical rotation was first described in detail by Ancel and Vintemberger[24] for *Rana fusca* eggs and probably occurs in most amphibians and some fish which display holoblastic cleavage. In *Xenopus*, cortical rotation occurs between 45 and 90 minutes after fertilization and is characterized by the 30° displacement of the inner cytoplasmic core relative to the outer cortex. The midline of rotation lies along an axis perpendicular to the animal-vegetal axis and coincides accurately with the prospective dorsal-ventral axis.[9,23]

Cortical rotation is dependent on the formation of a parallel array of microtubules at the shear zone between the inner cytoplasmic core and the outer cortex.[25](Fig. 2). The microtubules are dynamic, appearing at the onset of rotation and rapidly disappearing at the end of rotation, and are oriented with their plus ends pointing towards the future dorsal side.[26] The microtubules act as a scaffold upon which rotation occurs, allowing for the attachment of plus-end directed motors that drive movement of the cortex relative to the inner core.[27] Treatment of eggs with a variety of agents that prevent microtubule polymerization also block cortical rotation.[28,25] It remains unclear what spatial cues are used to initially orient the microtubule array, where the microtubules in the array originate, and how the microtubules in the array align. It has been suggested that astral microtubules emanating from the sperm aster may orient or even drive cortical rotation,[29,30] but the observation that a parallel microtubule array can form in prick activated eggs and isolated vegetal egg fragments argues that the sperm aster is not required for array formation. Still, it remains possible that the sperm aster may provide spatial cues for the orientation of the array in situ. In addition, the observation that rotation movements precede the appearance of a parallel array suggest that alignment of the microtubules may occur as a result of a feedback mechanism whereby microtubules orient rotation and rotation orients microtubules.[31,10]

Cortical rotation not only predicts the future dorsal-ventral axis, but it is also required for the specification of dorsal cell fates (Fig. 3). Inhibition of cortical rotation utilizing ultraviolet

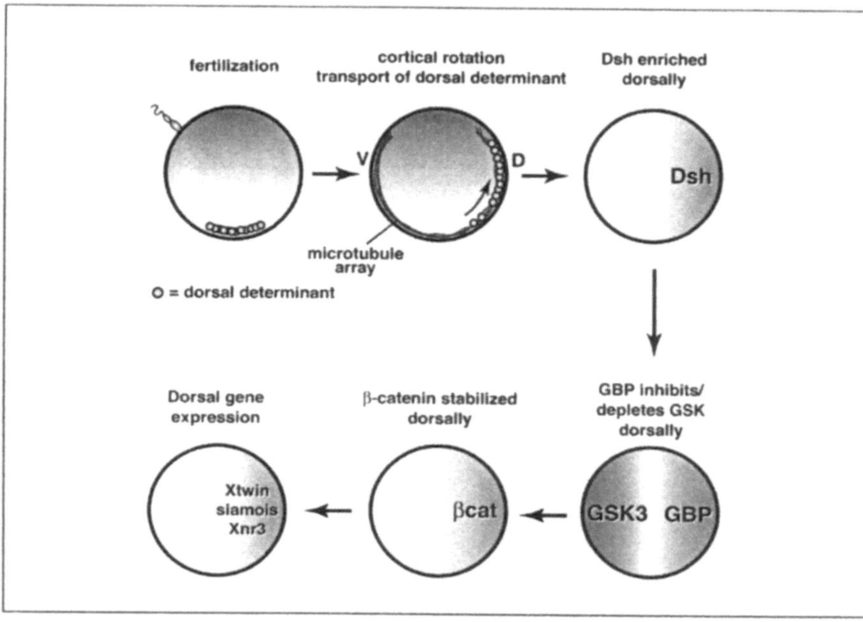


Figure 1. Simplified model of dorsal-ventral axis formation in Xenopus. Transport of the dorsal determinant from the vegetal pole to the future dorsal side is accompanied by the dorsal enrichment of Dsh. Accumulation of Dsh facilitates the interaction of GBP with GSK3. GBP blocks GSK3 function through inhibition of GSK3-mediated phosphorylation of β-catenin and depletion of GSK3 in dorsal blastomeres. As a result, β-catenin is stabilized and accumulates in the nucleus where it activates expression of dorsal-specific genes that specify dorsal cell fates. See text for details.

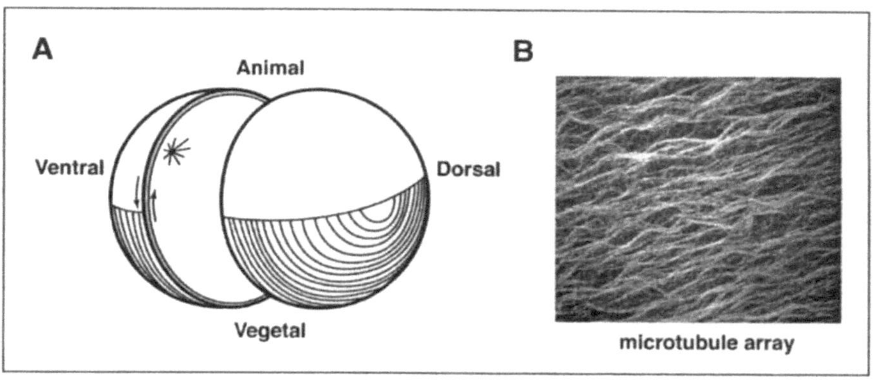


Figure 2. The microtubule array. (A) A parallel array of microtubules forms at the interface between the inner cytoplasmic core and the outer cortex. The orientation of the array is biased by the sperm aster and accurately predicts the location of the future dorsal-ventral axis. The microtubules serve as a scaffold for cortical rotation, the 30° rotation of the inner core relative to the outer cortex, and tracks for the transport of the dorsal determinant. (B) Confocal micrograph of the microtubule array during cortical rotation.

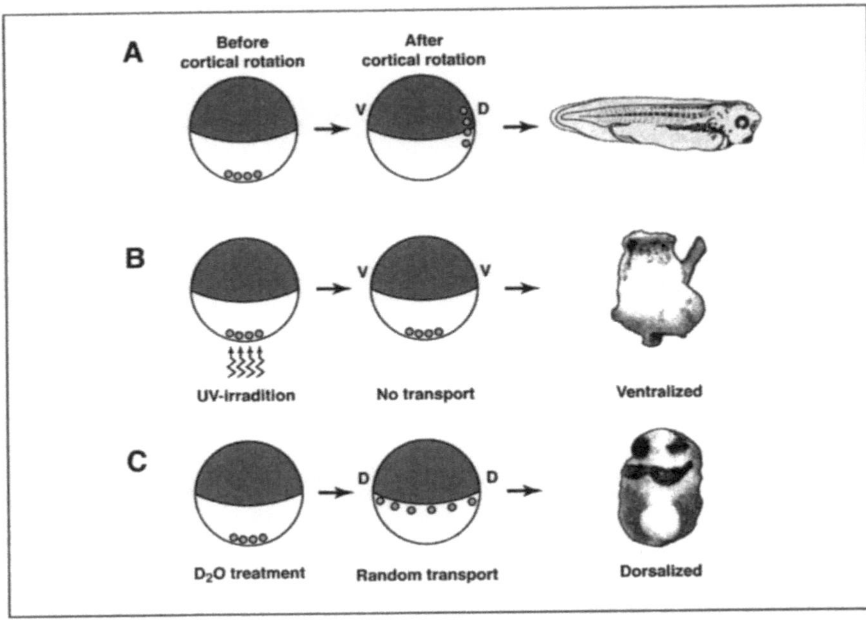


Figure 3. Transport of the dorsal determinant is required for the specification of dorsal cell fates. (A) In normal embryos, the dorsal determinant is transported from the vegetal pole to the future dorsal side. (B) UV-irradiation of the vegetal pole prior to the onset of cortical rotation prevents formation of the microtubule array and blocks transport of the dorsal determinant. Resulting embryos display a ventralized phenotype characterized by the lack of anterior and dorsal structures such as the head, notochord, somites, and neural tube. (C) Treatment of eggs with $D_2O$ precociously stabilizes microtubules in the vegetal cortex and results in the formation of a meshwork of microtubules. As a result, the dorsal determinant is transported in all directions and specificies dorsal cell fates throughout the entire equatorial zone. Treated embryos display a dorsalized phenotype characterized by excess cement gland, neural tissue, and notochord.

irradiation of the vegetal pole or inhibitors of microtubule polymerization blocks formation of Spemanns Organizer and subsequent development of dorsal structures including notochord, somites and neural tissue.[11] In the absence of dorsal structures, treated embryos develop a ventralized phenotype with ventral tissues of all three germ layers present.[32] Experimental manipulation of cortical rotation shows that the amount of rotation determines the extent of dorsal development.[22,23] Moreover, dorsal development can be rescued in eggs treated with microtubule depolymerizing agents by manually tipping treated eggs and allowing a gravity-induced rotation of the cortex.[22,28]

How does cortical rotation promote the development of dorsal cell fates? Current models propose that the microtubule array serves as tracks for the transport of a dorsal determinant from the vegetal pole to the future dorsal side of the embryo during cortical rotation. The dorsal determinant was initially defined embryologically as an activity that when transplanted to the ventral region of a recipient embryo can induce axis duplication.[33,7] Further transplantation and ablation studies demonstrated that the dorsal determinant is localized to the vegetal pole cortex at fertilization and moves to the prospective dorsal side during cortical rotation.[34,8,35] Moreover, cytoplasm at the equator shows the highest responsiveness to transplanted dorsal determinant suggesting that factors important for dorsal development are also localized to the equatorial region of the early embryo.[35] The dorsal determinant appears to be tightly associated with the egg cortex[35,36] and its movement towards the dorsal side mimics the transport of membrane bound organelles along the microtubule array towards the dorsal side.[37] Although the dorsal determinant is required to specify dorsal axial cell fates, it does not possess mesoderm inducing activity.[38] Instead, the dorsal determinant appears to stimulate production of a dorsal-modifying signal that, in combination with mes-endoderm inducers, promote the formation of the Spemann Organizer (reviewed in ref. 39). The region of the embryo possessing dorsal-modifying activity is called the Nieuwkoop center.[11]

What is the molecular nature of the dorsal determinant and how does its movement induce dorsal development? A growing body of evidence strongly supports the idea that the dorsal determinant functions as an activator of the canonical Wnt/β-catenin pathway (Fig. 4). A number of experiments have shown that overexpression of Wnt pathway components in ventral blastomeres can mimic the activity of the dorsal determinant and lead to the development of a complete secondary dorsal axis.[40,41,15,42,43] Importantly, loss-of-function experiments also demonstrate that the signaling through the Wnt pathway is required for establishment of the dorsal-ventral axis.[12,44-47,43,48,49]

The molecular nature of the dorsal determinant is unresolved at present, but the simplest explanation is that it is an upstream component of the Wnt pathway that becomes localized or regionally activated in response to cortical rotation. Potential suspects include a maternal Wnt ligand, a member of the Frizzled or LRP family of Wnt receptors, or an intracellular component of the pathway such as Dsh, CKIε, or GBP, all of which can mimic the dorsal determinant in gain-of-function experiments.[40-43,50-52] The role of each of these components in patterning the dorsal-ventral axis is summarized in Table 1 and discussed in detail below.

## Roles of Wnt and Frizzled in the Specification of Dorsal Cell Fates

Although several maternal Wnt genes have been identified (*Xwnt-5a*, *Xwnt-8b*, and *XWnt-11*),[53,2,54,55] it remains unclear whether any of these *Wnts* plays a role in axis specification. *XWnt-11* is the only maternally provided *Wnt* known that shows an asymmetric distribution in eggs and early embryos. *XWnt-11* transcripts are localized to the vegetal pole of unfertilized eggs[2] and *XWnt-11* protein accumulates on the dorsal side of the embryo during cleavage due to differential translation of its transcript.[56] However, neither *XWnt-11* nor *Xwnt-5A* can ectopically activate the Wnt/β-catenin pathway to induce formation of a complete secondary dorsal axis when overexpressed in ventral blastomeres suggesting that they do not play a role in specifying the endogenous dorsal-ventral axis.[2,54] Instead, *Xwnt-5A* and *XWnt-11* are implicated in regulating convergent extension movements during gastrulation[54,57-60] and may play a role in antagonizing the Wnt/β-catenin pathway on the ventral side of the embryo (see below and refs. 61,62). *Xwnt-8b*, on the other hand, does possess axis-inducing activity when overexpressed in ventral blastomeres.[55] However, *Xwnt-8b* transcripts are not localized vegetally where the dorsal determinant is found. In addition, overexpression of a dominant-negative form of Xwnt-8, a highly related *Xenopus* Wnt gene, in eggs and oocytes does not inhibit dorsal development suggesting that Xwnt-8b is not required for specification of dorsal cell fates.[63,64] Taken together, these data are consistent with the idea that the dorsal determinant may bypass the need for an extracellular Wnt ligand. It should be noted, however, that no study has stringently tested the requirement of maternal *Wnt* genes in axis specification through elimination of maternal transcripts and/or protein. Therefore, a definitive role for *Wnts* in dorsal development awaits further experimental analyses.

Several maternally expressed Frizzled receptors have been identified in Xenopus (*XFz-1*, *XFz-3*, *XFz-4*, and *XFz-7*)[65-68,48] raising the possibility that one or more of these receptors plays a role in axis specification. In support of this idea, antisense depletion of XFz-7 transcripts results in suppression of dorsal development and a reduction in expression of the

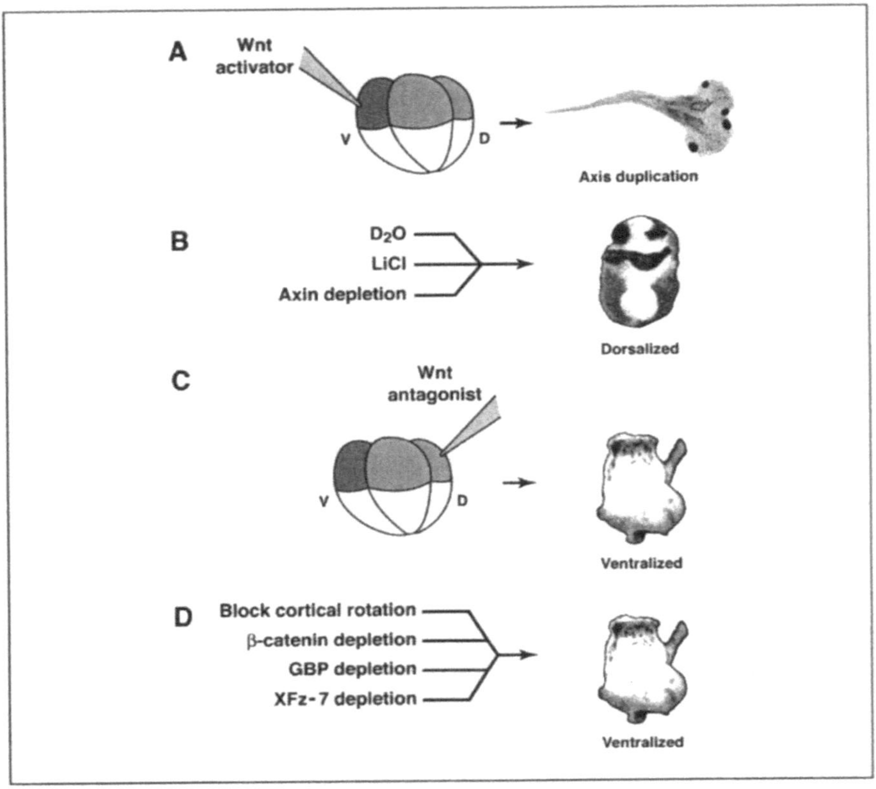


Figure 4. Perturbation of Wnt signaling results in defects in dorsal-ventral patterning. (A) Overexpression of activators of the Wnt pathway in ventral blastomeres results in axis duplication. (B) Treatments that promote inappropriate activation of the Wnt pathway throughout the early embryo result in a dorsalized phenotype. (C) Overexpression of antagonists of the Wnt pathway in dorsal blastomeres results in the loss of dorsal structures and a ventralized phenotype. (D) Inhibition of cortical rotation or depletion of XFz-7, GBP, or β-catenin blocks Wnt pathway activation and results in a ventralized phenotype.

dorsal-specific β-catenin target genes, *siamois* and *Xnr3*.[48] Morpholino-mediated knockdown of XFz-7 function also results in embryos with defects in dorsal-ventral patterning.[69] However, the phenotype of XFz-7 depleted embryos is not as severe as that of β-catenin depleted embryos. Thus, although XFz-7 is required for optimal activation of the Wnt pathway in dorsal cells its function may be partially redundant to other *XFz* genes or additional downstream components of the pathway. Important questions that remain to be answered regarding the role of XFz-7 in axis specification include determining where, when, and how XFz-7 is activated in the early embryo and determining how XFz-7 activates downstream components of the Wnt pathway. A related mystery is the intracellular localization of XFz-7. Is XFz-7 a component of the dorsal determinant localized to a specific class of transport vesicles? The requirement for XFz-7 in dorsal development does, however, imply that a Wnt ligand may also be necessary. This possibility and the requirement for other maternally expressed *XFz* genes in axis formation remain to be explored.

***Table 1. Wnt pathway genes and their proposed role in axis formation in* Xenopus**

| Gene | Expression/ Localization | Knockdown Phenotype[1] | Other Phenotypes and Functions[2] | Reference |
|---|---|---|---|---|
| Xwnt-8b<br>(V) | Maternal mRNA enriched<br>55<br>animally | | N.D.<br><br>Axis duplication | (D) Dorsalized |
| Xwnt-5A | Maternal | N.D. | (D) Gastrulation defects | 54 |
| XWnt-11 | Maternal mRNA localized to vegetal cortex; Protein becomes enriched dorsally | N.D. | (D) Gastrulation defects<br>(V) Partial axis duplication | 2,56,58 |
| XFz-1 | Maternal | N.D. | Can induce expression of siamois and Xnr3 in animal caps | 67 |
| XFz-3 | Maternal | N.D. | (D) Morphogenesis defects; Eye development | 65,111 |
| XFz-4 | Maternal | N.D. | N.D. | 68 |
| XFz-7 | Maternal | Ventralized | (D) Gastrulation defects<br>(V) Potentiates Wnt-induced axis duplication | 48,69,112,113 |
| Dsh | Maternal Protein becomes enriched dorsally | N.D. | (V) Axis duplication<br>(EE) Depletion has no effect on rate of β-catenin degradation | 42,66,70,73 |
| GBP | Maternal | Ventralized | (V) Axis duplication | 43 |
| CKIε | Maternal | N.D. | (V) Axis duplication; DN-CKI blocks Wnt-mediated axis duplication (EE) Inhibition increases rate of β-catenin degradation | 50, 51, 76 |
| Axin | Maternal | Dorsalized | (D) Ventralized<br>(V) DN-Axin induces axis duplication (EE) Required for β-catenin degradation | 49, 73, 90, 114 |
| XARP | Maternal | N.D. | (D) Ventralized<br>(V) DN-Axin induces axis duplication | 75 |
| APC | Maternal | N.D. | (V) Dorsalized (EE) Required for β-catenin degradation | 73,89 |
| GSK3 | Maternal Protein becomes depleted dorsally | N.D. | (D) Ventralized<br>(V) Kinase-dead GSK3 induces axis duplication<br>(EE) Required for β-catenin degradation | 44-46,73,77 |
| PP2A A | Maternal | N.D. | (D) Weak ventralization<br>(V) Blocks Wnt-induced axis duplication | 100 |

***continued on next page***

*Table 1. Continued*

| Gene | Expression/ Localization | Knockdown Phenotype[1] | Other Phenotypes and Functions[2] | Reference |
|---|---|---|---|---|
| PP2A C | Maternal | N.D. | (D) No effect (V) Blocks Wnt-induced axis duplication (EE) Required for b-catenin degradation | 100 |
| PP2A B56 | Maternal | N.D. | (D) Ventralized (V) Blocks Wnt-induced axis duplication (EE) Accelerates b-catenin degradation | 100 |
| bTrCP | Maternal; two transcripts enriched vegetally and one transcript localized animally | N.D. | (D) Ventralized (V) Blocks Wnt- and b-catenin-induced axis duplication; DN-bTrCP induces axis duplication | 104,105,106 |
| β-catenin | Maternal; protein becomes enriched dorsally | Ventralized | (D) No effect (V) Axis duplication | 10,12,13,15 16 |
| XTCF-3 | Maternal | N.D. | (D) XTCF-3 and DN-XTCF-3 ventralize (V) DN-XTCF-3 blocks β-catenin-induced axis duplication (EE) Blocks β-catenin degradation | 47,107,73 |
| ICAT | Maternal | N.D. | (D) Ventralized (V) Blocks Wnt- and _-catenin-induced axis duplication; DN-ICAT induces axis duplication | 108 |
| XCtBP | Maternal | N.D. | (D) No effect; A mutant activating form of XCtBP caused perturbation of head and notochord development | 107 |
| XGrg4 | Maternal | N.D. | (D) Ventralized; blocks expression of siamois and Xnr3 (V) Blocks Armadillo/TCF fusion induced axis duplication | 86,115 |
| XGrg5 | Maternal | N.D. | (D) Enhances activation of a siamois reporter gene (V) Potentiates Armadillo/TCF fusion induced axis duplication | 86,115 |
| CBP | Maternal | N.D. | (D) Inhibition of CBP by E1A causes ventralization | 84,116 |
| CamKII | Maternal Catalytic activity higher ventrally | N.D. | (D) Ventralization (V) DN-CamKII induces partial axis duplication | 62 |

*continued on next page*

## Transport and Asymmetric Accumulation of Dishevelled

The lack of compelling evidence implicating a Wnt ligand in axis specification has led to the proposal that the Wnt pathway is activated on the dorsal side of the embryo through a Wnt-independent mechanism. Consistent with this hypothesis, recent evidence demonstrates

***Table 1. Continued***

| Gene | Expression/ Localization | Knockdown Phenotype[1] | Other Phenotypes and Functions[2] | Reference |
|---|---|---|---|---|
| **Other Wnt pathway regulators with potential roles in axis formation** | | | | |
| Axam | N.D. | N.D. | (D) Ventralized<br>(V) Blocks Wnt- and Dsh-induced axis duplication | 117 |
| Duplin | N.D. | N.D. | (D) Ventralized (V) Blocks Wnt-, β-catenin- and siamois-induced axis duplication | 118,119 |
| IDAX | N.D. | N.D. | (D) Ventralized<br>(V) Blocks Wnt- and Dsh-induced axis duplication | 120 |
| LRP | N.D. | N.D. | (V) Axis duplication; DN-LRP blocks Wnt-induced axis duplication | 52 |
| Naked | N.D. | N.D. | (D) Ventralized; Blocks convergent extension movements<br>(V) Blocks Wnt-induced axis duplication | 121 |
| NLK | N.D. | N.D. | (D) Ventralized<br>(V) Blocks β-catenin-induced axis duplication | 122 |
| Par1 | N.D. | N.D. | (V) DN-Par1 blocks Wnt-induced axis duplication | 123 |
| PP2A PR61 | N.D. | N.D. | (D) Ventralized<br>(V) Blocks b-catenin-induced axis duplication | 124 |

1) N.D. = Not Determined
2) D = dorsal overexpression phenotype; V = ventral ovexpression phenotype; EE = activity in egg extracts;
DN = dominant negative

that Dsh possesses several characteristics in common with the endogenous dorsal determinant. As discussed earlier, the dorsal determinant is localized to the subcortical region at the vegetal pole prior of the unfertilized egg and is transported along the microtubule array to the future dorsal side during cortical rotation. Analysis of Dsh localization in the egg showed that endogenous Dsh displays a similar distribution in fertilized eggs. Prior to cortical rotation, Dsh associates with particles (0.6-1.2 μm diameter) found in the vegetal cortical region and similar Dsh particles become enriched in the cortex of the dorsal equatorial region following cortical rotation.[66] At later cleavage stages, steady-state levels of Dsh protein remain elevated dorsally with dorsal regions possessing approximately 2.4 fold higher levels then ventral regions.[66] Dsh is also differentially phosphorylated along the dorsal-ventral axis showing a 2.8 fold higher level of phosphorylation in dorsal regions compared to ventral regions.[70] Importantly, inhibition of cortical rotation by UV-irradiation of the vegetal pole prevented the dorsal enrichment of Dsh particles following cortical rotation and the dorsal accumulation of Dsh at later stages[66] thereby establishing a link between the asymmetric accumulation of Dsh and the spatially restricted activation of the Wnt pathway.

Examination of the dynamics of Dsh localization in the vegetal region of the fertilized egg using a GFP-tagged version of Dsh (Dsh-GFP) demonstrated that Dsh-GFP also associates with particles in the vegetal region of the fertilized egg.[66] Interestingly, Dsh-GFP particles are transported towards the prospective dorsal side of the embryo during cortical rotation. Movement of Dsh-GFP was abolished in embryos treated with agents that disrupt the microtubule array and was randomized in embryos treated with $D_2O$, a perturbation that causes the precocious and disorganized formation of the microtubule array.[71] Thus, the distribution and dynamics of both endogenous and ectopic Dsh mimic that of the dorsal determinant (Fig. 3). Since vesicles similar in size to Dsh-GFP particles are also known to be transported dorsally during cortical rotation[37] it has been proposed that Dsh may associate with a specific class of transport vesicles that are moved to the dorsal side via a kinesin-like motor.[66]

Although these data provide strong support for the idea that Dsh plays an important role in axis formation, attempts to show a requirement for Dsh in axis specification have thus far been unsuccessful. For example, expression of a dominant negative form of Dsh in dorsal cells does not prevent development of dorsal cell fates.[72] The caveat of these types of analyses is that they typically involve injection of mRNA encoding the dominant-negative form of the protein of interest into blastomeres at the 4-cell stage when a discernible difference between dorsal and ventral blastomeres is first manifest. Thus, if the protein of interest were required at a very early time, for example at the one-cell stage, then levels of the dominant-negative proteins would be insufficient to interfere with the function of the endogenous protein. A more stringent method for testing the affect of dominant-negative constructs on early development would be to express these constructs in oocytes thereby allowing enough time to produce sufficient amounts of the mutant protein to block the function of the endogenous protein. Knockdown experiments analyzing the role of Dsh utilizing anti-sense oligonucleotides and morpholinos have also been inconclusive due to the presence of a relatively stable pool of maternally provided Dsh protein. Neither depletion of Dsh transcripts in oocytes using antisense oligos (J. Heasman, J.R. Miller, and R.T. Moon, unpublished observations) nor inhibition of translation of Dsh mRNAs in oocytes using Dsh morpholinos (J.R. Miller, unpublished observations) eliminates maternal Dsh protein. Thus, although compelling circumstantial evidence implicates Dsh as a critical player in controlling the asymmetric activation of the Wnt pathway, the lack of genetic tools in *Xenopus* has prevented proving its involvement beyond a reasonable doubt. Such evidence will require the conception of new experimental tools or novel approaches to eliminate maternal protein function in early *Xenopus* embryos.

## Stabilization of β-Catenin on the Dorsal Side of the Embryo

Specification of dorsal cell fates requires the accumulation of newly synthesized β-catenin in dorsal blastomeres.[12,14,16] Central to understanding how β-catenin accumulates on the dorsal side of the embryo is determining how the destruction complex is asymmetrically regulated along the dorsal-ventral axis. The core destruction complex is composed of APC, Axin, and GSK3 and it is possible that any or all of these proteins are a target of Wnt pathway activation on the dorsal side of the embryo. A number of the usual suspects have been named as key antagonists of the destruction complex in dorsal blastomeres including Dsh, CKIε, and GBP. In addition, a novel role for XTCF-3 in promoting β-catenin stability has also emerged as a potential mechanism for protecting β-catenin from the destruction complex. Our current understanding of the role these proteins plays in promoting the stabilization and accumulation of β-catenin on the dorsal side is discussed below.

As stated above, Dsh becomes enriched on the dorsal side of the embryos through a process that is dependent on cortical rotation.[66] The observed transport of Dsh-GFP along the microtubule array suggests that transport of endogenous Dsh may contribute to its enrichment in dorsal blastomeres. It is thought that high levels of Dsh promote the stabilization of β-catenin in dorsal cells. In support of this idea, experimentally increasing levels of Dsh stabilizes β-catenin[66,73] and induces formation of a complete secondary axis when overexpressed in ventral blastomeres.[42] Furthermore, both Dsh and β-catenin accumulate in the cortex of the dorsal equatorial region immediately following cortical rotation suggesting that enrichment of Dsh results in an immediate stabilization of β-catenin on the dorsal side.[37,66]

Several important issues regarding the mechanism by which Dsh activates the Wnt pathway are unresolved. For example, it remains unclear how elevated levels of Dsh promote the stabilization of β-catenin on the dorsal side. One possible mechanism by which Dsh may stabilize β-catenin is that high levels of Dsh in the dorsal equatorial region might promote the formation of a transitional complex containing Dsh, GBP, Axin, and GSK3. Dsh would then hand GBP off to GSK resulting in the dissociation of the GBP/GSK3 complex.[74,75] Alternatively, Dsh may stimulate the activity of CKIε thereby enhancing interactions between β-catenin and XTCF-3 (see below). Another unsolved mystery is how XFz-7 regulates the activity of Dsh on the dorsal side of the embryo.

The Ser/Thr kinase CKIε is an important regulator of β-catenin stability. Overexpression of CKIε in ventral blastomeres stabilizes β-catenin and induces formation of a secondary axis indicating that CKIε can activate the maternal Wnt pathway in Xenopus.[50,51] In addition, expression of a kinase dead form of CKIε can block the dorsalizing activity of Xwnt-8[50,51] and inhibition of CKIε with CKI-7 can increase the rate of β-catenin degradation in egg extracts.[76] Epistasis experiments have placed CKIε downstream of Dsh and upstream of GSK3.[50] CKIε possesses several activities that may contribute to its ability to activate the Wnt pathway. CKIε can bind and phosphorylate Dsh[50] and XTCF-3,[76], and is present in a complex with Axin.[51] CKIε-mediated phosphorylation of Dsh promotes the binding of Dsh to GBP in egg extracts while phosphorylation of XTCF-3 increases the affinity of XTCF-3 for β-catenin[76] (see below). However, it is not known whether CKIε is required for dorsal development. Another unsolved mystery is whether Wnt signals directly stimulate CKIε activity. Finally, we do not know whether CKIε activity, localization and/or substrate interactions are spatially regulated along the dorsal-ventral axis.

A second crucial player involved in the regulation of β-catenin stability is GBP, a GSK3 binding protein capable of inhibiting GSK3 activity both in vitro and in vivo.[43] Overexpression of GBP in ventral blastomeres induces development of a complete secondary axis and can rescue dorsal development in UV-irradiated embryos showing that GBP expression is sufficient for the specification of dorsal cell fates.[43] Importantly, depletion of maternal GBP results in embryos displaying a ventralized phenotype demonstrating that GBP is also required for dorsal development. Examination of the mechanism by which GBP induces axis formation shows that expression of GBP abrogates the interaction of GSK3 with Axin.[74] In addition, GBP can inhibit the ability of GSK3 to phosphorylate a protein substrate (tau)[43] suggesting that it may also block the ability of GSK3 to phosphorylate β-catenin. Together, these data are consistent with the idea that GBP functions on the dorsal side of the embryo to diminish GSK3/Axin interactions thereby removing GSK3 from the destruction complex and preventing GSK3-mediated phosphorylation of β-catenin.

A second hypothesis put forward to explain how GBP regulates GSK3 along the dorsal-ventral axis proposes that GBP stimulates the depletion of GSK3 on the dorsal side of the embryo.[77] Examination of the steady state levels of GSK3 in dorsal and ventral regions of cleavage stage embryos showed that dorsal regions possess 20-30% less GSK3 than ventral regions. Inhibition of cortical rotation abolished dorsal depletion of GSK3 while dorsalization of embryos with $D_2O$ or expression of GBP enhanced depletion of GSK3. Thus, GBP-mediated dorsal depletion of GSK3 represents an additional mechanism by which asymmetric activation of the Wnt pathway might stimulate the stabilization of β-catenin on the dorsal side of the embryo. The importance of this mechanism for dorsal-ventral patterning awaits further analyses.

Although it is clear that GBP plays a central role in axis specification, many important questions regarding GBP function remain unanswered. For example, how is the activity of GBP regulated along the dorsal-ventral axis? One possibility is that GBP/GSK3 interactions are spatially regulated such that GBP preferentially binds and inhibits GSK3 in dorsal cells. Dorsally enhanced GBP/GSK3 interactions may be facilitated by the concerted action of Dsh and CKIε. Alternatively, the localization of GBP may be regulated either globally along the dorsal-ventral axis or locally within cells. In addition, although expression of GBP can cause a reduction in the levels of GSK3 it is not known whether GBP is required for the dorsal depletion of GSK3 observed in unperturbed embryos. Answers to these questions will provide important insights into the function of GBP in axis specification and shed light on the general mechanism of Wnt signal transduction in vertebrates.

In addition to the role XTCF-3 plays in regulating the expression of dorsal-specific target genes (see below), it may also function as a cytoplasmic regulator of β-catenin stability. In egg extracts, addition of XTCF-3 decreases binding of β-catenin to Axin and APC resulting in a significant reduction in the rate of β-catenin degradation.[76] Interestingly, approximately 70% of TCF protein is found in cytosolic fractions of stage 7.5 Xenopus embryos suggesting that competition between the destruction complex and XTCF-3 for β-catenin binding may be important for stabilizing β-catenin in dorsal blastomeres.

How might the ability of XTCF-3 to act as a cytoplasmic sink for β-catenin be regulated along the dorsal-ventral axis? A possible answer to this question comes from studies examining the regulation of XTCF-3/β-catenin interactions by CKIε and GSK3.[76] XTCF-3 binds and is phosphorylated by both CKIε and GSK3 in vitro. Functionally, phosphorylation of XTCF-3 by CKIε and GSK3 antagonistically effect the binding of XTCF-3 to β-catenin. Incubation of XTCF-3 with CKIε increases the affinity of XTCF-3 for β-catenin nearly four-fold, an affect that is blocked by addition of GSK3. Together these observations suggest that XTCF-3 may play an unexpected role in regulating β-catenin stability and provide evidence for a novel antagonistic relationship between CKIε and GSK3. In dorsal cells where CKIε activity is predicted to be high, CKIε might phosphorylate XTCF-3 thereby increasing the binding affinity of XTCF-3 for β-catenin. Conversely, on the ventral side where GSK3 is predicted to be high GSK3 might counteract the affect of CKIε on XTCF-3 preventing its association with β-catenin.

## β-Catenin, XTCF-3, and Activation of Dorsal-Specific Gene Expression

Antagonism of the destruction complex promotes the stabilization and nuclear accumulation of β-catenin on the dorsal side of the embryo.[14] In the nucleus, β-catenin interacts with XTCF-3, a maternally expressed member of the TCF/LEF family of architectural transcription factors, and stimulates transcription of dorsal-specific target genes.[47] To date, three direct targets of the β-catenin/XTCF-3 complex have been identified: *siamois*[78,79] and *Xtwin*[80], two highly conserved homeodomain containing transcription factors, and *Xnr3*[81], a member of the nodal-related family of signaling proteins. All three of these target genes possess consensus TCF/LEF binding sites upstream of the transcriptional start site which are bound by XTCF-3 in gel mobility shift assays. Mutation of these consensus sites abrogates expression demonstrating that XTCF-3 binding is required for transcriptional activation. *Xnr1*, another dorsally expressed nodal related gene, also possesses TCF/LEF sites in its promoter region[82] but the requirement of these sites for dorsal expression remains to be determined.

Expression of a β-catenin/LEF-1 fusion protein consisting of the carboxyl-terminal transactivation domain of β-catenin fused to the DNA binding domain of LEF-1 in ventral blastomeres results in development of a secondary axis.[83] Furthermore, replacement of the β-catenin transactivation domain with a heterologous transactivation domain also promotes dorsal development. Thus, TCF/LEF proteins appear to function simply by recruiting a transactivation domain and the basal transcription machinery to the promoter of Wnt responsive genes. This said, it appears that additional transcriptional co-activators, including CREB binding protein (CBP) and XGrg5 (a naturally truncated member of the Groucho family of transcriptional regulators), may play a role in stimulating expression of dorsal-specific target genes. Expression of CBP potentiates β-catenin-mediated activation of *siamois* and *Xnr3*[84] and expression of E1A, an inhibitor of CBP, can block Wnt- and β-catenin-mediated activation of siamois and Xnr3.[84,85] Moreover, dorsal expression of E1A causes a loss of dorsal structures supporting a role for CBP in the patterning of the endogenous dorsal-ventral axis.[84] Similarly, expression of XGrg5 in early embryos can enhance activation of a *siamois* reporter gene suggesting that it may play a role in potentiating the activity of β-catenin and XTCF-3 in dorsal nuclei.[86]

## The Destruction Complex and Ventral Inhibition of Wnt Signaling

Spatially restricted Wnt signaling in dorsal cells is mirrored by active inhibition of Wnt signaling in ventral cells normally required for specification of ventral cell fates. For example, experimental reduction of GSK3 or Axin activity on the ventral side of the embryo results in a dorsalized phenotype.[44-46,49] Likewise, XTCF-3 and Wnt/$Ca^{2+}$ signaling also play a role in actively repressing dorsal development on the ventral side of the embryo. Thus, specification of ventral cell fates requires spatially restricted inhibition of Wnt signaling. If Wnt signaling is inappropriately activated, dorsal cell fates are specified at the expense of ventral cell fates.

Wnt signaling is dependent on the presence or absence of β-catenin. On the ventral side of the embryo the proteins assigned the job of maintaining low levels of β-catenin are part of a dynamic, multi-protein machine called the destruction complex. As mentioned previously, key components of the destruction complex include APC, Axin, and GSK3 and the function of these proteins is to facilitate the phosphorylation of β-catenin by GSK3. Phosphorylated β-catenin binds βTrCP, which then stimulates addition of a poly-ubiquitin tag. The ubiquitination of β-catenin marks it for degradation by the proteasome. The function of the destruction complex and other Wnt antagonists during axis specification is summarized in Table 1 and discussed in detail below. More information on the destruction complex and its regulation by Wnt signaling can be found elsewhere.[87,88]

Although it is well established that APC functions to negatively regulate β-catenin in many systems, the role of APC in dorsal-ventral patterning in *Xenopus* is unsolved. Depletion of APC from egg extracts significantly reduces the degradation of β-catenin showing that APC is required for down-regulation of β-catenin levels.[73] However, overexpression of APC in ventral blastomeres results in development of a secondary axis suggesting that APC positively regulates β-catenin in early Xenopus embryos.[89] This apparent discrepancy in the function of APC in axis specification remains unresolved and resolution of this problem will require loss-of-function analyses in which the phenotypic consequence of APC depletion is determined.

Axin is a second key component of the destruction complex that binds APC, β-catenin, and GSK3 and stimulates the phosphorylation of β-catenin by GSK3. Overexpression of Axin or XARP, an Axin-related protein that has ~43% similarity to Axin, on the dorsal side can block formation of the endogenous dorsal axis.[90,91,75,49] In related analyses, ventral overexpression of a dominant-negative form of Axin or XARP unable to bind APC results in axis duplication.[90,75] In egg extracts, addition of Axin accelerates β-catenin degradation and depletion of Axin blocks degradation[73] consistent with the idea that Axin plays a critical role in maintaining low levels of β-catenin on the ventral side of the embryo.

Depletion of maternal Axin mRNA utilizing antisense oligonucleotides results in embryos displaying a severe hyperdorsalized phenotype.[49] This phenotype is very similar to that seen following incubation of eggs with $D_2O$[71] (see Fig. 2) or treatment of cleavage stage embryos with LiCl, an inhibitor of GSK3.[92-94] The dramatic phenotype induced by depletion of maternal Axin mRNA occurred even though total levels of Axin protein were only reduced by 40% at the blastula stage.[49] The antisense oligos used in this study did not significantly affect the levels of XARP suggesting that XARP function is not sufficient to downregulate β-catenin on the ventral side. Depletion of Axin resulted in stabilization of β-catenin and its accumulation in both dorsal and ventral nuclei. Unexpectedly, injection of Axin mRNA into ventral, but not dorsal, blastomeres at the 4-cell stage rescued the knockdown phenotype. These data are puzzling since, as mentioned above, dorsal overexpression of Axin in wild type embryos results in suppression of dorsal development.[90,91,49] The reason for this apparent discrepancy is unclear at this time and will require further studies to resolve. Despite this surprise, the observation that depletion of maternal Axin hyperdorsalizes embryos demonstrates that Axin is required for the appropriate patterning of the dorsal-ventral axis in *Xenopus*.

In mammalian cells, Axin function is regulated by phosphorylation. Axin is a substrate of GSK3 and treatment of cells with LiCl decreased levels of Axin suggesting that GSK3 promotes the stabilization of Axin.[95] A related observation is that Wnt stimulation of cells causes dephosphorylation of Axin and a decrease in the total levels of Axin.[96] In *Xenopus*, however, little is known about how Axin might be regulated along the dorsal-ventral axis. Levels of Axin protein appear to be similar in dorsal and ventral regions at the 4- to 8-cell stage[49], but could show dorsal-ventral differences at later cleavage stages. In addition, although it is known that Axin is present in both cytoplasmic and membrane fractions of early embryos[49], the intracellular localization of endogenous Axin and its potential regulation along the dorsal-ventral axis has not been explored. Finally, the question of whether interactions between Axin and its many binding partners (e.g., APC, GSK3, Dsh, and β-catenin) are spatially regulated has not been addressed.

Wnt signaling is dependent on the regulation of GSK3-mediated phosphorylation of β-catenin.[97] In Xenopus, overexpression of GSK3 in dorsal blastomeres results in ventralization whereas overexpression of a kinase-dead mutant of GSK3 in ventral blastomeres ectopically stabilizes β-catenin and causes axis duplication.[44-46,14] GSK3 can directly phosphorylate β-catenin in vitro and can promote the phosphorylation of β-catenin in vivo.[98] Moreover, mutation of the putative GSK3 phosphorylation sites in β-catenin greatly enhances the stability and signaling activity of β-catenin.[98] These same sites are mutated in a variety of human cancers underscoring the importance of GSK3-mediated phosphorylation for the regulation of β-catenin.[87,88] Together these data provide substantial support for the idea that GSK3 stimulates β-catenin degradation on the ventral side of the embryo.

How is GSK3 regulated? As described above, GBP appears to play a central role in regulating GSK3 function along the dorsal-ventral axis. It is thought that GBP downregulates GSK3 in dorsal cells through one or all of the following mechanisms. GBP might inhibit GSK-mediated phosphorylation of β-catenin, promote the dissociation of GSK3 from the destruction complex, enhance the depletion of GSK3, and/or inhibit GSK-mediated phosphorylation of XTCF-3. One major issue that remains unresolved is whether activation of the Wnt pathway in dorsal cells involves the downregulation of GSK3 catalytic activity. The total catalytic activity of GSK3 in dorsal and ventral regions is not significantly different casting doubt on the idea that GSK3 activity is suppressed dorsally.[77] However, a caveat of these experiments is that the activity of GSK3 was measured by its ability to phosphorylate a peptide substrate. Since GBP may not block the ability of GSK3 to phosphorylate all peptide substrates[74] the lack of a difference in GSK3 activity along the dorsal-ventral axis may not accurately reflect the specific activity of GSK3 towards protein substrates. In addition, it is not known whether dorsal-ventral differences exist in the activity of Axin-associated GSK3, the pool of GSK3 required for regulation of β-catenin.

A central theme in signal transduction is that where there are kinases, a phophastase will not be far behind. Axis formation in *Xenopus* is no exception and recent studies have shown that the activity of the destruction complex may be regulated by Protein Phosphatase 2A (PP2A), a heterotrimeric Ser/Thr protein phosphatase.[99,100] Dorsal overexpression of the B56α regulatory subunit of PP2A results in ventralization of embryos showing that ectopic activation of PP2A can block endogenous dorsal Wnt signaling.[100] Moreover, ventral expression of B56α, or either the A or C subunit of PP2A can block secondary axis formation induced by Xwnt-8.[100] Co-immunoprecipitation experiments in cultured cells[101] and *Xenopus*[100] demonstrate that PP2A A, C, and B56α subunits are present in a complex with Axin suggesting that PP2A is a component of the destruction complex. Consistent with this idea, expression of B56α in egg extracts or addition of purified PP2A C subunit to egg extracts increases the rate of β-catenin degradation.[100] Conversely, addition of either okadaic acid or microcystin-LR, two general inhibitors of Ser/Thr protein phosphatases, to egg extracts significantly reduces the rate of β-catenin degradation.[100] Finally, addition of PP2A C subunit, but not PP1 C subunit, was able to fully restore β-catenin degradation to microcystin-LR depleted extracts showing that PP2A is likely to be the only Ser/Thr protein phosphatase required for β-catenin degradation in early *Xenopus* embryos.[100] These data provide compelling evidence that PP2A plays a critical role in regulating Wnt pathway activity along the dorsal-ventral axis. Important questions that remain unanswered include determining which components of the Wnt pathway are substrates for PP2A, determining the subunit composition and spatial distribution of PP2A during early development, and examining whether the activity of PP2A is regulated along the dorsal-ventral axis, and the function of PP2A in axis specification, however, may be more complex than described above. This follows from the observations that expression of the Cα subunit of PP2A can synergize with Dsh to induce axis duplication.[102] In addition, overexpression of the SV40 small t antigen and a B'ε subunit of PP2A, both negative regulators of the C subunit, blocked the ability of Dsh to induce axis duplication and a *siamois* reporter gene. Interestingly, the Cα subunit did not appear to affect β-catenin stability but instead acted on XTCF-3, converting it from a repressor to an activator. These data are consistent with the idea that PP2A might function antagonistically to GSK3 in regulating XTCF-3 function.

In addition to these observations, other alternative roles for PP2A have been proposed. PP2A can dephosphorylate Axin in vitro, an affect that mimics Wnt3a induced dephosphorylation seen in cultured cells.[96] In related experiments, Wnt mediated dephosphorylation of Axin reduced binding of Axin to β-catenin suggesting that PP2A might play a role in activating Wnt signaling by decreasing Axin/β-catenin interactions. PP2A can also dephosphorylate APC in vitro.[103]

These contradictory data regarding PP2A function may be resolved if distinct PP2A heterotrimeric complexes function at multiple levels in the Wnt pathway. For example, one regulatory subunit, such as a member of the B56 family, might target PP2A to one set of substrates leading to inhibition of signaling, while a second regulatory subunit might target PP2A to an alternative substrate resulting in activation of signaling. Numerous kinases have been shown to take part in Wnt signaling and, therefore, it is not unlikely that PP2A acts on distinct substrates at multiple levels of the Wnt pathway. Confirmation of this idea will require defining the composition of different PP2A complexes involved in Wnt signaling and then determining the substrate specificity of each of these complexes.

Phosphorylation of β-catenin marks it for destruction by the proteasome. The intermediary in this process is an E3 ubiquitin ligase that recognizes β-catenin only when phosphorylated and catalyzes the covalent addition of ubiquitin. βTrCP is a component of the E3 ligase identified in *Xenopus* that directly binds phosphorylated β-catenin.[104] Overexpression studies show that βTrCP can inhibit dorsal development and block the activity of Xwnt-8.[105,106] Conversely, overexpression of a dominant-negative mutant of βTrCP in ventral blastomeres results in stabilization of β-catenin and axis duplication.[105,106,104] These observations explain how GSK3-mediated phosphorylation of β-catenin promotes the rapid degradation of β-catenin by the proteasome and how inhibition of phosphorylation allows β-catenin to avoid the proteasome.

## Ventral Repression of Dorsal-Specific Genes

Mutation of the TCF/LEF sites in the *siamois* promoter blocks β-catenin-dependent activation of the reporter gene on the dorsal side but the same mutant promoter is still active on the ventral side of the embryo where levels of nuclear β-catenin are very low.[78] These data suggest that XTCF-3 functions as an activator in dorsal cells when complexed to β-catenin and a repressor in ventral cells when β-catenin is absent. The mechanism by which XTCF-3 acts as a repressor appears to involve its interaction with additional co-repressors including XGrg4, a member of the Groucho family of transcriptional regulators,[86] and XCtBP.[107] Dorsal injection of XGrg4 results in repression of the endogenous siamois gene and a ventralized phenotype[86] showing that expression of TCF target genes may result from competition between β-catenin and XGrg4.

ICAT is an additional maternally expressed protein that may also inhibit the transcriptional activity of β-catenin on the ventral side of the embryo.[108] ICAT, a small protein of 81 amino acids that shows no significant sequence similarity to known proteins, binds to armadillo repeats 10-12 of β-catenin. Functional studies show that ICAT can block the interaction between β-catenin and XTCF-3. Overexpression of ICAT on the dorsal side leads to a loss of dorsal cell fates consistent with the idea that ICAT prevents β-catenin/XTCF-3 interactions. In related experiments, overexpression of a dominant-negative form of ICAT in ventral blastomeres induced axis duplication. Together these data suggest that ICAT normally functions on the ventral side of the embryo to prevent β-catenin/XTCF-3 interactions indirectly inhibiting expression of dorsal target genes.

## Ventral Inhibition of Wnt/β-Catenin Signaling by the Wnt/$Ca^{2+}$ Pathway

Inhibition of the Wnt/β-catenin pathway in ventral cells may also involve the activity of the Wnt/$Ca^{2+}$ pathway and signaling via $Ca^{2+}$/calmodulin-dependent protein kinase II (CamKII).[62] Measurements of CamKII activity along the dorsal-ventral axis showed elevated activity in ventral regions of cleavage stage embryos. Overexpression of an activated form of CamKII in dorsal cells suppresses dorsal cell fates whereas overexpression of a kinase-dead form of CamKII in ventral cells suppresses ventral cell fates suggesting that CamKII plays a role in regulating axis specification. The function of CamKII in specifying ventral cell fates may be stimulated by an endogenous Wnt ligand since expression of a dominant negative form of XWnt-11 in ventral cells suppresses the spatial difference in CamKII activity. Both Xwnt-5A and XWnt-11 are expressed maternally and are capable of activating the Wnt/$Ca^{2+}$ pathway and CamKII in Xenopus[109,62] raising the possibility that one or both of these genes plays a role in activating CamKII. The observation that XWnt-11 protein accumulates dorsally[56] and appears to be required for gastrulation movements[110] shifts the focus to Xwnt-5A. The targets of CamKII also remain a mystery. It is possible that CamKII might directly impinge on the canonical Wnt/β-catenin pathway or may influence dorsal-ventral patterning through a parallel pathway.

## Other Potential Regulators of Wnt Signaling during Axis Formation

Over the past several years, studies from a variety of systems have identified a long list of putative regulators of the canonical Wnt/β-catenin pathway. The activity of a number of these factors in regulating Wnt signaling during axis specification has been determined utilizing overexpression analyses. The activity of these factors in *Xenopus* is listed in Table 1. Since expression data is lacking for many of these factors their precise function in axis formation remains unclear. Thus, determining whether these genes play critical roles in dorsal-ventral patterning in *Xenopus* awaits further analyses.

## A Working Model for the Function of Wnt Signaling in the Patterning of the Dorsal-Ventral Axis in *Xenopus*

How might things work? The model illustrated in Fig. 5 is a proposed mechanism for the role of Wnt signaling in the establishment of the dorsal-ventral axis in *Xenopus* based on the literature discussed above. It should be noted that although the model presented here is consistent with current literature, so are many others.

The *Xenopus* egg possesses an inherent animal-vegetal asymmetry that is manifest at the molecular level by the asymmetric distribution of specific mRNAs and proteins to either the animal or vegetal cortical regions (Fig. 5A). The dorsal determinant is among these factors and is localized to the vegetal pole in tight association with the cortex. Fertilization occurs anywhere in the animal hemisphere and triggers formation of a parallel array of microtubules in the vegetal region at the interface of the outer cortex and inner cytoplasmic core. The sperm aster influences the orientation of the microtubule array but is not required for its formation. The polarized polymerization of the microtubules in the array (plus ends point dorsally), however, is critical for patterning of the dorsal-ventral axis as it forms a new axis of polarity in the egg. The polarity of the microtubule determines not only the direction of cortical rotation but also serves as tracks for the transport of the dorsal determinant from the vegetal pole to the future dorsal side of the embryo.

The molecular nature of the dorsal determinant remains a mystery although the localization and movement of Dsh-GFP suggest that Dsh may be one component (Fig. 5A). Dsh, through an association with a kinesin-like motor, might be transported dorsally along the microtubule array. Movement of Dsh from the vegetal pole to the prospective dorsal side would lead to the accumulation of Dsh in the cortical region on the dorsal side and establishment of an asymmetry in the Wnt pathway. The requirement of XFz-7 for the specification of dorsal cell fates suggests that signaling through XFz-7 and the axis inducing Wnt, Xwnt-8b, might influence the activity of Dsh. One possibility is that XFz-7, Xwnt-8b, and Dsh may associate with a specific class of vesicles that are transported dorsally during cortical rotation. Alternatively, XFz-7 might function in a parallel pathway that also leads to activation of dorsal-specific target genes perhaps through activation of CKIε.

The accumulation of Dsh on the dorsal side would then promote stabilization of β-catenin through its interactions with CKIε and GBP (Fig. 5A). CKIε is predicted to play two roles in promoting the stabilization of β-catenin. First, CKIε phosphorylates Dsh leading to increased association of Dsh with GBP. Dsh also binds Axin, and through this interaction Dsh might then deliver GBP to GSK3 in the destruction complex. Binding of GBP to GSK3 might inhibit the ability of GSK3 to phosphorylate β-catenin, promote dissociation of GSK3 from the destruction complex, and/or stimulate the depletion of GSK3 on the dorsal side. Second, CKIε might phosphorylate XTCF-3, promoting the binding of XTCF-3 to β-catenin. Since β-catenin binding to XTCF-3, Axin, and APC is exclusive, association of β-catenin with XTCF-3 would protect β-catenin from the destruction complex. Furthermore, binding of β-catenin to XTCF-3 would promote the nuclear accumulation of the β-catenin/XTCF-3 complex. In the nucleus,

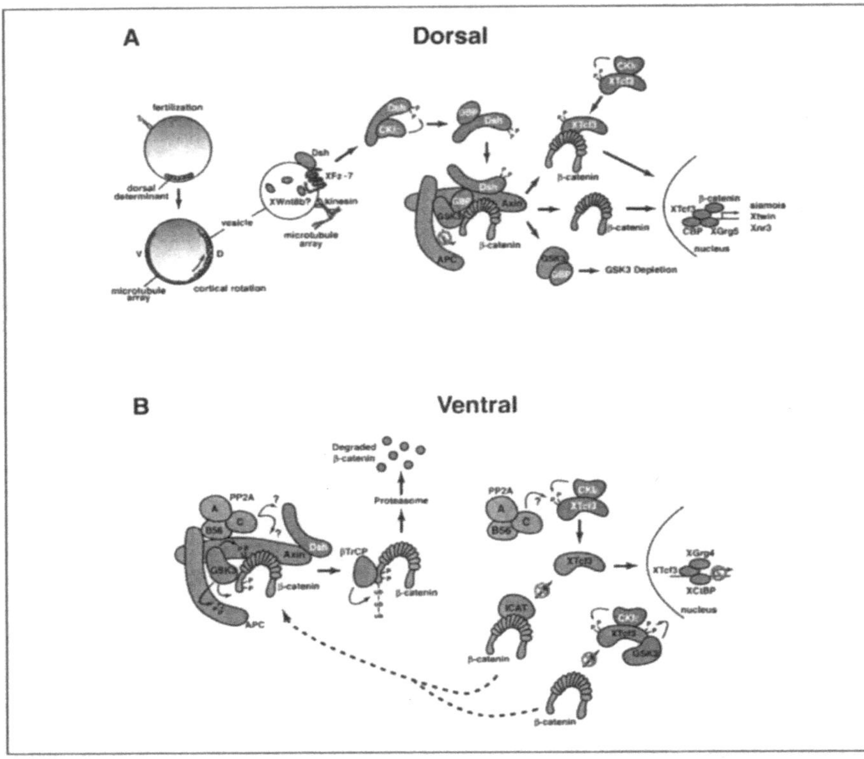


Figure 5. Model for the role of Wnt signaling in the formation of the dorsal-ventral axis in Xenopus. (A) Activation of the Wnt pathway on the dorsal side of the embryo is proposed to be dependent on the transport of Dsh from the vegetal pole to the future dorsal side during cortical rotation. Transport of Dsh may be facilitated through its association with a specific class of vesicles that may also contain Xwnt-8b and XFz-7. On the dorsal side, XFz-7 and Dsh function together with CKIε and GBP to antagonize the destruction complex and stabilize β-catenin. GBP might inhibit GSK3-mediated phosphorylation of β-catenin by physically blocking GSK3/β-catenin interactions, promoting dissociation of GSK3 from the destruction complex, and/or enhancing the depletion of GSK3. Binding of β-catenin to XTCF-3 may also play a role in protecting β-catenin from degradation. The interaction of β-catenin with XTCF-3 might be enhanced by phosphorylation of XTCF-3 by CKIε. Stabilization of β-catenin leads to its accumulation in the nucleus where together with XTCF-3 it activates dorsal-specific target genes. (B) On the ventral side of the embryo, the destruction complex facilitates the rapid degradation of β-catenin by the proteasome. GSK3-mediated phosphorylation of β-catenin leads to its interaction with βTrCP and subsequent ubiquitination. Poly-ubiquitination serves as a tag on β-catenin, marking it for degradation by the proteasome. XTCF-3/β-catenin interactions are also blocked on the ventral side by the concerted action of ICAT, GSK3, and PP2A. In the absence of β-catenin in the nucleus XTCF-3 functions as a transcriptional repressor blocking expression of dorsal target genes. See text for details.

the β-catenin/XTCF-3 complex would bind sequence elements present in the upstream regions of dorsal-specific target genes and stimulate transcription.

On the ventral side of the embryo, the destruction complex is active and promotes the rapid degradation of β-catenin (Fig. 6B). Axin and APC bind β-catenin and facilitate its phosphorylation by GSK3. APC and Axin are also substrates of GSK3 and phosphorylation of each of these proteins might lead to enhanced binding of β-catenin. Phosphorylated β-catenin is then bound by β-TRCP and is ubiquitinated, a tag that marks β-catenin for destruction by the proteasome. PP2A also stimulates β-catenin degradation on the ventral side although its substrates are unknown. One possibility is that PP2A may dephosphorylate XTCF-3 thereby decreasing its affinity for β-catenin. The low level of β-catenin present in ventral blastomeres precludes its interaction with XTCF-3. Moreover, ICAT might bind free β-catenin and prevent its interaction with XTCF-3. Acting without β-catenin, XTCF-3 functions together with XCtBP and XGrg4 to repress transcription of dorsal-specific target genes. Signaling through the Wnt/$Ca^{2+}$ pathway and CamKII also promotes ventral development. The targets of CamKII may include components of the canonical Wnt/β-catenin pathway or additional unidentified effectors.

## Major Unresolved Issues

Although recent years have seen a wealth of new information regarding the role of Wnt signaling in dorsal-ventral patterning, many important question remain unanswered. Is a maternal Wnt ligand(s) required for axis specification? How does XFz-7 promote dorsal development? Is Dsh required for dorsal development and if so how does dorsal enrichment of Dsh promote the stabilization of β-catenin? How is the activity of CKIε, GBP, PP2A, and GSK3 regulated along the dorsal-ventral axis? Is APC required for axis specification? Answers to these questions will require detailed examination of protein-protein interactions, intracellular localization, and activity of Wnt pathway components in dorsal and ventral regions of the early embryo. Although answers may not come easy, the ingenuity and perseverance of researchers in the *Xenopus* field will assure many future successes. These data will not only provide information about the role of Wnt signaling in dorsal-ventral patterning in Xenopus but will also provide critical insights into the general mechanism of Wnt signal transduction. In addition, since aberrant activation of Wnt signaling is implicated in a variety of human cancers, future research on Wnt signaling is also likely to yield valuable information with high relevance to the pathogenesis and treatment of human disease.

### *Acknowledgements*

I would like to thank Caroline D'Cunha for comments on this manuscript. Owing to rapid progress in the field and limited space for discussion, I apologize to those researchers whose data may have been omitted from this review.

## References

1. Rebagliati MR, Weeks DL, Harvey RP et al. Identification and cloning of localized maternal RNAs from Xenopus eggs. Cell 1985; 42(3):769-77.
2. Ku M, Melton DA. Xwnt-11: A maternally expressed Xenopus wnt gene. Development 1993; 119(4):1161-1173.
3. Lustig KD, Kroll KL, Sun EE et al. Expression cloning of a Xenopus T-related gene (xombi) involved in mesodermal patterning and blastopore lip formation. Development 1996; 122(12):4001-4012.
4. Stennard F, Carnac G, Gurdon JB. The Xenopus T-box gene, antipodean, encodes a vegetally localised maternal mrna and can trigger mesoderm formation. Development 1996; 122(12):4179-4188.
5. Zhang J, King ML. Xenopus VegT RNA is localized to the vegetal cortex during oogenesis and encodes a novel T-box transcription factor involved in mesodermal patterning. Development 1996; 122(12):4119-4129.
6. Horb ME, Thomsen GH. A vegetally localized T-box transcription factor in Xenopus eggs specifies mesoderm and endoderm and is essential for embryonic mesoderm formation. Development 1997; 124(9):1689-1698.

7. Fujisue M, Kobayakawa Y, Yamana K. Occurrence of dorsal axis-inducing activity around the vegetal pole of an uncleaved Xenopus egg and displacement to the equatorial region by cortical rotation. Development 1993; 118:163-170.
8. Sakai M. The vegetal determinants required for the Spemann Organizer move equatorially during the first cell cycle. Development 1996; 122(7):2207-14.
9. Vincent JP, Oster GF, Gerhart JC. Kinematics of gray crescent formation in Xenopus eggs: The displacement of subcortical cytoplasm relative to the egg surface. Dev Biol 1986; 113(2):484-500.
10. Larabell CA, Rowning BA, Wells J et al. Confocal microscopy analysis of living Xenopus eggs and the mechanism of cortical rotation. Development 1996; 122(4):1281-1289.
11. Gerhart J, Danilchik M, Doniach T et al. Cortical rotation of the Xenopus egg: Consequences for the anteroposterior pattern of embryonic dorsal development. Development 1989; 107(Suppl):37-51.
12. Heasman J, Crawford A, Goldstone K et al. Overexpression of cadherins and underexpression of beta-catenin inhibit dorsal mesoderm induction in early Xenopus embryos. Cell 1994; 79(5):791-803.
13. Schneider S, Steinbeisser H, Warga RM et al. Beta-catenin translocation into nuclei demarcates the dorsalizing centers of frog and fish embryos. Mech Dev 1996; 57:191-198.
14. Larabell CA, Torres M, Rowning BA et al. Establishment of the dorso-ventral axis in Xenopus embryos is presaged by early asymmetries in beta-catenin that are modulated by the Wnt signaling pathway. J Cell Biol 1997; 136(5):1123-1136.
15. Funayama N, Fagotto F, McCrea P et al. Embryonic axis induction by the armadillo repeat domain of beta-catenin: Evidence for intracellular signaling. J Cell Biol 1995; 128:959-968.
16. Heasman J, Kofron M, Wylie C. Beta-catenin signaling activity dissected in the early Xenopus embryo: A novel antisense approach. Dev Biol 2000; 222(1):124-134.
17. Heasman J. Patterning the Xenopus blastula. Development 1997; 124(21):4179-4191.
18. De Robertis EM, Larrain J, Oelgeschlager M et al. The establishment of Spemann's Organizer and patterning of the vertebrate embryo. Nat Rev Genet 2000; 1(3):171-181.
19. Gamse J, Sive H. Vertebrate anteroposterior patterning: The Xenopus neurectoderm as a paradigm. Bioessays 2000; 22(11):976-986.
20. Kimelman D, Griffin KJ. Vertebrate mesendoderm induction and patterning. Curr Opin Genet Dev 2000; 10(4):350-356.
21. Bouwmeester T. The spemann-mangold organizer: The control of fate specification and morphogenetic rearrangements during gastrulation in Xenopus. Int J Dev Biol 2001; 45(1):251-258.
22. Scharf SR, Gerhart JC. Determination of the dorsal-ventral axis in eggs of Xenopus laevis: Complete rescue of uv-impaired eggs by oblique orientation before first cleavage. Dev Biol 1980; 79(1):181-198.
23. Vincent JP, Gerhart JC. Subcortical rotation in Xenopus eggs: An early step in embryonic axis specification. Dev Biol 1987; 123(2):526-539.
24. Ancel P, Vintemberger P. Recherches sur le determinisme de la smetrie bilaterale dans l'oeuf des amphibiens. Bull Biol Fr Belg 1948; 31 (Suppl.):1-182.
25. Elinson RP, Rowning B. A transient array of parallel microtubules in frog eggs: Potential tracks for a cytoplasmic rotation that specifies the dorso-ventral axis. Dev Biol 1988; 128(1):185-197.
26. Houliston E, Elinson RP. Patterns of microtubule polymerization relating to cortical rotation in Xenopus laevis eggs. Development 1991; 112(1):107-117.
27. Houliston E, Le Guellec R, Kress M et al. The kinesin-related protein eg5 associates with both interphase and spindle microtubules during Xenopus early development. Dev Biol 1994; 164(1):147-159.
28. Scharf SR, Gerhart JC. Axis determination in eggs of Xenopus laevis: A critical period before first cleavage, identified by the common effects of cold, pressure and ultraviolet irradiation. Dev Biol 1983; 99(1):75-87.
29. Gerhart J, Ubbels G, Black S et al. A reinvestigation of the role of the grey crescent in axis formation in Xenopus laevis. Nature 1981; 292(5823):511-516.
30. Ubbels GA, Hara K, Koster CH et al. Evidence for a functional role of the cytoskeleton in determination of the dorsoventral axis in Xenopus laevis eggs. J Embryol Exp Morphol 1983; 77:15-37.
31. Vincent JP, Scharf SR, Gerhart JC. Subcortical rotation in Xenopus eggs: A preliminary study of its mechanochemical basis. Cell Motil Cytoskeleton 1987; 8(2):143-154.

32. Cooke J, Smith JC. The midblastula cell cycle transition and the character of mesoderm in U.V.-induced nonaxial Xenopus development. Development 1987; 99(2):197-210.
33. Yuge M, Kobayakawa Y, Fujisue M et al. A cytoplasmic determinant for dorsal axis formation in an early embryo of Xenopus laevis. Development 1990; 110(4):1051-1056.
34. Kikkawa M, Takano K, Shinagawa A. Location and behavior of dorsal determinants during first cell cycle in Xenopus eggs. Development 1996; 122(12):3687-3696.
35. Kageura H. Activation of dorsal development by contact between the cortical dorsal determinant and the equatorial core cytoplasm in eggs of Xenopus laevis. Development 1997; 124(8):1543-1551.
36. Marikawa Y, Li Y, Elinson RP. Dorsal determinants in the Xenopus egg are firmly associated with the vegetal cortex and behave like activators of the Wnt pathway. Dev Biol 1997; 191(1):69-79.
37. Rowning BA, Wells J, Wu M et al. Microtubule-mediated transport of organelles and localization of beta-catenin to the future dorsal side of Xenopus eggs. Proc Natl Acad Sci USA 1997; 94(4):1224-1229.
38. Holowacz T, Elinson RP. Properties of the dorsal activity found in the vegetal cortical cytoplasm of Xenopus eggs. Development 1995; 121(9):2789-2798.
39. Harland R, Gerhart J. Formation and function of Spemann's Organizer. Annu Rev Cell Dev Biol 1997; 13:611-667.
40. Smith WC, Harland RM. Injected XWnt-8 RNA acts early in Xenopus embryos to promote formation of a vegetal dorsalizing center. Cell 1991; 67(4):753-765.
41. Sokol S, Christian JL, Moon RT et al. Injected Wnt RNA induces a complete body axis in Xenopus embryos. Cell 1991; 67(4):741-752.
42. Sokol SY, Klingensmith J, Perrimon N et al. Dorsalizing and neuralizing properties of XDsh, a maternally expressed Xenopus homolog of dishevelled. Development 1995; 121(6):1637-1647.
43. Yost C, Farr GH, 3rd, Pierce SB et al. GBP, an inhibitor of GSK-3, is implicated in Xenopus development and oncogenesis. Cell 1998; 93(6):1031-1041.
44. Dominguez I, Itoh K, Sokol SY. Role of glycogen synthase kinase 3 beta as a negative regulator of dorsoventral axis formation in Xenopus embryos. Proc Natl Acad Sci USA 1995; 92(18):8498-8502.
45. He X, Saint-Jeannet JP, Woodgett JR et al. Glycogen synthase kinase-3 and dorsoventral patterning in Xenopus embryos. Nature 1995; 374(6523):617-622.
46. Pierce SB, Kimelman D. Regulation of Spemann Organizer formation by the intracellular kinase XGSK-3. Development 1995; 121(3):755-765.
47. Molenaar M, van de Wetering M, Oosterwegel M et al. XTCF-3 transcription factor mediates beta-catenin-induced axis formation in Xenopus embryos. Cell 1996; 86(3):391-399.
48. Sumanas S, Strege P, Heasman J et al. The putative Wnt receptor Xenopus Frizzled-7 functions upstream of beta-catenin in vertebrate dorsoventral mesoderm patterning. Development 2000; 127(9):1981-1990.
49. Kofron M, Klein P, Zhang F et al. The role of maternal Axin in patterning the Xenopus embryo. Dev Biol 2001; 237(1):183-201.
50. Peters JM, McKay RM, McKay JP et al. Casein kinase I transduces Wnt signals. Nature 1999; 401(6751):345-350.
51. Sakanaka C, Leong P, Xu L et al. Casein kinase I epsilon in the Wnt pathway: Regulation of beta-catenin function. Proc Natl Acad Sci USA 1999; 96(22):12548-12552.
52. Tamai K, Semenov M, Kato Y et al. LDL-receptor-related proteins in Wnt signal transduction. Nature 2000; 407(6803):530-535.
53. Christian JL, Gavin BJ, McMahon AP et al. Isolation of cDNAs partially encoding four Xenopus Wnt-1/Int-1-related proteins and characterization of their transient expression during embryonic development. Dev Biol 1991; 143(2):230-234.
54. Moon RT, Campbell RM, Christian JL et al. Xwnt-5a: A maternal Wnt that affects morphogenetic movements after overexpression in embryos of Xenopus laevis. Development 1993; 119(1):97-111.
55. Cui Y, Brown JD, Moon RT et al. Xwnt-8b: A maternally expressed Xenopus Wnt gene with a potential role in establishing the dorsoventral axis. Development 1995; 121(7):2177-2186.
56. Schroeder KE, Condic ML, Eisenberg LM et al. Spatially regulated translation in embryos: Asymmetric expression of maternal Wnt-11 along the dorsal-ventral axis in Xenopus. Dev Biol 1999; 214(2):288-297.

57. Heisenberg CP, Tada M, Rauch GJ et al. Silberblick/Wnt-11 mediates convergent extension movements during zebrafish gastrulation. Nature 2000; 405(6782):76-81.
58. Tada M, Smith JC. XWnt-11 is a target of Xenopus brachyury: Regulation of gastrulation movements via Dishevelled, but not through the canonical Wnt pathway. Development 2000; 127(10):2227-2238.
59. Kuhl M, Geis K, Sheldahl LC et al. Antagonistic regulation of convergent extension movements in Xenopus by Wnt/beta-catenin and Wnt/$Ca^{2+}$ signaling. Mech Dev 2001; 106(1-2):61-76.
60. Wallingford JB, Vogeli KM, Harland RM. Regulation of convergent extension in Xenopus by Wnt5a and Frizzled-8 is independent of the canonical Wnt pathway. Int J Dev Biol 2001; 45(1):225-227.
61. Torres MA, Yang-Snyder JA, Purcell SM et al. Activities of the Wnt-1 class of secreted signaling factors are antagonized by the Wnt-5a class and by a dominant negative cadherin in early Xenopus development. J Cell Biol 1996; 133(5):1123-1137.
62. Kuhl M, Sheldahl LC, Malbon CC et al. Ca(2+)/calmodulin-dependent protein kinase II is stimulated by Wnt and Frizzled homologs and promotes ventral cell fates in Xenopus. J Biol Chem 2000; 275(17):12701-12711.
63. Hoppler S, Brown JD, Moon RT. Expression of a dominant-negative Wnt blocks induction of myod in Xenopus embryos. Genes Dev 1996; 10(21):2805-2817.
64. Moon RT, Brown JD, Torres M. Wnts modulate cell fate and behavior during vertebrate development. Trends Genet 1997; 13(4):157-162.
65. Shi DL, Goisset C, Boucaut JC. Expression of XFz3, a Xenopus Frizzled family member, is restricted to the early nervous system. Mech Dev 1998; 70(1-2):35-47.
66. Miller JR, Rowning BA, Larabell CA et al. Establishment of the dorsal-ventral axis in Xenopus embryos coincides with the dorsal enrichment of Dishevelled that is dependent on cortical rotation. J Cell Biol 1999; 146(2):427-437.
67. Brown JD, Hallagan SE, McGrew LL et al. The maternal Xenopus beta-catenin signaling pathway, activated by Frizzled homologs, induces goosecoid in a cell non-autonomous manner. Dev Growth Differ 2000; 42(4):347-357.
68. Shi DL, Boucaut JC. Xenopus Frizzled 4 is a maternal mrna and its zygotic expression is localized to the neuroectoderm and trunk lateral plate mesoderm. Mech Dev 2000; 94(1-2):243-245.
69. Sumanas S, Ekker SC. Xenopus Frizzled-7 morphant displays defects in dorsoventral patterning and convergent extension movements during gastrulation. Genesis 2001; 30(3):119-122.
70. Rothbacher U, Laurent MN, Deardorff MA et al. Dishevelled phosphorylation, subcellular localization and multimerization regulate its role in early embryogenesis. Embo J 2000; 19(5):1010-1022.
71. Scharf SR, Rowning B, Wu M et al. Hyperdorsoanterior embryos from Xenopus eggs treated with $D_2O$. Dev Biol 1989; 134(1):175-188.
72. Sokol SY. Analysis of Dishevelled signalling pathways during Xenopus development. Curr Biol 1996; 6(11):1456-67.
73. Salic A, Lee E, Mayer L et al. Control of beta-catenin stability: Reconstitution of the cytoplasmic step of the Wnt pathway in Xenopus egg extracts. Molecular Cell 2000; 5:523-532.
74. Farr III GH, Ferkey DM, Yost C et al. Interaction among GSK-3, GBP, Axin and Apc in Xenopus axis specification. J Cell Biol 2000; 148:691-701.
75. Itoh K, Antipova A, Ratcliffe MJ et al. Interaction of Dishevelled and Xenopus Axin-related protein is required for Wnt signal transduction. Mol Cell Biol 2000; 20(6):2228-2238.
76. Lee E, Salic A, Kirschner MW. Physiological regulation of beta-catenin stability by TCF-3 and CK1epsilon. J Cell Biol 2001; 154(5):983-993.
77. Dominguez I, Green JB. Dorsal downregulation of GSK3beta by a non-Wnt-like mechanism is an early molecular consequence of cortical rotation in early Xenopus embryos. Development 2000; 127(4):861-8.
78. Brannon M, Gomperts M, Sumoy L et al. A beta-catenin/XTCF-3 complex binds to the siamois promoter to regulate dorsal axis specification in Xenopus. Genes Dev 1997; 11(18):2359-2370.
79. Fan MJ, Gruning W, Walz G et al. Wnt signaling and transcriptional control of siamois in Xenopus embryos. Proc Natl Acad Sci USA 1998; 95(10):5626-5631.

80. Laurent MN, Blitz IL, Hashimoto C et al. The Xenopus homeobox gene twin mediates Wnt induction of goosecoid in establishment of Spemann's Organizer. Development 1997; 124(23):4905-4916.
81. McKendry R, Hsu SC, Harland RM et al. LEF-1/TCF proteins mediate Wnt-inducible transcription from the Xenopus nodal-related 3 promoter. Dev Biol 1997; 192(2):420-31.
82. Hyde CE, Old RW. Regulation of the early expression of the Xenopus nodal-related 1 gene, xnr1. Development 2000; 127(6):1221-1229.
83. Vleminckx K, Kemler R, Hecht A. The c-terminal transactivation domain of beta-catenin is necessary and sufficient for signaling by the LEF-1/beta-catenin complex in Xenopus laevis. Mech Dev 1999; 81(1-2):65-74.
84. Hecht A, Vleminckx K, Stemmler MP et al. The p300/CBP acetyltransferases function as transcriptional coactivators of beta-catenin in vertebrates. Embo J 2000; 19(8):1839-1850.
85. Takemaru KI, Moon RT. The transcriptional coactivator CBP interacts with beta-catenin to activate gene expression. J Cell Biol 2000; 149(2):249-254.
86. Roose J, Molenaar M, Peterson J et al. The Xenopus Wnt effector XTCF-3 interacts with Groucho-related transcriptional repressors. Nature 1998; 395(6702):608-612.
87. Miller JR, Hocking AM, Brown JD et al. Mechanism and function of signal transduction by the Wnt/beta-catenin and Wnt/$Ca^{2+}$ pathways. Oncogene 1999; 18(55):7860-7872.
88. Polakis P. Wnt signaling and cancer. Genes Dev 2000; 14(15):1837-1851.
89. Vleminckx K, Wong E, Guger K et al. Adenomatous polyposis coli tumor suppressor protein has signaling activity in Xenopus laevis embryos resulting in the induction of an ectopic dorsoanterior axis. J Cell Biol 1997; 136(2):411-420.
90. Zeng L, Fagotto F, Zhang T et al. The mouse fused locus encodes axin, an inhibitor of the Wnt signaling pathway that regulates embryonic axis formation. Cell 1997; 90(1):181-192.
91. Fagotto F, Jho E, Zeng L et al. Domains of Axin involved in protein-protein interactions, Wnt pathway inhibition, and intracellular localization. J Cell Biol 1999; 145(4):741-756.
92. Kao KR, Elinson RP. The entire mesodermal mantle behaves as Spemann's Organizer in dorsoanterior enhanced Xenopus laevis embryos. Dev Biol 1988; 127(1):64-77.
93. Kao KR, Elinson RP. Dorsalization of mesoderm induction by lithium. Dev Biol 1989; 132(1):81-90.
94. Hedgepeth CM, Conrad LJ, Zhang J et al. Activation of the Wnt signaling pathway: A molecular mechanism for lithium action. Dev Biol 1997; 185(1):82-91.
95. Yamamoto H, Kishida S, Kishida M et al. Phosphorylation of Axin, a Wnt signal negative regulator, by Glycogen Synthase Kinase-3beta regulates its stability. J Biol Chem 1999; 274(16):10681-10684.
96. Willert K, Shibamoto S, Nusse R. Wnt-induced dephosphorylation of Axin releases beta-catenin from the Axin complex. Genes Dev 1999; 13(14):1768-1773.
97. Ferkey DM, Kimelman D. GSK-3: New thoughts on an old enzyme. Dev Biol 2000; 225(2):471-479.
98. Yost C, Torres M, Miller JR et al. The axis-inducing activity, stability, and subcellular distribution of beta-catenin is regulated in Xenopus embryos by Glycogen Synthase Kinase 3. Genes Dev 1996; in press.
99. Seeling JM, Miller JR, Gil R et al. Regulation of beta-catenin signaling by the B56 subunit of Protein Phosphatase 2A. Science 1999; 283(5410):2089-2091.
100. Li X, Yost HJ, Virshup DM et al. Protein Phosphatase 2A and its B56 regulatory subunit inhibit Wnt signaling in Xenopus. Embo J 2001; 20(15):4122-41231.
101. Hsu W, Zeng L, Costantini F. Identification of a domain of Axin that binds to the serine/threonine Protein Phosphatase 2A and a self-binding domain. J Biol Chem 1999; 274(6):3439-3445.
102. Ratcliffe MJ, Itoh K, Sokol SY. A positive role for the PP2A catalytic subunit in Wnt signal transduction. J Biol Chem 2000; 275(46):35680-35683.
103. Ikeda S, Kishida M, Matsuura Y et al. GSK-3beta-dependent phosphorylation of Adenomatous Polyposis Coli gene product can be modulated by beta-catenin and Protein Phosphatase 2A complexed with Axin. Oncogene 2000; 19(4):537-545.
104. Liu C, Kato Y, Zhang Z et al. Beta-TrCP couples beta-catenin phosphorylation-degradation and regulates Xenopus axis formation. Proc Natl Acad Sci USA 1999; 96(11):6273-6278.

105. Marikawa Y, Elinson RP. Beta-trcp is a negative regulator of wnt/beta-catenin signaling pathway and dorsal axis formation in xenopus embryos. Mech Dev 1998; 77(1):75-80.
106. Lagna G, Carnevali F, Marchioni M et al. Negative regulation of axis formation and Wnt signaling in Xenopus embryos by the F-box/WD40 protein beta TrCP. Mech Dev 1999; 80(1):101-106.
107. Brannon M, Brown JD, Bates R et al. XCtBP is a XTCF-3 co-repressor with roles throughout Xenopus development. Development 1999; 126(14):3159-31570.
108. Tago K, Nakamura T, Nishita M et al. Inhibition of Wnt signaling by Icat, a novel beta-catenin-interacting protein. Genes Dev 2000; 14(14):1741-1749.
109. Sheldahl LC, Park M, Malbon CC et al. Protein Kinase C is differentially stimulated by Wnt and Frizzled homologs in a G-protein-dependent manner. Curr Biol 1999; 9(13):695-698.
110. Smith JC, Conlon FL, Saka Y et al. XWnt-11 and the regulation of gastrulation in Xenopus. Philos Trans R Soc Lond B Biol Sci 2000; 355(1399):923-930.
111. Rasmussen JT, Deardorff MA, Tan C et al. Regulation of eye development by Frizzled signaling in Xenopus. Proc Natl Acad Sci U S A 2001; 98(7):3861-3866.
112. Djiane A, Riou J, Umbhauer M et al. Role of Frizzled 7 in the regulation of convergent extension movements during gastrulation in Xenopus laevis. Development 2000; 127(14):3091-3100.
113. Medina A, Reintsch W, Steinbeisser H. Xenopus Frizzled 7 can act in canonical and non-canonical Wnt signaling pathways: Implications on early patterning and morphogenesis. Mech Dev 2000; 92(2):227-237.
114. Hedgepeth CM, Deardorff MA, Klein PS. Xenopus Axin interacts with Glycogen Synthase Kinase-3 beta and is expressed in the anterior midbrain. Mech Dev 1999; 80(2):147-151.
115. Molenaar M, Brian E, Roose J et al. Differential expression of the Groucho-related genes 4 and 5 during early development of Xenopus laevis. Mech Dev 2000; 91(1-2):311-315.
116. Fujii G, Tsuchiya R, Itoh Y et al. Molecular cloning and expression of Xenopus p300/CBP. Biochim Biophys Acta 1998; 1443(1-2):41-54.
117. Kadoya T, Kishida S, Fukui A et al. Inhibition of Wnt signaling pathway by a novel Axin-binding protein. J Biol Chem 2000; 275(47):37030-37037.
118. Sakamoto I, Kishida S, Fukui A et al. A novel beta-catenin-binding protein inhibits beta-catenin-dependent TCF activation and axis formation. J Biol Chem 2000; 275(42):32871-8.
119. Kobayashi M, Kishida S, Fukui A et al. Nuclear localization of duplin, a beta-catenin-binding protein, is essential for its inhibitory activity on the Wnt signaling pathway. J Biol Chem 2001;13:13.
120. Hino S, Kishida S, Michiue T et al. Inhibition of the Wnt signaling pathway by Idax, a novel Dvl-binding protein. Mol Cell Biol 2001; 21(1):330-342.
121. Zeng W, Wharton KA, Jr., Mack JA et al. Naked cuticle encodes an inducible antagonist of Wnt signalling. Nature 2000; 403(6771):789-795.
122. Ishitani T, Ninomiya-Tsuji J, Nagai S et al. The Tak1-Nlk-Mapk-related pathway antagonizes signalling between beta-catenin and transcription factor TCF. Nature 1999; 399(6738):798-802.
123. Sun TQ, Lu B, Feng JJ et al. Par-1 is a Dishevelled-associated kinase and a positive regulator of Wnt signalling. Nat Cell Biol 2001; 3(7):628-636.
124. Yamamoto H, Hinoi T, Michiue T et al. Inhibition of the Wnt signaling pathway by the PR61 subunit of Protein Phosphatase 2A. J Biol Chem 2001; 276:29;26875-26882.

# CHAPTER 5

# The Role of Wnt Signaling in Vertebrate Head Induction and the Organizer-Gradient Model Dualism

**Clemens Kiecker and Christof Niehrs**

## Abstract

The prevailing model for anteroposterior (AP) axis formation during vertebrate embryogenesis implies distinct organizer regions inducing head and trunk structures, respectively. A gradient of posteriorizing activity (transformer) which regulates AP patterning of the neuraxis has been suggested on the other hand by classical studies using amphibian embryos. Here, we will review the roles of Wnt signaling during head formation in various vertebrate model organisms. Early in vertebrate embryogenesis, organizer (and thus head) formation depends on a Wnt-type β-catenin-mediated signal. During gastrulation, posteriorizing Wnt/β-catenin signaling antagonizes the head organizer whose distinguishing feature is the secretion of Wnt inhibitors. The interplay between head organizer-derived Wnt inhibitors and posteriorizing Wnts establishes a gradient of Wnt/β-catenin activity that regulates AP patterning of the vertebrate gastrula. Thus, the concepts of head and trunk organizer and of the transformer gradient are reconciled in a dualistic view of AP axis formation.

## Introduction

In vertebrates, the two main body axes, anteroposterior (AP) and dorsoventral (DV), are progressively established during early embryogenesis. The anterior extremity of a vertebrate is defined by its head, a complex structure formed from derivatives of all germ layers. The vertebrate head is characterized by an elaborate segmented brain, sensory organs and structural components such as skull bones, teeth and facial muscles. These features distinguish vertebrates even from their closest nonvertebrate relatives, the cephalochordates.[1] Here, we will review the different and sometimes opposing roles played by growth factors of the Wnt family during head formation with special emphasis on the emerging notion that inhibition of Wnt/β-catenin signaling is crucially required for anterior specification. As we will discuss below, the seemingly different organizer and gradient models for AP axis formation are reconciled in light of the emerging underlying molecular mechanism.

*Wnt Signaling in Development*, edited by Michael Kühl.

## The Vertebrate Organizer

Fundamental progress in understanding embryonic axis formation was made in the 1920s when the embryologists Hilde Mangold and Hans Spemann demonstrated the extraordinary inductive potency of the amphibian dorsal blastopore lip by performing transplantation experiments on newt embryos. They explanted the comparably small dorsal lip region from a donor gastrula and grafted it to the ventral side of a host. This treatment induced a complete secondary embryonic axis, including a well-patterned head, within the host embryo. Only few structures of the secondary axis were derived from the transplanted tissue while most of the ectopic tissues were derived from the host. Mangold and Spemann concluded that the transplanted cells had induced and organized the surrounding host tissue nonautonomously to become part of the secondary axis.[2]

Tissues corresponding to Spemann's organizer have since been identified in chick, fish, rabbit and mouse.[3-7] The organizer regions of all these vertebrates show comparable tissue fates and are able to induce twinning when transplanted to ectopic locations in host embryos.[3-5,7-10] Thus, the vertebrate organizer appears to be an evolutionary conserved structure with a primary function in embryonic axis formation.

## Classical Models for Anteroposterior Axis Formation

At least three different tissues derived from the organizer are distinguishable at neurula stages (from anterior to posterior): (1) the anterior endoderm (AE) which will give rise to the liver at later stages, (2) the prechordal mesendoderm (PME) which will differentiate into head mesenchyme, head and eye muscles and foregut and (3) the chordamesoderm (CM) which will give rise to the notochord. Together, these organizer derivatives are referred to as axial mesendoderm.

In the 1930s, Spemann's colleague Otto Mangold dissected the archenteron roof (containing the axial mesendoderm) of early newt neurulae into four transversal segments and implanted these pieces into the blastocoel of early gastrulae. As in the Mangold-Spemann experiment, the grafts were able to induce ectopic structures in the host embryos. Their inductive potential, however, seemed to be more restricted compared with early organizer grafts: The anteriormost grafts (containing AE) only induced ectopic balancers and parts of the oral apparatus while grafts containing PME and anterior CM frequently induced ectopic head structures such as balancers, eyes, otic vesicles, fore- and midbrain; grafts containing medial CM induced ectopic trunks including hindbrain and spinal cord tissue and the posteriormost type of grafts (containing posterior CM) induced ectopic tails (Fig. 1A). Mangold concluded that derivatives of the organizer differ not only in fate but also in inducing potential, with anterior organizer inducing head structures and posterior organizer inducing trunk and tail structures.[11] Similarily, Einsteck assays using early gastrula lips resulted in the formation of complete secondary axes including heads while late gastrula lips only induced secondary trunks (Fig. 1B). Taken together, these results suggested that the organizer consists of distinct head- and trunk-organizing regions and that the region between PME and anterior CM harbors the strongest head-inducing potential.

An alternative model for AP axis formation was introduced in the 1950s by the 'Dutch school' of embryologists represented by Pieter Nieuwkoop and colleagues.[12] Nieuwkoop implanted folds of competent ectoderm into prospective neural plates of amphibian embryos at different AP levels. Analysis of the grafts revealed that folds implanted at the forebrain level had differentiated exclusively into anterior neural structures (such as nasal pits, pineal gland, eyes and forebrain) while folds implanted at the hindbrain level had differentiated into hindbrain at their base and into forebrain distally. Folds implanted at the spinal cord level had differentiated into spinal cord at their base, into hindbrain medially and into forebrain distally (Fig. 2A). Comparable results were obtained using flattened explants of ectoderm juxtaposed with

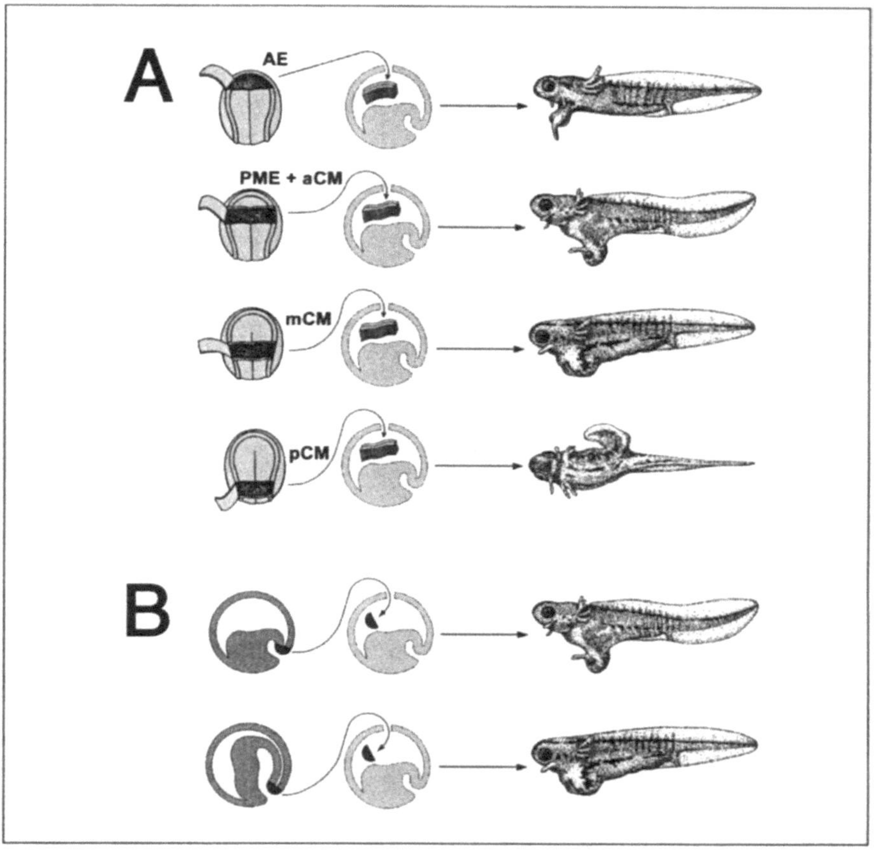


Figure 1. Regionally specific induction by Spemann's organizer. (A) Mangold's experiment demonstrating regionally specific induction by axial mesendoderm.[11] Transversal segments of the archenteron roof of newt neurulae (left) were implanted into the blastocoel of early gastrulae (middle). At tadpole stage, grafted embryos displayed characteristic regionally specific inductions of balancers, heads, trunks and tails, respectively (right). AE, anterior endoderm; aCM, anterior chordamesoderm; mCM, medial chordamesoderm; pCM, posterior chordamesoderm; PME, prechordal mesendoderm. (B) Regionally specific induction by early and late blastopore lips. Dorsal lip tissue was explanted (left) at early or late gastrula stages and grafted into the blastocoel of early newt gastrulae (middle). At tadpole stage, grafted embryos displayed head and trunk duplications, respectively (right). Modified from ref. 160, with kind permission of Sinauer Associates.

fragments of CM.[13] Based on these observations Nieuwkoop proposed the sequential activity of two signals (Fig. 2B): First, a signal from the dorsal mesoderm (organizer) induces the overlying ectoderm to become neuroectoderm of anterior character (activation step). Second, a signal which is absent from the anterior of the embryo and becomes increasingly stronger posteriorly confers progressively posterior identity to the corresponding part of the neural plate (transformation step).

Similarly, a model based on the activity of two opposing activities was devised by the 'Finnish school' of embryologists. According to this model, AP axial patterning is regulated by the concentration ratio between an archencephalic (forebrain-type) and a spinocaudal

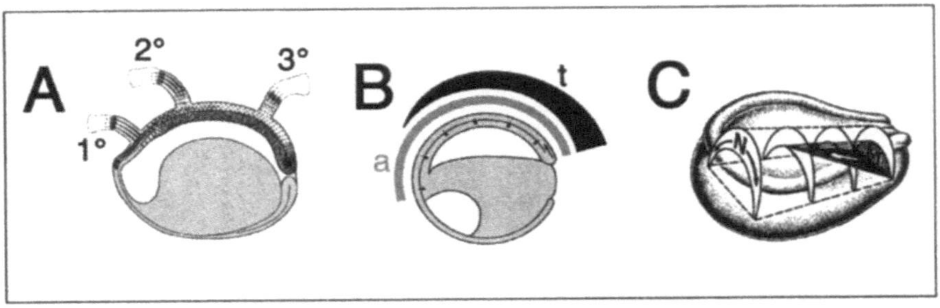


Figure 2. Classical models for AP patterning. In all schemes, anterior points to the left, dorsal to the top. (A) Sagittal section of an early neurula newt embryo with ectodermal folds implanted at three different AP levels of the neural plate (1°,2°,3°). Different patterns indicate different AP types of neural tissue (for simplicity, ventral tissues are shown in grey).[12] (B) Nieuwkoop's interpretation of A: An activating signal (a, dark grey) from dorsal mesoderm induces anterior neuroectoderm dorsally along the entire AP axis. Subsequently, posterior identity is imposed by a transforming signal (t, black) which increases posteriorly. The activating signal acts vertically between tissue layers (arrowheads), the transforming signal may act vertically or by planar propagation (arrow).[137] (C) The two-inducer model proposed by Saxén and Toivonen:[14] Two opposing gradients formed by the archencephalic inducer (N) and the spinocaudal inducer (arrow) determine cell fate along the AP axis. Panels A, C are modified from ref. 138; with kind permission of: The International Journal of Developmental Biology (UBC Press, Bilbao).

(posteriorizing/mesodermalizing) inducer (Fig. 2C).[14,15] It is of note that both Nieuwkoops two-step and the two-inducer model predict gradients of posteriorizing factors. As we will discuss below, organizer and gradient models merely reflect different perspectives of the same phenomenon.

## Wnt/β-Catenin Signaling Antagonizes the Vertebrate Head Organizer

Gain- and loss-of-function approaches in *Xenopus* and zebrafish have revealed that Wnt/β-catenin-type signaling is necessary and sufficient for organizer induction and thus, indirectly, for head formation during an early phase of embryogenesis (reviewed in chapter 4; refs. 16,17). The competence of *Xenopus* and zebrafish embryos to respond to Wnt signals changes dramatically after the midblastula transition (MBT) when zygotic transcription starts. Post-MBT Wnt/β-catenin signaling does not have organizer-inducing properties anymore but it antagonizes the head organizer and confers posteriorizing activity: Overexpression of various Wnts, the Wnt transducers β-catenin or XTCF-3, or treatment with the artificial Wnt pathway activator $Li^+$ after MBT lead to repression of anterior and concomittant induction of posterior neural markers and suppression of head development (Fig. 3).[18-32] Wnt/β-catenin signaling is able to counteract *Xenopus* head organizer activity at least until the end of gastrulation as revealed by experiments with embryos carrying an inducible *XWnt-8* transgene.[33] A posteriorizing role for Wnts is also suggested by studies on transgenic mice ectopically expressing chick *Wnt-8C* and by $Li^+$ treatment of gastrulating chick embryos.[34,35]

Wnt/β-catenin signaling not only suppresses anterior but is also required for posterior development: Overexpression of Wnt antagonists in *Xenopus* and zebrafish, respectively, results in the formation of embryos with enlarged head structures and shortened trunks (Fig. 3).[36-46] Similarly, both overexpression of GSK3β in *Xenopus* anterior ectoderm and depletion of β-catenin from this region using morpholino antisense oligos elicits neural anteriorization.[47,48] Mice mutant for *Wnt-3A, Wnt-5A, LEF-1* and *TCF-1* -encoding nuclear transducers of Wnt signaling-, and *LRP6* -encoding a Wnt coreceptor- show posterior truncations, providing genetic evidence for the requirement for Wnts in posteriorization.[49-52] Pronounced trunk defects

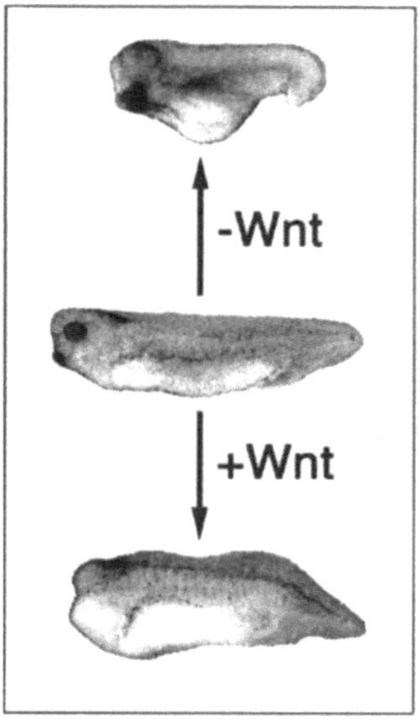


Figure 3. Wnts posteriorize *Xenopus* embryos. Activation of the Wnt/β-catenin pathway (+Wnt) leads to headless tadpoles (lower panel); its inhibition (-Wnt) results in enlargement of anterior and concomittant reduction of trunk structures (upper panel).

are also observed in zebrafish lacking *Wnt-8* function.[53] Interestingly, allelic combinations of mouse *Wnt-3A*⁻ and *Vestigial tail*, a hypomorphic mutation of *Wnt-3A*, display dose-dependent posterior truncations suggesting a requirement for increasing levels of Wnt signaling to specify increasingly more posterior fates.[54] Likewise, posterior neural fates are progressively deleted in zebrafish by increasing doses of morpholino oligos.[55]

Thus, Wnts elicit phenotypically opposing effects with early signaling promoting and late signaling counteracting head formation. This complicates the phenotypic interpretation of global gain- and loss-of-function experiments and advocates the importance of manipulating gene expression in defined time windows. The change of cellular competence to respond to Wnt signals (from organizer induction to head antagonism) does not seem to be caused by the employment of a different (noncanonical) Wnt pathway but by a change of nuclear cofactors.[28,31]

## Tissues with Posteriorizing Activity Express Wnts

Nieuwkoop and colleagues have described graded transforming activity residing within posterior axial mesoderm (CM) and this has been confirmed by tissue recombination experiments in *Xenopus*, mouse and chick.[56-59] However, studies in *Xenopus*, chick and zebrafish have also revealed a posteriorizing function of nonaxial tissues.[60-64] All tissues implicated in posteriorization express at least one member of the Wnt family during gastrulation (Table 1).

## The Two Inhibitor Model for Head Induction

How does the notion of separate head and trunk organizers relate to the posteriorizing activity of post-MBT Wnt signaling? The vertebrate organizer expresses a large number of secreted growth factors and growth factor antagonists which continue to be expressed in the axial mesendoderm in different patterns. It is well established that inhibitors of bone morphogenetic proteins (BMPs—a TGFβ subfamily) which are expressed all along the AP axis (Table 2) mediate the dorsalizing and neural inducing activity of Spemann's organizer in amphibians (reviewed in ref. 65). Ectopic expression of BMP antagonists on the ventral side of *Xenopus* embryos induces secondary trunks lacking anterior neural structures up to the hindbrain

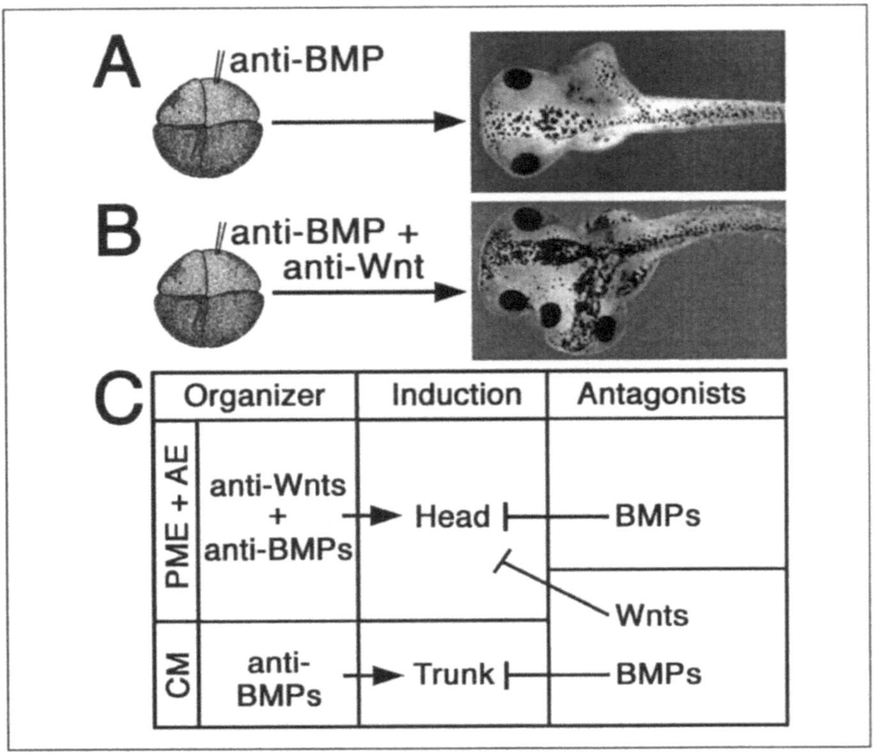


Figure 4. The two inhibitor model for head induction. (A) Ectopic BMP antagonism results in trunk duplication, (B) a combination of BMP and Wnt antagonism induces heads. (C) Diagram representing the two inhibitor model.[66]

suggesting that trunk organizer function is mediated by BMP inhibition (Fig. 4A). A distinguishing feature of the anterior derivatives of the organizer (AE and PME) is the expression of Wnt inhibitors (Table 2; reviewed in refs. 66,67). Coexpression of BMP and Wnt antagonists induces complete secondary axes including heads with anterior neural structures (Fig. 4B).[38,42] Based on these observations, the 'two inhibitor model' for head induction has been proposed: The head organizer functions by simultaneously antagonizing BMP and Wnt signaling while the trunk organizer only antagonizes BMP signaling (Fig. 4C).[38,66]

In amphibians, heterotopic grafting studies have located the strongest head-inducing potential to PME and anterior CM and ablation studies have proven a requirement for this part of the organizer in head formation.[11,68] Grafts of corresponding regions in chick and zebrafish also result in anterior neural induction, emphasizing the instructive role of anterior axial mesendoderm, a major source of Wnt inhibitors, in head development (Fig. 5).[9,69]

Supporting the two inhibitor model, inhibition of Wnt signaling is necessary in vivo for formation of anterior neural structures in *Xenopus*, zebrafish and mouse: Inhibition of the secreted Wnt antagonist Dickkopf1 (Dkk-1) in *Xenopus* using neutralizing antibodies as well as genetic inactivation of the Wnt antagonists TCF-3 and *axin1* in the zebrafish *headless* and *masterblind* mutants, respectively, all result in microcephalic embryos.[27,42,70,71] A targeted knockout of the *Dkk-1* gene in mouse leads to embryos completely lacking rostral head structures.[72]

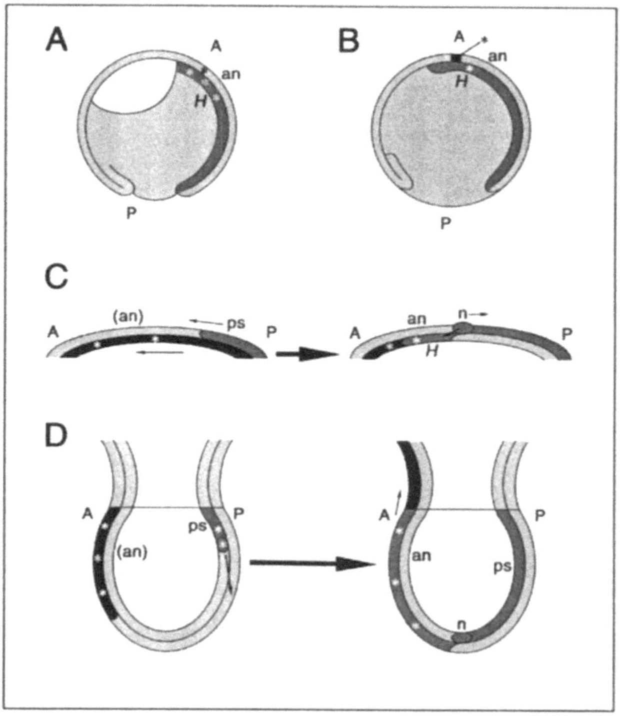


Figure 5. Comparative scheme of vertebrate gastrulae. Simplified schemes of midsagittal sections through (A) *Xenopus*, (B) zebrafish, (C) chick and (D) mouse gastrulae. Axial mesendodermal tissues are shown in dark grey; expression of Wnt inhibitors is indicated by asterisks. A, (prospective) anterior; P, (prospective) posterior; H, regions for which strong anterior- (head-) inducing potential has been demonstrated in transplantations; an, anterior neuroectoderm; (an), prospective anterior neuroectoderm; n, node; ps, primitive streak. (A) In the *Xenopus* mid- to late gastrula, axial mesendoderm (AE, PME, CM, from anterior to posterior) underlies the neural plate. Strongest head-inducing potential resides in the region between PME and anterior CM. Wnt inhibitors are expressed in AE, PME, anterior CM and anterior neuroectoderm. (B) At comparable stages, zebrafish and *Xenopus* gastrulae are strikingly similar. In zebrafish, neuroectoderm is bordered anteriorly by a single row of cells (row-1 cells, black) which are required for anterior neural development and secrete a Wnt inhibitor (see Note Added in Proof). (C) The early chick embryo is a bilayered structure consisting of the extraembryonic hypoblast (black) and the overlying epiblast (grey) which will give rise to the embryo proper. Early in gastrulation (left), the primitive streak starts to extend from the posterior side of the embryo while hypoblast cells move anteriorly (arrows). At later stages (right), mesendodermal cells migrate through the node (arrow) and displace the hypoblast anteriorly. The node regresses posteriorly during this process (arrow). Both hypoblast and anterior axial mesendoderm secrete Wnt inhibitors. (D) In the cup-shaped early mouse gastrula (left), extraembryonic anterior visceral endoderm (AVE, black) juxtaposes the prospective anterior neural plate while the primitive streak forms posteriorly. Both are sources of Wnt inhibitors. Later in gastrulation (right), AE and PME have displaced the AVE (arrow) and continue to secrete Wnt inhibitors.

Interestingly, a very recent study has revealed that instructive signaling by insulin-like growth factors (IGFs) is necessary and sufficient for anterior neural induction in *Xenopus*.[73] In line with the two inhibitor model, IGF signaling has been demonstrated to counteract the Wnt/β-catenin pathway. Epistatically, this inhibition seems to occur between GSK3β

*Table 1. Wnt expression in vertebrate gastrulae*

| | Chick | Mouse | *Xenopus* | Zebrafish |
|---|---|---|---|---|
| **Wnt-2B/13** | | *primitive streak and mesoderm*[139] | | |
| **Wnt-3** | | *primitive streak and mesoderm*[140] | | |
| **Wnt-3A** | *primitive streak, posterior lateral plate*[74,76] | *CM*[49] | *CM*[23,32] | |
| **Wnt-5A**[1] | | *graded in posterior half of embryo, including primitive streak*[50] | *broad in ectoderm and mesoderm*[141] | = Wnt-5; *axial mesoderm*[143] |
| **Wnt-5B**[1] | | | [142] | |
| **Wnt-6** | | | [144] | |
| **Wnt-7B** | | | *ectoderm*[25] | |
| **Wnt-8** | *nonaxial mesoderm*[145] | *posterior epiblast and mesoderm*[118] | *ventrolateral mesoderm*[18] | *nonaxial mesendoderm, neuroectoderm* (Wnt-8B)[20] |
| **Wnt-8B** | | | *anterior neuroectoderm*[135] | |
| **Wnt-11**[1] | *newly formed mesendoderm*[146] | | *surrounding the blastopore*[147] | *blastoderm margin, paraxial head mesoderm, neuro-ectoderm*[107] |

[1] Note that Wnt-5A and -11 are unlikely to be posteriorizing because they do not appear to signal via β-catenin but employ a different, noncanonical Wnt pathway (see chapter 1).

and β-catenin and one of the next important steps will certainly be to analyze the underlying biochemical mechanism. Furthermore, it remains to be investigated whether Wnt inhibition is the major effect of IGF signaling in head induction or whether IGFs perform other independent functions beyond Wnt inhibition—a likely possibility given that, in contrast to known Wnt inhibitors, IGFs are able to ectopically induce anterior structures such as eyes and cement glands.

Similar to the head, the vertebrate heart is derived from anterior regions of the gastrulating embryo. Although the definitive heart ends up in a ventral position, its precursor -the cardiogenic mesoderm- is located adjacent to the PME, in a region with low levels of Wnt/β-catenin signaling. Three recent studies using chick and *Xenopus* embryos have highlighted a central role of Wnt inhibition in heart specification, lending further support to the two inhibitor model (reviewed in chapter 11).[74-76] Notably, ectopic expression of *Wnt-8C* in transgenic mouse embryos results in reduced heart and foregut structures suggesting that low levels of Wnt signaling are required for anterior specification of all germ layers.[34]

***Table 2. Regional-specific expression of Secreted Organizer Effectors in* Xenopus**

| Gene | Function | Expression in AE | PME | CM | References |
|---|---|---|---|---|---|
| *Chordin* | BMP antagonists | - | ? | + | 149,150 |
| *Follistatin, Follistatin-related protein* | | - | - | + | 151-153 |
| *Noggin* | | - | + | + | 154,155 |
| *Crescent/Frzb2* | Wnt antagonists | + | + | - | 115,116 |
| *Dickkopf1* | | + | + | - | 42 |
| *Frzb1* | | ? | + | - | 39,40 |
| *Cerberus* | BMP, Nodal, Wnt inhibitor | + | - | - | 156,157 |
| *Antivin* | Nodal inhibitor | - | + | + | 158 |
| *Cyp26* | RA-degrading enzyme | - | + | - | 159 |

## Anterior Organizing Centers Express Wnt Inhibitors

As discussed above, anterior axial mesendoderm functions as head organizer in amphibians, zebrafish and chick (Fig. 5A-C).[9,11,69] In the mouse, however, grafts of both late and early gastrula organizer regions only induce secondary embryonic axes lacking anterior neural structures.[7,77] The murine anterior visceral endoderm (AVE), an extraembryonic structure derived from the distal tip of the early embryo, shifts anteriorly before the onset of gastrulation until it underlies the prospective anterior neural plate (Fig. 5D).[78-80] Remarkably, the AVE has been found to express many organizer genes involved in anterior specification and its surgical ablation results in anterior defects.[78,79] Analyses of chimeric mice in which the function of AVE-expressed genes is specifically disrupted in extraembryonic tissues has supported the notion that the AVE is essential for head formation (for reviews see refs. 79,81). Yet, AVE alone is not able to ectopically induce anterior neural markers when grafted heterotopically.[77] Recombination of ectodermal explants with AVE does not result in anterior neural induction although posterior markers become suppressed, suggesting a rather permissive role for AVE in anterior development.[82] It is likely that this suppression of posterior fates is mediated by Wnt antagonism and, in line with this hypothesis, the AVE secretes the Wnt antagonists Cerberus-like (Cer-l) and Dkk-1.[83-86] As mentioned above, *Dkk-1*$^{-/-}$ mice are headless but the loss of Cer-l function does not elicit head defects.[72,87-91] Thus, if Wnt inhibition is an essential function of AVE it may be mediated redundantly.

An AVE-like role has been suggested for the anterior hypoblast of the chick embryo which is also a source of several Wnt inhibitors (Fig. 5C). This tissue is not able to stably induce forebrain identity in naïve chick epiblast, but it transiently induces early neural markers and directs epiblast movements.[92,93] A model has been proposed in which the role of the anterior hypoblast (and the murine AVE) is to protect presumptive anterior neuroectoderm from posteriorizing activities.[82,93] Yet, despite emerging similarities between the mammalian AVE and the chick anterior hypoblast (gene expression, pregastrulation movements), there are fundamental differences between these tissues with regard to their inducing abilities as revealed by heterospecies transplantations: Rabbit anterior hypoblast is able to stably induce anterior neural fates upon transplantation under chick epiblast, compared to only transient induction of some pre-neural markers by chick anterior hypoblast.[92,93] The molecular basis for these differences remains elusive.

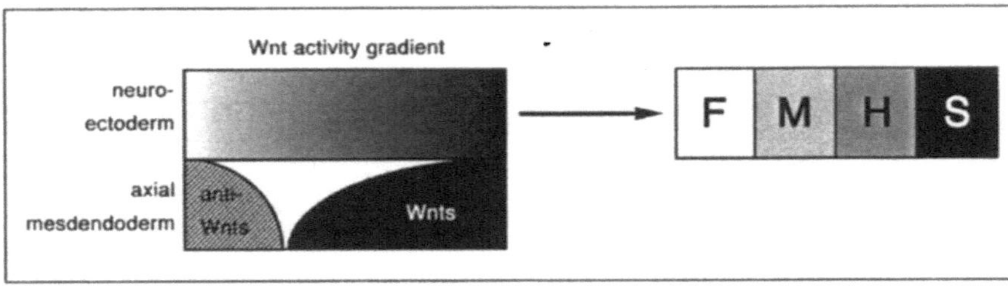


Figure 6. Simplified model for AP patterning of neuroectoderm by a Wnt activity gradient. Wnts (black) and Wnt inhibitors are expressed in the axial mesendoderm underlying the neuroectoderm. The expression of Wnts and Wnt inhibitors in the neuroectoderm is not shown for simplicity. Their antagonistic activities result in the formation of a Wnt activity gradient that patterns the AP neuraxis into forebrain (F), midbrain (M), hindbrain (H) and spinal cord (S) regions. Note that formation of the posterior nervous system also requires other factors (FGFs, RA) which are not shown here.

If the role of the AVE in head induction is permissive, which structures do actually provide instructive head-inducing signals in the mouse embryo? The earlier observation that mouse organizer grafts fail to induce complete embryonic axes including anterior neural structures has been challenged recently: Transplantation of mouse or rabbit organizers into chick host embryos results in formation of complete secondary axes including forebrain structures, indicating that, in principle, the mammalian organizer harbors the capacity for anterior neural induction.[94] Very recently, Tam and colleagues have described induction of complete secondary axes in mouse embryos by mid-gastrula stage organizers.[10] Thus, the early gastrula organizer (EGO; see ref. 77) may not yet have acquired full head-inducing potential while the late gastrula organizer (see ref. 7) may have lost its head-inducing potential already. A similar situation is encountered in chick: Neither the anterior primitive streak of the early chick gastrula nor the node after the emigration of the head process are able to induce complete secondary embryonic axes.[95,96] Finally, it has been demonstrated by ablation that anterior axial mesendoderm is not only sufficient but also required for forebrain development in mice.[97]

Ablation of a single row of cells from the ectoderm that borders the neural plate anteriorly in the zebrafish gastrula (row-1 cells) results in head defects (Fig. 5B), indicating the presence of another early anterior-inducing center.[98] Notably, a Wnt inhibitor of the sFRP (secreted Frizzled-related protein) class has been identified as mediator of row-1 function (see Note Added in Proof). Taken together, all of the tissues implicated in anterior neural induction in various species are sources of secreted Wnt inhibitors.

## A Transforming Gradient of Wnt/β-Catenin Activity Regulates AP Neural Patterning

While Spemann and coworkers suggested distinct head- and trunk-inducing regions within the embryo—and this view is in agreement with the two-inhibitor model for vertebrate head induction—the AP axis of the embryo is patterned by a posteriorizing gradient of a transformer according to the activation-transformation model proposed by Nieuwkoop and colleagues. Clearly, the two-inhibitor model—in its simple form—is not sufficient to explain regional AP patterning of the neuraxis like Nieuwkoop's model. To reconcile the two-inhibitor and the activation-transformation model we have investigated recently whether Wnts may constitute the transforming signal.[32] We found that in *Xenopus* (1) Wnts posteriorize neuroectoderm dose-dependently, (2) Wnt/β-catenin signaling is required for AP ectodermal patterning during gastrulation and (3) Wnts are able to signal directly and over a distance within neuroectoderm, conferring polar AP neural pattern to neuralized ectodermal explants and in vivo. Importantly, an endogenous AP gradient of Wnt/β-catenin signaling was detected in the presumptive neural plate of the *Xenopus* gastrula. These data support a model in which a posteriorizing activity gradient of Wnt/β-catenin signaling is established during gastrulation by the interplay of head organizer-derived Wnt inhibitors and posteriorly expressed Wnts (Fig. 6). Furthermore, they suggest that Wnts may act as morphogens in vertebrates. However, a recent study describes nonautonomous induction of posterior neural markers following overexpression of β-catenin in *Xenopus* ectodermal explants.[29] Differences in the experimental setup may explain this apparent discrepancy. In particular, we have used an earlier readout in our experimental approach.

In vivo, the expression domains of regional neural marker genes change following modulation of Wnt signaling: Forced activation of the Wnt/β-catenin pathway after the MBT results in a loss of anterior neural fates while, conversely, an expansion of anterior at the expense of posterior neural markers is observed upon overexpression of Wnt antagonists. Yet, the expression domains of these markers always respect certain boundaries and do never expand throughout the entire neural plate.[32] This suggests that the competence of neural cells to respond to Wnts is regionally restricted. An organizer-independent ectodermal AP prepattern which depends on differential competence of the epiblast has been described in zebrafish.[99] Besides Wnts, other signaling molecules such as fibroblast growth factors (FGFs) and retinoic acid (RA) have been implicated in posteriorization which may regulate cellular competence (see refs. 100-102 for reviews). FGFs are potent posteriorizing agents (e. g. see ref. 103), yet, Wnt antagonists rescue FGF-induced posteriorization in *Xenopus*, suggesting that this effect is mediated—at least in part—indirectly through Wnts.[23,27] Interestingly, a recent study in chick suggests that FGFs are required during gastrulation to maintain a population of neural progenitors within Hensen's node.[104] The authors propose a model in which FGFs are not a transforming agent but define a niche for neural stem cells in the posteriorly regressing organizer where cells remain exposed to the true transformer (discussed in ref. 105). Obviously, Wnts are good candidates for this signal. Taken together, AP neural patterning is regulated by Wnt/β-catenin signaling while RA and FGFs may also contribute to this process, maybe by regulating the competence of neural cells to respond to posteriorizing Wnt signals.

## Wnt Signaling and Gastrulation Movements

Zebrafish *silberblick* mutants display mild cyclopia and other head midline defects.[106] Interestingly, a mutation in the *Wnt-11* locus has been identified as the cause of this phenotype.[107] Yet, the *silberblick* head defects are unlikely to result from a failure in head induction or AP patterning: Several recent studies in *Xenopus* and *zebrafish* have provided strong evidence for a noncanonical, β-catenin-independent Wnt pathway (involving Wnt-11 and Dishevelled) regulating the convergent-and-extension movements characteristic of vertebrate gastrulation (see Chapter 1).[107-114] Impaired elongation of axial mesendoderm which is a source of midline patterning signals is the most probable explanation for the *silberblick* phenotype. Overexpression of the Wnt inhibitor Crescent/Frzb2 (Crs) in *Xenopus* elicits a comparable phenotype, suggesting that Crs inhibits Wnt-11-like signals.[115,116] Remarkably, the neural plate is anteriorized in Crs-expressing embryos.[116] Thus, Crs may inhibit a spectrum of different Wnts implicated in canonical and noncanonical signaling. In conclusion, head defects may arise due to inhibition of noncanonical Wnt signaling.

## Later Roles of Wnts

At later stages of vertebrate development, Wnts and Wnt inhibitors are expressed in complex patterns and perform various roles depending on their timing and location of expression. Almost all Wnts and many Wnt modulators are differentially expressed in the embryonic central nervous system.[50,117-119] Not much is known concerning the precise functions of Wnt signaling at these stages. They have been implicated in pattern refinement and subregionalisation, as retrograde signals in axon remodeling and synaptic differentiation and in modulating apoptosis (refs. 120,121). Wnt-1 which is expressed in the mesencephalon is required for maintaining the midbrain-hindbrain boundary (isthmic) organizer and for the development of the cerebellum, an isthmus-derived structure (reviewed in refs. 122,123). A local Wnt-3A signal from the developing cerebral cortex promotes cell proliferation during hippocampus formation.[124,125]

Gain- and loss-of-function analyses in all vertebrate model organisms have revealed that Wnt signals are crucial for formation and specification of the neural crest, a specialized cell population that is induced at the lateral edges of the neural plate and becomes localized to the dorsalmost region of the neural tube.[24,25,126-129] Neural crest cells undergo an epithelial-mesenchymal transition, migrate to many different targets within the developing embryo and give rise to a variety of different cell types (reviewed in ref. 130). The cranial neural crest populates the branchial arches, gives rise to cranial ganglia and craniofacial cartilage and bone and is therefore of central importance for later events in head formation.[130-133] Many Wnts are predominantly expressed in dorsal regions of the neural tube (e. g. in *Xenopus*: XWnt-1—midbrain, dorsal neural tube; XWnt-2B—dorsal forebrain/midbrain boundary; XWnt-3A—dorsal neural tube; XWnt-7B—dorsal neural tube; XWnt-8B—dorsal di- and mesencephalon, forebrain/midbrain boundary).[25,134-136] Mouse *Wnt-1*$^{-/-}$;*Wnt-3A*$^{-/-}$ double mutants display, besides many other defects, a dramatic reduction of neural crest and severe craniofacial abnormalities.[126] Interestingly, *Wnt-5A*$^{-/-}$ mutant mice also have abnormally shaped heads. Wnt-5A—which is considered not to activate the canonical Wnt/β-catenin pathway because it belongs to a distinct class of Wnts—is expressed in the outgrowing first branchial arch and in outgrowing facial primordia. It has been suggested that a Wnt-5A pathway is generally involved in regulating the outgrowth of extending structures.[50]

Taken together, Wnts play various roles during later stages of head development which may be positive as described for midbrain, hippocampus and neural crest development. However, the large number of Wnts and Wnt modulators expressed in a highly regionalized manner highlights the need for further detailed investigations.

## Conclusions and Outlook

At least three phases of Wnt signaling can be distinguished during vertebrate head formation. First, early β-catenin-dependent signaling plays a central role in organizer induction which is a prerequisite for head formation. Second, Wnt/β-catenin signaling antagonizes the head organizer during gastrulation. Different anterior-organizing centers have been identified in vertebrate gastrulae and all of them secrete Wnt antagonists. The interplay of these antagonists with posteriorizing Wnts may establish a tranforming activity gradient of Wnt/β-catenin signaling which regulates AP neural patterning. Based on these molecular data, we propose that Spemann's concept of separate head and trunk organizers and Nieuwkoop's transforming gradient do not contradict each other but represent two different views of AP axis formation: Like an electron will appear as a wave or a particle, AP patterning appears as regulated by organizers or gradients, depending on the experimental approach.

While Wnt/β-catenin signaling regulates AP patterning, noncanonical Wnt-11-type signaling promotes axial elongation and interference with this pathway may affect head formation indirectly. In the third and least characterized phase of Wnt signaling, multiple Wnts are expressed in a highly regionalized fashion. Wnts play important roles in the regional patterning of the brain and in the specification of neural crest which is essential for later steps of craniofacial development.

In the future, it will be crucial to modulate Wnt signaling during defined intervals and in restricted regions in order to specifically target distinct Wnt functions. First steps in this direction have been made by using inducible transgenes and tissue transplantation approaches.[28,29,31-33] Furthermore, the identification of transcriptional targets of Wnt signaling will be of great interest. In particular, the identification of genes activated by later (e. g. posteriorizing) Wnt/β-catenin signaling is in its very beginnings.[26]

## *Note Added in Proof*

While this chapter was reviewed, a study using chick embryos has been published which lends support to the idea that a gradient of Wnt/β-catenin signaling regulates AP patterning of the early neural plate (Nordström U, Jessell TM, Edlund T. Progressive induction of caudal neural character by graded Wnt signaling. Nat Neurosci 2002; 5:525-532). Furthermore, a Frzb-like Wnt inhibitor (Tlc) was identified which mediates the anteriorizing function of row-1 cells in zebrafish (Houart C, Caneparo L, Heisenberg C et al. Establishment of the telencephalon during gastrulation by local antagonism of Wnt signaling. Neuron 2002; 35:255-265). This and another study suggest that in the late zebrafish gastrula Wnt signaling regulates patterning between fore- and midbrain (Kim SH, Shin J, Park HC et al. Specification of an anterior neuroectoderm patterning by Frizzled8a-mediated Wnt8b signalling during late gastrulation in zebrafish. Development 2002; 129:4443-4455). Recently, Wnt6 expression has been described in chick (Schubert FR, Mootoosamy RC, Walters EH et al. Wnt6 marks sites of epithelial transformations in the chick embry. Mech Dev 2002; 114:143-148) and has been proposed to act as a direct inducer of neural crest (Garcia-Castro MI, Marcelle C, Bronner-Fraser M. Ectodermal Wnt function as a neural crest inducer. Science 2002; 297:848-851).

## *Acknowledgements*

We apologize to all researchers whose work we could not cite due to space constraints. We thank Dr. Gary Davidson for helpful comments on the manuscript.

## References

1. Shimeld SM, Holland PW. Vertebrate innovations. Proc Natl Acad Sci USA 2000; 97:4449-4452.
2. Spemann H, Mangold H. Über Induktion von Embryonalanlagen durch Implantation artfremder Organisatoren. Arch Mikrosk Anat Entwicklungsmechan 1924; 100:599-638.
3. Waddington CH. Induction by the primitive streak and its derivatives in the chick. J Exp Zool 1933; 10:38-46.
4. Oppenheimer JM. Transplantation experiments on developing teleosts (Fundulus and Perca). J Exp Zool 1936; 72:409-437.
5. Waddington C. Organizers in mammalian development. Nature 1936; 138:125.
6. Blum M, Gaunt SJ, Cho KW et al. Gastrulation in the mouse: the role of the homeobox gene goosecoid. Cell 1992; 69:1097-1106.
7. Beddington RS. Induction of a second neural axis by the mouse node. Development 1994; 120:613-620.
8. Shih J, Fraser SE. Characterizing the zebrafish organizer: microsurgical analysis at the early-shield stage. Development 1996; 122:1313-1322.
9. Saúde L, Woolley K, Martin P et al. Axis-inducing activities and cell fates of the zebrafish organizer. Development 2000; 127:3407-3417.
10. Kinder SJ, Tsang TE, Wakamiya M et al. The organizer of the mouse gastrula is composed of a dynamic population of progenitor cells for the axial mesoderm. Development 2001; 128:3623-3634.
11. Mangold O. Über die Induktionsfähigkeit der verschiedenen Bezirke der Neurula von Urodelen. Naturwissenschaften 1933; 21:761-766.
12. Nieuwkoop PD, Botterenbrood EC, Kremer A et al. Activation and organization of the central nervous system in amphibians. J Exp Zool 1952; 120:1-108.
13. Nieuwkoop PD, Nigtevecht GV. Neural activation and transformation in explants of competent ectoderm under the influence of fragments of anterior notochord in urodeles. J Embryol Exp Morph 1954; 2:175-193.

14. Saxén L, Toivonen S. The two-gradient hypothesis in primary induction: The combined effect of two types of inducers mixed in different ratios. J Embryol Exp Morph 1961; 9:514-533.
15. Toivonen S, Saxén L. Morphogenetic interaction of presumptive neural and mesodermal cells mixed in different ratios. Science 1968; 159:539-540.
16. Schier AF. Axis formation and patterning in zebrafish. Curr Opin Genet Dev 2001; 11:393-404.
17. Solnica-Krezel L, Driever W. The role of the homeodomain protein Bozozok in zebrafish axis formation. Int J Dev Biol 2001; 45:299-310.
18. Christian JL, Moon RT. Interactions between Xwnt-8 and Spemann organizer signaling pathways generate dorsoventral pattern in the embryonic mesoderm of Xenopus. Genes Dev 1993; 7:13-28.
19. Kelly GM, Erezyilmaz DF, Moon RT. Induction of a secondary embryonic axis in zebrafish occurs following the overexpression of β-catenin. Mech Dev 1995; 53:261-273.
20. Kelly GM, Greenstein P, Erezyilmaz DF et al. Zebrafish wnt8 and wnt8b share a common activity but are involved in distinct developmental pathways. Development 1995; 121:1787-1799.
21. McGrew LL, Lai CJ, Moon RT. Specification of the anteroposterior neural axis through synergistic interaction of the Wnt signaling cascade with noggin and follistatin. Dev Biol 1995; 172:337-342.
22. Fredieu JR, Cui Y, Maier D et al. Xwnt-8 and lithium can act upon either dorsal mesodermal or neurectodermal cells to cause a loss of forebrain in Xenopus embryos. Dev Biol 1997; 186:100-114.
23. McGrew LL, Hoppler S, Moon RT. Wnt and FGF pathways cooperatively pattern anteroposterior neural ectoderm in Xenopus. Mech Dev 1997; 69:105-114.
24. Saint-Jeannet JP, He X, Varmus HE et al. Regulation of dorsal fate in the neuraxis by Wnt-1 and Wnt-3a. Proc Natl Acad Sci USA 1997; 94:13713-13718.
25. Chang C, Hemmati-Brivanlou A. Neural crest induction by Xwnt7B in Xenopus. Dev Biol 1998; 194:129-134.
26. McGrew LL, Takemaru K, Bates R et al. Direct regulation of the Xenopus engrailed-2 promoter by the Wnt signaling pathway, and a molecular screen for Wnt-responsive genes, confirm a role for Wnt signaling during neural patterning in Xenopus. Mech Dev 1999; 87:21-32.
27. Kazanskaya O, Glinka A, Niehrs C. The role of Xenopus dickkopf1 in prechordal plate specification and neural patterning. Development 2000; 127:4981-4992.
28. Darken RS, Wilson PA. Axis induction by wnt signaling: Target promoter responsiveness regulates competence. Dev Biol 2001; 234:42-54.
29. Domingos PM, Itasaki N, Jones CM et al. The Wnt/β-Catenin Pathway Posteriorizes Neural Tissue in Xenopus by an Indirect Mechanism Requiring FGF Signaling. Dev Biol 2001; 239:148-160.
30. Gamse JT, Sive H. Early anteroposterior division of the presumptive neurectoderm in Xenopus. Mech Dev 2001; 104:21-36.
31. Hamilton FS, Wheeler GN, Hoppler S. Difference in XTCF-3 dependency accounts for change in response to β-catenin-mediated Wnt signaling in Xenopus blastula. Development 2001; 128:2063-2073.
32. Kiecker C, Niehrs C. A morphogen gradient of Wnt/β-catenin signaling regulates anteroposterior neural patterning in Xenopus. Development 2001; 128:4189-4201.
33. Wheeler GN, Hamilton FS, Hoppler S. Inducible gene expression in transgenic Xenopus embryos. Curr Biol 2000; 10:849-852.
34. Pöpperl H, Schmidt C, Wilson V et al. Misexpression of Cwnt8C in the mouse induces an ectopic embryonic axis and causes a truncation of the anterior neuroectoderm. Development 1997; 124:2997-3005.
35. Roeser T, Stein S, Kessel M. Nuclear β-catenin and the development of bilateral symmetry in normal and LiCl-exposed chick embryos. Development 1999; 126:2955-2965.
36. Hoppler S, Brown JD, Moon RT. Expression of a dominant-negative Wnt blocks induction of MyoD in Xenopus embryos. Genes Dev 1996; 10: 2805-2817.
37. Pierce SB, Kimelman D. Overexpression of Xgsk-3 disrupts anterior ectodermal patterning in Xenopus. Dev Biol 1996; 175: 256-264.
38. Glinka A, Wu W, Onichtchouk D et al. Head induction by simultaneous repression of Bmp and Wnt signaling in Xenopus. Nature 1997; 389: 517-519.
39. Leyns L, Bouwmeester T, Kim SH et al. Frzb-1 is a secreted antagonist of Wnt signaling expressed in the Spemann organizer. Cell 1997; 88: 747-756.

40. Wang S, Krinks M, Lin K et al. Frzb, a secreted protein expressed in the Spemann organizer, binds and inhibits Wnt-8. Cell 1997; 88:757-766.
41. Deardorff MA, Tan C, Conrad LJ et al. Frizzled-8 is expressed in the Spemann organizer and plays a role in early morphogenesis. Development 1998; 125:2687-2700.
42. Glinka A, Wu W, Delius H et al. Dickkopf-1 is a member of a new family of secreted proteins and functions in head induction. Nature 1998; 391:357-362.
43. Hsieh JC, Kodjabachian L, Rebbert ML et al. A new secreted protein that binds to Wnt proteins and inhibits their activities. Nature 1999; 398:431-436.
44. Fekany-Lee K, Gonzalez E, Miller-Bertoglio V et al. The homeobox gene bozozok promotes anterior neuroectoderm formation in zebrafish through negative regulation of BMP2/4 and Wnt pathways. Development 2000; 127:2333-2345.
45. Hashimoto H, Itoh M, Yamanaka Y et al. Zebrafish Dkk1 functions in forebrain specification and axial mesendoderm formation. Dev Biol 2000; 217:138-152.
46. Shinya M, Eschbach C, Clark M et al. Zebrafish Dkk1, induced by the pre-MBT Wnt signaling, is secreted from the prechordal plate and patterns the anterior neural plate. Mech Dev 2000; 98:3-17.
47. Itoh K, Tang TL, Neel BG et al. Specific modulation of ectodermal cell fates in Xenopus embryos by glycogen synthase kinase. Development 1995; 121:3979-3988.
48. Heasman J, Kofron M, Wylie C. β-catenin signaling activity dissected in the early Xenopus embryo: a novel antisense approach. Dev Biol 2000; 222:124-134.
49. Takada S, Stark KL, Shea MJ et al. Wnt-3a regulates somite and tailbud formation in the mouse embryo. Genes Dev 1994; 8:174-189.
50. Yamaguchi TP, Bradley A, McMahon AP et al. A Wnt5a pathway underlies outgrowth of multiple structures in the vertebrate embryo. Development 1999; 126:1211-1223.
51. Galceran J, Farinas I, Depew MJ et al. Wnt3a$^{-/-}$-like phenotype and limb deficiency in LEF-1$^{-/-}$ TCF-1$^{-/-}$ mice. Genes Dev 1999; 13:709-717.
52. Pinson KI, Brennan J, Monkley S et al. An LDL-receptor-related protein mediates Wnt signaling in mice. Nature 2000; 407:535-538.
53. Lekven AC, Thorpe CJ, Waxman JS et al. Zebrafish wnt8 encodes two wnt8 proteins on a bicistronic transcript and is required for mesoderm and neurectoderm patterning. Dev Cell 2001; 1:103-114.
54. Greco TL, Takada S, Newhouse MM et al. Analysis of the vestigial tail mutation demonstrates that Wnt-3a gene dosage regulates mouse axial development. Genes Dev 1996; 10:313-324.
55. Erter CE, Wilm TP, Basler N et al. Wnt8 is required in lateral mesendodermal precursors for neural posteriorization in vivo. Development 2001; 128:3571-3583.
56. Hemmati-Brivanlou A, Stewart RM, Harland RM. Region-specific neural induction of an engrailed protein by anterior notochord in Xenopus. Science 1990; 250:800-802.
57. Doniach T, Musci TJ. Induction of anteroposterior neural pattern in Xenopus: evidence for a quantitative mechanism. Mech Dev 1995; 53:403-413.
58. Ang SL, Rossant J. Anterior mesendoderm induces mouse Engrailed genes in explant cultures. Development 1993; 118:139-149.
59. Rowan AM, Stern CD, Storey KG. Axial mesendoderm refines rostrocaudal pattern in the chick nervous system. Development 1999; 126:2921-2934.
60. Bang AG, Papalopulu N, Kintner C et al. Expression of Pax-3 is initiated in the early neural plate by posteriorizing signals produced by the organizer and by posterior non-axial mesoderm. Development 1997; 124:2075-2085.
61. Muhr J, Jessell TM, Edlund T. Assignment of early caudal identity to neural plate cells by a signal from caudal paraxial mesoderm. Neuron 1997; 19:487-502.
62. Woo K, Fraser SE. Specification of the zebrafish nervous system by nonaxial signals. Science 1997; 277:254-257.
63. Ensini M, Tsuchida TN, Belting HG et al. The control of rostrocaudal pattern in the developing spinal cord: specification of motor neuron subtype identity is initiated by signals from paraxial mesoderm. Development 1998; 125:969-982.
64. Bang AG, Papalopulu N, Goulding MD et al. Expression of Pax-3 in the lateral neural plate is dependent on a Wnt-mediated signal from posterior nonaxial mesoderm. Dev Biol 1999; 212:366-380.

65. De Robertis EM, Larraín J, Oelgeschläger M et al. The establishment of Spemann's organizer and patterning of the vertebrate embryo. Nat Rev Genet 2000; 1:171-181.
66. Niehrs C. Head in the WNT: the molecular nature of Spemann's head organizer. Trends Genet 1999; 15:314-319.
67. Kiecker C, Niehrs C. The role of prechordal mesendoderm in neural patterning. Curr Opin Neurobiol 2001; 11:27-33.
68. Schneider VA, Mercola M. Spatially distinct head and heart inducers within the Xenopus organizer region. Curr Biol 1999; 9:800-809.
69. Pera EM, Kessel M. Patterning of the chick forebrain anlage by the prechordal plate. Development 1997; 124:4153-4162.
70. Kim CH, Oda T, Itoh M et al. Repressor activity of Headless/TCF-3 is essential for vertebrate head formation. Nature 2000; 407:913-916.
71. Heisenberg CP, Houart C, Take-Uchi M et al. A mutation in the Gsk3-binding domain of zebrafish Masterblind/Axin1 leads to a fate transformation of telencephalon and eyes to diencephalon. Genes Dev 2001; 15:1427-1434.
72. Mukhopadhyay M, Shtrom S, Rodriguez-Esteban C et al. Dickkopf1 is required for embryonic head induction and limb morphogenesis in the mouse. Dev Cell 2001; 1:423-434.
73. Pera EM, Wessely O, Li SY et al. Neural and Head Induction by Insulin-like Growth Factor Signals. Dev Cell 2001; 1:655-665.
74. Marvin MJ, Di Rocco G, Gardiner A et al. Inhibition of Wnt activity induces heart formation from posterior mesoderm. Genes Dev 2001; 15:316-327.
75. Schneider VA, Mercola M. Wnt antagonism initiates cardiogenesis in Xenopus laevis. Genes Dev 2001; 15:304-315.
76. Tzahor E, Lassar AB. Wnt signals from the neural tube block ectopic cardiogenesis. Genes Dev 2001; 15:255-260.
77. Tam PP, Steiner KA. Anterior patterning by synergistic activity of the early gastrula organizer and the anterior germ layer tissues of the mouse embryo. Development 1999; 126:5171-5179.
78. Thomas P, Beddington R. Anterior primitive endoderm may be responsible for patterning the anterior neural plate in the mouse embryo. Curr Biol 1996; 6:1487-1496.
79. Beddington RS, Robertson EJ. Axis development and early asymmetry in mammals. Cell 1999; 96:195-209.
80. Perea-Gomez A, Rhinn M, Ang SL. Role of the anterior visceral endoderm in restricting posterior signals in the mouse embryo. Int J Dev Biol 2001; 45:311-320.
81. De Souza FS, Niehrs C. Anterior endoderm and head induction in early vertebrate embryos. Cell Tissue Res 2000; 300:207-217.
82. Kimura C, Yoshinaga K, Tian E et al. Visceral endoderm mediates forebrain development by suppressing posteriorizing signals. Dev Biol 2000; 225:304-321.
83. Belo JA, Bouwmeester T, Leyns L et al. Cerberus-like is a secreted factor with neuralizing activity expressed in the anterior primitive endoderm of the mouse gastrula. Mech Dev 1997; 68:45-57.
84. Biben C, Stanley E, Fabri L et al. Murine cerberus homologue mCer-1: a candidate anterior patterning molecule. Dev Biol 1998; 194:135-151.
85. Shawlot W, Deng JM, Behringer RR. Expression of the mouse cerberus-related gene, Cerr1, suggests a role in anterior neural induction and somitogenesis. Proc Natl Acad Sci USA 1998; 95:6198-6203.
86. Pearce JJ, Penny G, Rossant J. A mouse cerberus/Dan-related gene family. Dev Biol 1999; 209:98-110.
87. Simpson EH, Johnson DK, Hunsicker P et al. The mouse Cer1 (Cerberus related or homologue) gene is not required for anterior pattern formation. Dev Biol 1999; 213:202-206.
88. Belo JA, Bachiller D, Agius E et al. Cerberus-like is a secreted BMP and nodal antagonist not essential for mouse development. Genesis 2000; 26:265-270.
89. Shawlot W, Min Deng J, Wakamiya M et al. The cerberus-related gene, Cerr1, is not essential for mouse head formation. Genesis 2000; 26:253-258.
90. Stanley EG, Biben C, Allison J et al. Targeted insertion of a lacZ reporter gene into the mouse Cer1 locus reveals complex and dynamic expression during embryogenesis. Genesis 2000; 26:259-264.

91. Borges AC, Marques S, Belo JA. The BMP antagonists cerberus-like and noggin do not interact during mouse forebrain development. Int J Dev Biol 2001; 45:441-444.
92. Knoetgen H, Viebahn C, Kessel M. Head induction in the chick by primitive endoderm of mammalian, but not avian origin. Development 1999; 126:815-825.
93. Foley AC, Skromne I, Stern CD. Reconciling different models of forebrain induction and patterning: a dual role for the hypoblast. Development 2000; 127:3839-3854.
94. Knoetgen H, Teichmann U, Wittler L et al. Anterior neural induction by nodes from rabbits and mice. Dev Biol 2000; 225:370-380.
95. Lemaire L, Roeser T, Izpisua-Belmonte JC et al. Segregating expression domains of two goosecoid genes during the transition from gastrulation to neurulation in chick embryos. Development 1997; 124:1443-1452.
96. Storey KG, Crossley JM, De Robertis EM et al. Neural induction and regionalisation in the chick embryo. Development 1992; 114:729-741.
97. Camus A, Davidson BP, Billiards S et al. The morphogenetic role of midline mesendoderm and ectoderm in the development of the forebrain and the midbrain of the mouse embryo. Development 2000; 127:1799-1813.
98. Houart C, Westerfield M, Wilson SW. A small population of anterior cells patterns the forebrain during zebrafish gastrulation. Nature 1998; 391:788-792.
99. Koshida S, Shinya M, Mizuno T et al. Initial anteroposterior pattern of the zebrafish central nervous system is determined by differential competence of the epiblast. Development 1998; 125:1957-1966.
100. Doniach T. Basic FGF as an inducer of anteroposterior neural pattern. Cell 1995; 83:1067-1070.
101. Mason I. Neural induction: do fibroblast growth factors strike a cord? Curr Biol 1996; 6:672-675.
102. Gavalas A, Krumlauf R. Retinoid signaling and hindbrain patterning. Curr Opin Genet Dev 2000; 10:380-386.
103. Pownall ME, Tucker AS, Slack JM et al. eFGF, Xcad3 and Hox genes form a molecular pathway that establishes the anteroposterior axis in Xenopus. Development 1996; 122:3881-3892.
104. Mathis L, Kulesa PM, Fraser SE. FGF receptor signaling is required to maintain neural progenitors during Hensen's node progression. Nat Cell Biol 2001; 3:559-566.
105. Krull CE, Krumlauf R. Building from the bottom up. Nat Cell Biol 2001; 3:E138-E139.
106. Heisenberg CP, Brand M, Jiang YJ et al. Genes involved in forebrain development in the zebrafish, Danio rerio. Development 1996; 123:191-203.
107. Heisenberg CP, Tada M, Rauch G-J et al. Silberblick/Wnt11 mediates convergent extension movements during zebrafish gastrulation. Nature 2000; 405:76-81.
108. Sokol S. A role for Wnts in morpho-genesis and tissue polarity. Nat Cell Biol 2000; 2:E124-E125.
109. Tada M, Smith JC. Xwnt11 is a target of Xenopus Brachyury: regulation of gastrulation movements via Dishevelled, but not through the canonical Wnt pathway. Development 2000; 127:2227-2238.
110. Wallingford JB, Rowning BA, Vogeli KM et al. Dishevelled controls cell polarity during Xenopus gastrulation. Nature 2000; 405:81-85.
111. Kühl M, Geis K, Sheldahl LC et al. Antagonistic regulation of convergent extension movements in Xenopus by Wnt/β-catenin and Wnt/$Ca^{2+}$ signaling. Mech Dev 2001; 106:61-76.
112. Topczewski J, Sepich DS, Myers DC et al. The zebrafish glypican knypek controls cell polarity during gastrulation movements of convergent extension. Dev Cell 2001; 1:251-264.
113. Wallingford JB, Harland RM. Xenopus Dishevelled signaling regulates both neural and mesodermal convergent extension: parallel forces elongating the body axis. Development 2001; 128:2581-92.
114. Wallingford JB, Vogeli KM, Harland RM. Regulation of convergent extension in Xenopus by Wnt5a and Frizzled-8 is independent of the canonical Wnt pathway.
115. Bradley L, Sun B, Collins-Racie L et al. Different activities of the frizzled-related proteins frzb2 and sizzled2 during Xenopus anteroposterior patterning. Dev Biol 2000; 227:118-132.
116. Pera EM, De Robertis EM. A direct screen for secreted proteins in Xenopus embryos identifies distinct activities for the Wnt antagonists Crescent and Frzb-1. Mech Dev 2000; 96:183-195.
117. Parr BA, Shea MJ, Vassileva G et al. Mouse Wnt genes exhibit discrete domains of expression in the early embryonic CNS and limb buds. Development 1993; 119:247-261.

118. Bouillet P, Oulad-Abdelghani M, Ward SJ et al. A new mouse member of the Wnt gene family, mWnt-8, is expressed during early embryogenesis and is ectopically induced by retinoic acid. Mech Dev 1996; 58:141-152.
119. Kim AS, Lowenstein DH, Pleasure SJ. Wnt receptors and Wnt inhibitors are expressed in gradients in the developing telencephalon. Mech Dev 2001; 103:167-172.
120. Patapoutian A, Reichardt LF. Roles of Wnt proteins in neural development and maintenance. Curr Opin Neurobiol 2000; 10:392-399.
121. Wilson SW, Rubenstein JL. Induction and dorsoventral patterning of the telencephalon. Neuron 2000; 28:641-651.
122. Rhinn M, Brand M. The midbrain-hindbrain boundary organizer. Curr Opin Neurobiol 2001; 11:34-42.
123. Wurst W, Bally-Cuif L. Neural plate patterning: upstream and downstream of the isthmic organizer. Nat Rev Neurosci 2001; 2:99-108.
124. Galceran J, Miyashita-Lin EM, Devaney E et al. Hippocampus development and generation of dentate gyrus granule cells is regulated by LEF-1. Development 2000; 127:469-482.
125. Lee SM, Tole S, Grove E et al. A local Wnt-3a signal is required for development of the mammalian hippocampus. Development 2000; 127:457-467.
126. Ikeya M, Lee SM, Johnson JE et al. Wnt signaling required for expansion of neural crest and CNS progenitors. Nature 1997; 389:966-970.
127. Dorsky RI, Moon RT, Raible DW. Control of neural crest cell fate by the Wnt signaling pathway. Nature 1998; 396:370-373.
128. Deardorff MA, Tan C, Saint-Jeannet JP et al. A role for frizzled 3 in neural crest development. Development 2001; 128:3655-3663.
129. Tan C, Deardorff MA, Saint-Jeannet JP et al. Kermit, a frizzled interacting protein, regulates frizzled 3 signaling in neural crest development. Development 2001; 128:3665-3674.
130. LaBonne C, Bronner-Fraser M. Molecular mechanisms of neural crest formation. Annu Rev Cell Dev Biol 1999; 15:81-112.
131. Köntges G, Lumsden A. Rhombencephalic neural crest segmentation is preserved throughout craniofacial ontogeny. Development 1996; 122:3229-3242.
132. Krull CE. Segmental organization of neural crest migration. Mech Dev 2001; 105:37-45.
133. Wilkie AO, Morriss-Kay GM. Genetics of craniofacial development and malformation. Nat Rev Genet 2001; 2:458-468.
134. Wolda SL, Moody CJ, Moon RT. Overlapping expression of Xwnt-3A and Xwnt-1 in neural tissue of Xenopus laevis embryos. Dev Biol 1993; 155:46-57.
135. Cui Y, Brown JD, Moon RT et al. Xwnt-8b: a maternally expressed Xenopus Wnt gene with a potential role in establishing the dorsoventral axis. Development 1995; 121:2177-2186.
136. Landesman Y, Sokol SY. Xwnt-2b is a novel axis-inducing Xenopus Wnt, which is expressed in embryonic brain. Mech Dev 1997; 63:199-209.
137. Lumsden A, Krumlauf R. Patterning the vertebrate neuraxis. Science 1996; 274:1109-1115.
138. Saxén L. Neural induction. Int J Dev Biol 1989; 33:21-48.
139. Zakin LD, Mazan S, Maury M et al. Structure and expression of Wnt13, a novel mouse Wnt2 related gene. Mech Dev 1998; 73:107-116.
140. Liu P, Wakamiya M, Shea MJ et al. Requirement for Wnt3 in vertebrate axis formation. Nat Genet 1999; 22:361-365.
141. Moon RT, Campbell RM, Christian JL et al. Xwnt-5A: a maternal Wnt that affects morphogenetic movements after overexpression in embryos of Xenopus laevis. Development 1993; 119:97-111.
142. Kuiken GA, Bertens PJ, Peterson-Maduro J et al. The promoter of the Xwnt-5C gene contains octamer and AP-2 motifs functional in Xenopus embryos. Nucleic Acids Res 1994; 22:1675-1680.
143. Krauss S, Korzh V, Fjose A et al. Expression of four zebrafish wnt-related genes during embryogenesis. Development 1992; 116:249-259.
144. Wolda SL, Moon RT. Cloning and developmental expression in Xenopus laevis of seven additional members of the Wnt family. Oncogene 1992; 7:1941-1947.
145. Hume CR, Dodd J. Cwnt-8C: a novel Wnt gene with a potential role in primitive streak formation and hindbrain organization. Development 1993; 119:1147-1160.

146. Eisenberg CA, Gourdie RG, Eisenberg LM. Wnt-11 is expressed in early avian mesoderm and required for the differentiation of the quail mesoderm cell line QCE-6. Development 1997; 124:525-536.
147. Ku M, Melton DA. Xwnt-11: a maternally expressed Xenopus wnt gene. Development 1993; 119:1161-1173.
148. Heisenberg CP, Tada M, Rauch GJ et al. Silberblick/Wnt11 mediates convergent extension movements during zebrafish gastrulation. Nature 2000; 405:76-81.
149. Sasai Y, Lu B, Steinbeisser H et al. Xenopus chordin: a novel dorsalizing factor activated by organizer-specific homeobox genes. Cell 1994; 79:779-790.
150. Piccolo S, Sasai Y, Lu B et al. Dorsoventral patterning in Xenopus: inhibition of ventral signals by direct binding of chordin to BMP-4. Cell 1996; 86:589-598.
151. Hemmati-Brivanlou A, Kelly OG, Melton DA. Follistatin, an antagonist of activin, is expressed in the Spemann organizer and displays direct neuralizing activity. Cell 1994; 77:283-295.
152. Fainsod A, Deissler K, Yelin R et al. The dorsalizing and neural inducing gene follistatin is an antagonist of BMP-4. Mech Dev 1997; 63:39-50.
153. Okabayashi K, Shoji H, Onuma Y et al. cDNA cloning and distribution of the Xenopus follistatin-related protein. Biochem Biophys Res Commun 1999; 254:42-48.
154. Smith WC, Harland RM. Expression cloning of noggin, a new dorsalizing factor localized to the Spemann organizer in Xenopus embryos. Cell 1992; 70:829-840.
155. Zimmerman LB, De Jesus-Escobar JM, Harland RM. The Spemann organizer signal noggin binds and inactivates bone morphogenetic protein 4. Cell 1996; 86:599-606.
156. Bouwmeester T, Kim S, Sasai Y et al. Cerberus is a head-inducing secreted factor expressed in the anterior endoderm of Spemann's organizer. Nature 1996; 382:595-601.
157. Piccolo S, Agius E, Leyns L et al. The head inducer Cerberus is a multifunctional antagonist of Nodal, BMP and Wnt signals. Nature 1999; 397:707-710.
158. Cheng AM, Thisse B, Thisse C et al. The lefty-related factor Xatv acts as a feedback inhibitor of nodal signaling in mesoderm induction and L-R axis development in Xenopus. Development 2000; 127:1049-1061.
159. Hollemann T, Chen Y, Grunz H et al. Regionalized metabolic activity establishes boundaries of retinoic acid signaling. EMBO J 1998; 17:7361-7372.
160. Gilbert SF. Developmental Biology, 6th ed. Sinauer Associates, Inc. Publishers, 2000; Sunderland, MA.

# Chapter 6

# Epithelial Planar Cell Polarity in Drosophila

**Marek Mlodzik**

## Abstract

In multicellular organisms epithelia form highly organized and polarized structures. Epithelial apical-basolateral polarity enables the tissues to perform functions like vectorial transport of fluid (e.g., in several glands, the gut or kidney) or directed secretion of specialized components to either their apical or basal side. However, the function of many tissues requires an additional axis of polarity within an epithelium, which is a uniform polarity of single cells or multicellular units within the plane of the epithelium. This type of polarity is commonly referred to as epithelial planar cell polarity (Fig. 1), or as often used in *Drosophila*, tissue polarity. The establishment and maintenance of such polarity is an important feature of many epithelia during development and disease.

## Epithelial Planar Cell Polarity

Typical examples of epithelial planar cell polarity, or planar cell polarity (PCP) for short, include aspects of skin development and function in vertebrates, e.g., the ordered appearance of scales in fish or feathers in birds. Planar cell polarization is also evident in many internal organs. Such examples include the neurosensory epithelium of the inner ear, where the stereocilial bundles are aligned for normal sensitivity to sound, and the oviduct, where the orientation of the cilia allows for directional transport of the egg.

Representative PCP examples in invertebrates include most of the exocuticle in insects, where the single hairs produced by each cuticular cell and the sensory organs like the mechanosensory bristles are aligned within either the anterior-posterior axis on the main body or the proximal-distal axis on the appendages[1-3] (Fig. 2A,B). In addition, in insect compound eyes PCP is evident in the uniform ordered appearance of the facet unit eyes, or ommatidia[4] (Fig. 2C). It is a very interesting biological problem, how cells that are many, often hundreds, of cell diameters apart can adopt the same polarity within the epithelial plane.

The determination of apical-basolateral polarity is usually a direct consequence of an interaction of the cell with its environment. This can be mediated by an interaction with the basal extracellular matrix, or by general cell adhesion properties (reviewed in ref. 5). The establishment of planar cell polarity is, however, more complex, as there are no obvious permanent clues along the anterior-posterior, proximal-distal or dorsal-ventral axes. It is thus thought that the signal(s) that control PCP operate over a long range. Moreover, the specific cellular responses to PCP signaling are very diverse in different tissues, ranging from the organization of cytoskeletal elements in single cells to the general organization of multicellular units both internally and with respect to their surrounding environment. The structures of the Drosophila exocuticle are particularly well suited to the study of the establishment of PCP, due to the ease of observation and experimental accessibility.

*Wnt Signaling in Development*, edited by Michael Kühl.

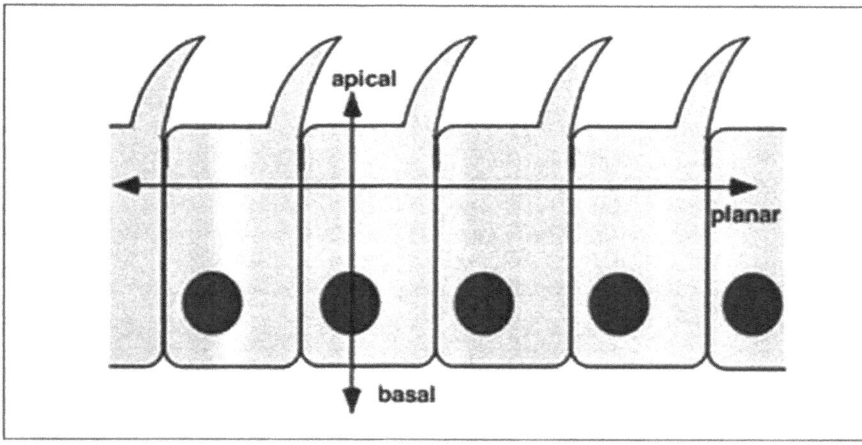


Figure 1. Schematic drawing of epithelial planar cell polarity. In an epithelial cell sheets planar polarity is perpendicular to the apical-basal polarity axis.

Strikingly, a growing number of aspects of PCP signaling appears also required for cellular shape changes throughout development in many different animals. Recently, the process of convergence and extension in vertebrate development has been shown to share molecular features with planar polarization and *Drosophila* PCP signaling. [6-10] Similarly, the process of dorsal closure in *Drosophila* embryogenesis has overlapping requirements with planar polarity signaling (reviewed in ref. 11). Thus the study of PCP determination serves as a general model system for linking signaling pathways to cellular organization and shape changes.

## Planar Cell Polarity in *Drosophila*

*Drosophila* provides a unique model system to study the establishment of PCP (or as also called in flies, tissue polarity) as all adult structures are polarized within the plane (reviewed in refs. 1,3,12,13). This is most obvious in the wings (Fig. 2 A), the dorsal thorax/notum (Fig. 2 B) and the eyes (Fig. 2 C). In the wing, each cell orients itself proximal-to-distally, generating an actin-based hair at the distal vertex. On the thorax, like most parts of the main body, orientation of cells is in the anterior-posterior axis, apparent, in addition to the actin based hairs, in the orientation of the bristles, the peripheral nervous system (PNS) sensory organs. In the eye, planar polarity is reflected in the regular arrangement of the ommatidia in the dorsal-ventral axis. These distinct appearances of polarization in these different tissues underlie a profound difference in the mechanistic aspects of PCP establishment (see below).

Mutations in genes required for planar polarity establishment in general result in random orientation of hairs, bristles or ommatidia (reviewed in refs. 1,3,13; see also refs. 14-16) (Fig. 3). Several of the PCP (or tissue polarity) genes have been identified genetically over the past three decades. These are, when affecting all tissues, referred to as "primary" PCP genes and comprise a growing number of genes, including *frizzled* (fz), *dishevelled* (dsh), *prickle-spiny legs* (pk), *rhoA*, *strabismus* (*stbm*; also called *Van Gogh*), *flamingo* (*fmi*; also called *starry night*) and *diego* (dgo) (Refs. 14-23). See list of all reported genes with original references in Table 1. In addition, several other genes have been identified that appear required for PCP establishment in specific tissues. For example, *fuzzy*, *inturned* and *multiple wing hairs* (mwh) only affect polarity of the cellular hairs on the wing[24,25], whereas mutations in *nemo* and *roulette* appear only to affect ommatidial polarity. [26] Taken together, these observations suggested that the genes affecting

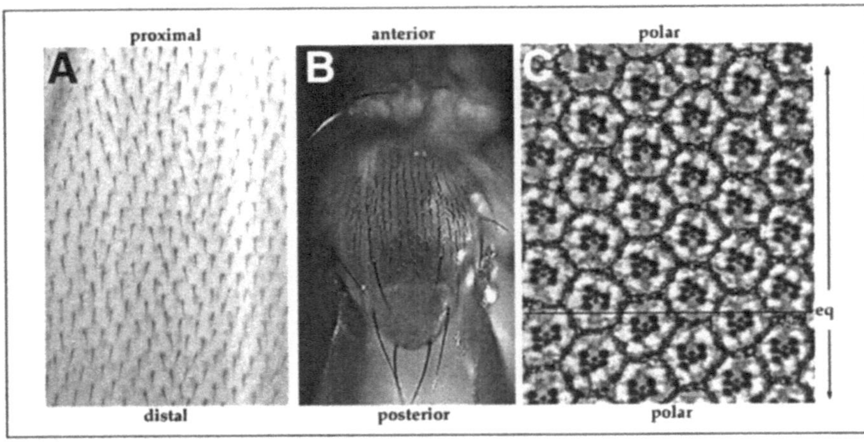


Figure 2. Examples of epithelial planar cell polarity in Drosophila. (A) wing (B), dorsal thorax, and (C) eye. Note that the axis of polarity is different in the three examples, and that each tissue displays different aspects of tissue polarization. Whereas single cells are polarized in the wing as evident by the ordered appearance of the hairs (A), groups of cells are reflecting polarization in the notum and the eye.

PCP in all tissues are components of a common signaling pathway and mechanism, responsible for reading and relaying a common PCP signal (see below). In contrast, the tissue specific genes like *fuzzy* and *roulette* act as effectors of this signaling pathway during the execution stage of PCP establishment in the respective tissues.

Although the polarizing signal has not yet been identified it is likely to be a member of the Wnt family of growth factors as the primary receptor is Frizzled (Fz, see below). Moreover in zebrafish, Wnt-11 has been identified as a key factor regulating convergence and extension[7], a process that is regulated largely by the same Fz/PCP pathway as polarization in *Drosophila*. However, as the axes of PCP orientation are distinct in different tissues, it is thought that the source of the signal is defined independently in each tissue/organ. Whereas the presumptive signal polarizes the wing (or other appendages) from proximal to distal, the trunk is polarized in the anterior-posterior axis, and the eye is polarized in the dorsal-ventral axis. Thus, in each tissue PCP generation can be subdivided in three steps: (1) Definition of the source of a polarizing signal; (2) reception and interpretation of the signal in single cells or groups of cells; and (3) organization of the cells (groups of cells) in response to the signal. Although steps 1 and 3 appear to be tissue specific, the second step shares most (if not all) factors and is a generally conserved signaling pathway and mechanism. This part is also conserved evolutionarily and is thought to generally regulate cellular polarization also in vertebrates, as evident in the convergent and extension process during vertebrate gastrulation[6-8,10,27] and inner ear stereo-cilia polarization (see below).

## Wnt/Frizzled Signaling and Planar Cell Polarity Establishment

As mentioned above, the signal(s) activating the PCP pathway in *Drosophila* are not yet known, but likely to be a Wnt family member, possibly a Wnt-11 as in zebrafish.[7] However, the receptor is encoded by *frizzled* (*fz*), the founding member of the Frizzled receptor family of Wnt receptors,[28] and is required in all tissues.[1] Molecular cloning of *fz* and its primary sequence predicted a seven-transmembrane receptor-like molecule[29], and more recently, members of the Fz protein family were identified as the receptors for the Wnt family of secreted

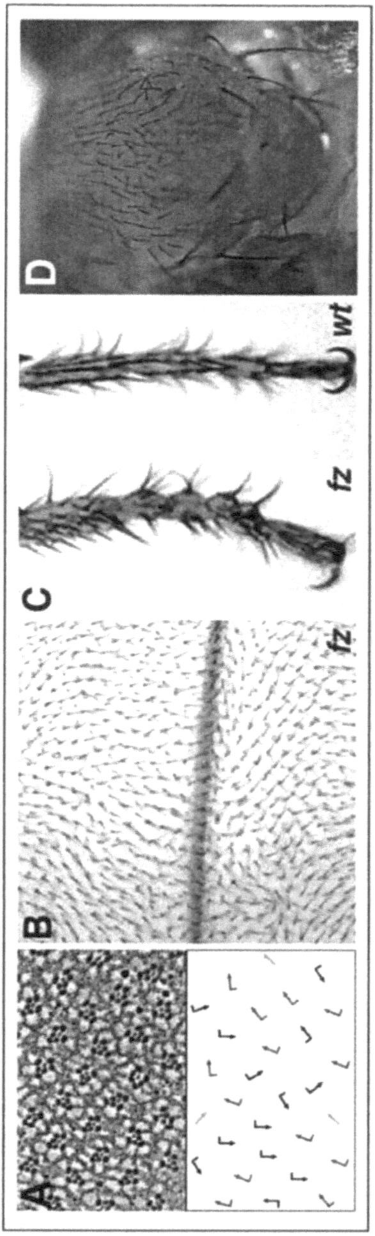


Figure 3. The planar cell polarity (PCP) phenotypes. All panels show fz- tissues (and a wt control in C). (A) Ommatidial arrangement in a dorsal area of the eye, with schematic presentation below (black arrow: correct alignment, red arrow: misrotated ommatidium, green arrow: wrong or no chiral form, compare to Fig. 2 C). (B) Wing phenotype (distal is to the right; compare to Fig. 6.2 A for wt). (C) Leg phenotype with wt control. Note that defects extend also to the arrangement of the joints. (D) Notum phenotype. Note the misaligned bristles on the dorsal thorax (compare to Fig. 2 B for wt).

growth factors[28] and implicated in the Wg/Wnt signaling pathway.[30-34] Besides *fz*, *dishevelled* (dsh) is the best characterized polarity gene.[15,35,36] Genetic experiments and epistasis analysis has placed dsh downstream of *fz*, both in canonical Wnt signaling as well as in Fz/PCP signaling (reviewed in ref. 35). Although Fz and Dsh are part of both, the canonical Wnt/Wg signaling pathway and Fz/PCP signaling, the other known components of canonical Wnt signaling do not display planar polarity defects,[37,38] indicating that a distinct pathway is functioning downstream of Fz and Dsh in PCP establishment.

Dsh is a conserved cytoplasmic protein found in all species from *C.elegans* to humans. All Dsh proteins identified to date share three highly conserved domains found also in other proteins: a N-terminal DIX (Dishevelled-Axin) domain, a central PDZ domain and a C-terminal DEP (Dishevelled, Egl-10, Pleckstrin) domain (reviewed in ref. 35). All three domains have been implicated as protein-protein interaction modules, and thus Dsh might serve as an adapter or scaffolding molecule.

A combination of genetic and biochemical studies has recently demonstrated that a Fz/PCP signaling pathway downstream of Fz and Dsh consists of the small GTPases RhoA and Rac, the Rho-associated kinase (*dROK*) and the STE20-like kinase Misshapen (Msn) (Refs. 14,39-42). Distinct downstream signaling components are then associated with the Fz/PCP pathway depending on the tissue. This divergence is most obvious in the final response to Fz/PCP signaling in the distinct tissues. Whereas a transcriptional response is important in the eye, cytoskeletal reorganization is the main response in wing cells (see below).

In the eye, genetic interactions suggest that JNK/p38-type MAPK modules act downstream of Rac and Msn[38,40,41,43] (summarized in Fig. 4). Consistently, in biochemical assays,

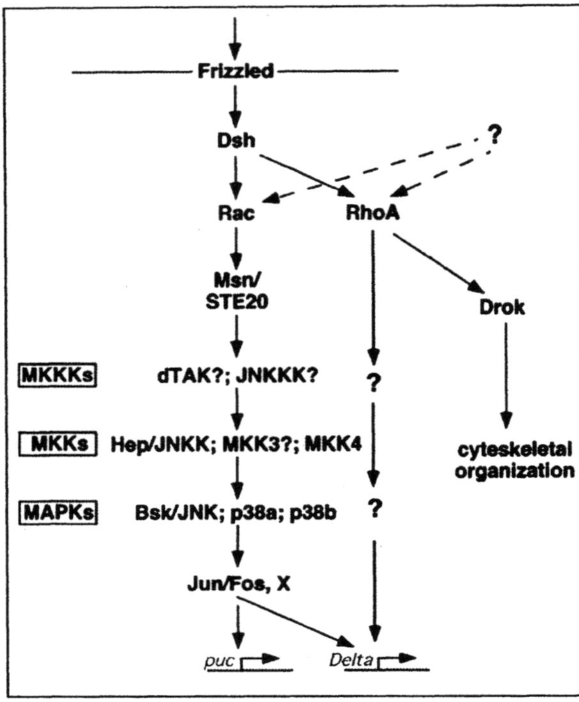


Figure 4. The Frizzled/PCP pathway as it is used in the Drosophila eye. Note that the nuclear signaling event leading to the transcriptional activation of Delta is specific to eye PCP establishment. The members of the kinase cascade acting downstream of Misshapen appear largely redundant, the effectors of RhoA in nuclear signaling are not known.

*Drosophila* Dsh (and its human homologues) act as potent activators of JNK signaling in a Msn dependent way.[38,41,44] Nevertheless, the role of the JNK/p38 kinases in Fz/PCP signaling must be redundant, as the respective single mutants do not show polarity phenotypes in simple loss-of-function analyses and only when more than one of these related kinases is affected, a significant effect is seen.[41,43] It was also recently shown that the AP-1 transcription factor (consisting of Jun and Fos) is one of the necessary nuclear factors downstream of Fz and the JNK/p38 modules.[43,45] Similarly, the role of Rac1 and Rac2 in the pathway is supported by genetic studies.[38-40]

In the wing, where reorganization of cytoskeletal elements is the main response to Fz/PCP signaling the downstream signaling components include the Rho-associated kinase *dROK*, the nonmuscle myosin regulatory light chain (*MRLC/sqh*) and Myosin II (*zipper*) (Ref. 42). The wing cell specific aspects of Fz/PCP signaling are summarized in Fig. 5.

It remains unclear how Fz is molecularly linked to Dsh, and how Dsh subsequently transduces the signal to further downstream components. Although Fz recruits Dsh to the membrane,[37,46] there is no reported direct molecular interaction between them. Similarly, it is not yet known what factor(s) Dsh binds to activate its downstream planar polarity effectors like RhoA and Rac, and/or Msn/STE20. However, a recent study in *Xenopus* has identified a Formin homology domain protein, Daam1, as a bridging factor between Dsh and Rho[27]. Daam1 binds both Dsh and Rho via distinct protein regions and is required for Fz mediated Rho activation.

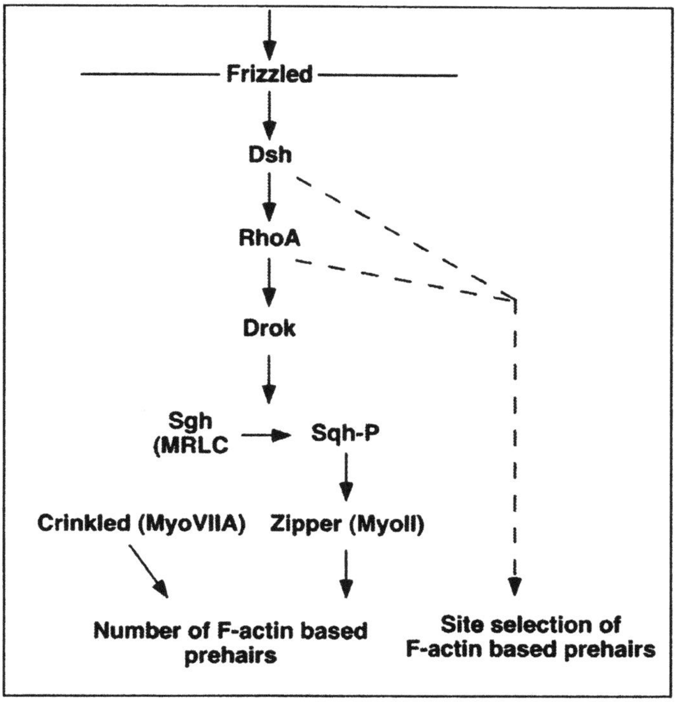


Figure 5. The Frizzled/PCP pathway as it has been established in wing cell polarization (adapted from Winter et al. 2000; ref. 42). Most of the PCP signaling is thought to lead directly to cytoskeletal organization and the determination of the site of the assembly of the distal actin spike giving rise to the wing hair. A similar pathway might act in the context of convergent extension associated cell shape changes in gastrulation in vertebrates

Thus, Daam1 is a very good candidate to recruit Rho to the sites of Dsh localization. The confirmation of this finding in *Drosophila* is not yet available.

## Frizzled Signaling and Its Link to the Other PCP Genes

Several other primary PCP genes have been genetically and molecularly characterized (Table 1). Their link to Fz signaling remains however unclear. *Strabismus/Van Gogh* (stbm) also shows similar phenotypic features as *fz* and *dsh*. [20,23] Although, its molecular features suggest a transmembrane protein, there is no functional homology with other proteins. The *flamingo/starry night* (*fmi*) gene also shows similar phenotypic features as *fz* and *dsh* and encodes an atypical cadherin with a seven-pass transmembrane receptor-like region[17,21] (Table 1). Fmi is differentially localized at cell-cell boundaries in the wing in a *fz* dependent manner (see below). A new PCP gene, *diego* (*dgo*), was identified recently in a gain-of-function screen.[18] *diego* encodes a cytoplasmic protein with 6 Ankyrin repeats and it co-localizes with Fmi at membranes. Moreover, *fmi* and *dgo* interact genetically and overexpression of Dgo leads to Fmi mislocalization. Thus it was proposed that Fmi and Dgo act together in PCP establishment.[18] Unlike *fz*, *dsh*, *stbm* and *fmi*, *dgo* however does not affect all tissues. Although it shows typical PCP phenotypes in the wing and the eye, it does not affect polarity on the notum in loss-of-function analyses[18] (Table 1). Last but not least, the *prickle-spiny legs* locus (*pk*) is another 'old' primary

***Table 1. Primary polarity genes involved in PCP establishment in* Drosophila**

| Gene | Tissues Gene Is Required in | Molecular Features | References |
|---|---|---|---|
| frizzled (fz) | all adult cuticular tissues | seven-pass transmembrane receptor, binds Wg/Wnt ligands | 22, 29 |
| disheveled (dsh) | all adult cuticular tissues | cytoplasmic protein containing DIX, PDZ, DEP domains, recruited to membrane by Fz | 15, 37, 38 |
| prickle (pk) (incl. spiny legs isoform) | all adult cuticular tissues | cytoplasmic protein with 3 LIM domains and PET domain | 19 |
| strabismus (stbm) (a.k.a. Van Gogh) | all adult cuticular tissues | novel putative four-pass transmembrane protein | 20, 23 |
| flamingo (fmi)/ starry night (stan) | all adult cuticular tissues | atypical cadherin with seven-pass transmembrane region | 21 |
| diego | eye, wing | cytoplasmic ankyrin repeat protein, colocalizes with Fmi | 18 |
| RhoA | eye, wing* | small GTPase | 14 |
| *dROCK* | eye, wing* | Rho-associated kinase, affects a subset of PCP aspects in wing (multiple wing hairs) and eye (ommatidial rotation) | 42 |
| misshapen (msn) | eye, wing, notum* | STE20-like S/T protein kinase | 41 |

* other tissues were not tested in homozygous clones, embryonic lethal

PCP gene. It was cloned recently and encodes a protein containing three LIM domains and an additional conserved PET (Prickle-Espinas-Testin) domain.[19] As it shares 4 domains likely to be involved in protein-protein interactions, it was proposed that it may serve a scaffolding function.[19] The roles of *stbm* and *pk* in PCP generation and establishment and their relationship to Fz signaling or the other PCP genes remains, however, obscure due to the lack of any reported molecular or genetic interactions.[14,19]

Strikingly, it was recently proposed that a similar, or the same, Fz/PCP signaling pathway regulates polarized cell shape changes during vertebrate development in the process of convergence and extension during gastrulation.[7,10,27] Interestingly, the conservation extends beyond the Fz/Dsh signaling aspect and includes vertebrate homologs of several of the other PCP genes. In particular, the Xenopus Stbm homolog *xStbm* has been shown to affect the related convergent and extension polarization process.[6,8] In addition, several members of the Protocadherin gene superfamily have been shown to be affected in vertebrate hearing loss mutations related to aberrant stereo-cilia polarization.[47,48] These recent observations suggest that Fz/planar polarity signaling and the associated other PCP genes comprise a conserved

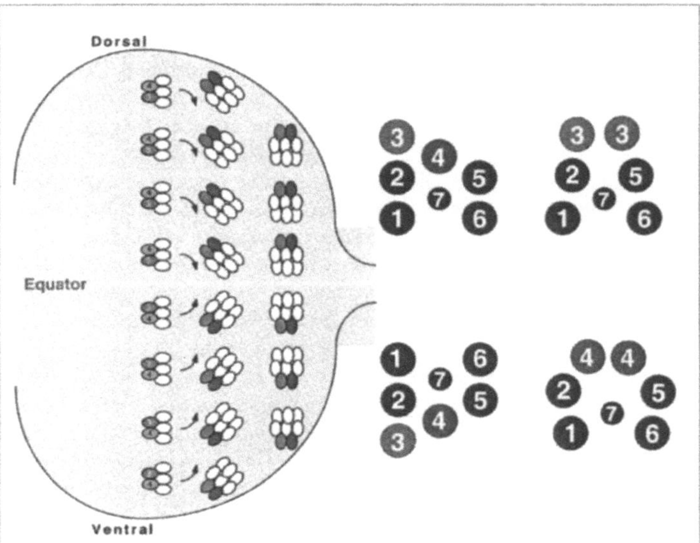


Figure 6. Establishment of planar polarity in the eye. Left side: Cartoon of the logic of polarity generation during eye development. Initially ommatidial preclusters are organized in the antro-posterior axis and are symmetrical. Subsequently they rotate 90° towards the equator (the D/V-midline) and at the end of this process chirality established by the positions of the R3/R4 cells. Right side: Schematic presentation of chiral organization of the respective ommatidia. In addition to the two chiral forms, symmetrical clusters with R3/R3 or R4/R4 pairs can be found in PCP mutants (compare to Figs. 2 C and 3 A).

mechanism and genetic cassette of genes employed in polarized cell shape changes throughout the animal kingdom.

## Tissue Specific Responses to PCP Signaling

As mentioned above the Fz/PCP pathway and the other "primary" polarity genes lead to tissue specific responses in the different organs. The distinct cellular responses in the relevant tissues are summarized below:

### *a) Planar Polarization in the Wing*

In the wing every cell generates its orientation and polarization more or less independently of each other and reorganizes its cytoskeleton to form an actin spike that points towards the distal vertex of each cell. This actin spike gives rise to the "wing hairs" on the adult wing, one in each cell pointing distally (Fig. 2 A). Thus, the main requirement for PCP signaling is to directly regulate the cytoskeletal organization (reviewed in ref. 2). In such a scenario the asymmetric subcellular localization (or enrichment) of a component is sufficient for the process of polarization. Accordingly, asymmetric localization within the proximal-distal axis has been observed for several components of PCP signaling and determination. In particular, Fmi, Fz, Dsh and Diego are enriched in the proximal-distal axis, either on both sides (Fmi, Dgo) or just the distal side (Fz, Dsh).[18,21,49-51] Strikingly, the localization of each of these depends on the presence of all the other ones, suggesting a complicated interdependent signaling circuitry or a protein complex. In primary PCP gene mutants most cells still form a single actin hair, however at the wrong site within the cell and/or with a wrong orientation. The failure to cause an asymmetric enrichment of some factors can lead to a "multiple wing hair" phenotype,[24,25] where one cell fails to focus the machinery generating the actin spike.

Although asymmetric localization appears to be a prerequisite, it is not yet clear how this is linked to the actual polarization of the cytoskeleton field and Fz, the presumed receptor for the signal (reviewed in refs. 13,52). First hints at a link to the subsequent execution of the polarization resulting in the formation of one F-actin based hair comes from a recent study.[42] These linking components include the previously mentioned Drok and also MRLC/Sqh and *MyoII/Zip* genes.[42] In addition, this process requires the *fuzzy* (*fy*), *inturned* (*in*) and *multiple wing hair* (*mwh*) genes.[24,25] These are likely to function in a parallel pathway to Fz/Dsh/Drok signaling and their function is not clear with respect to the Fz/PCP pathway.

### *b) Planar Polarization on the Notum*

In contrast to the wing, in the eye (see below) or the notum (dorsal thorax; Fig. 2 B) groups of cells make a coordinated decision, and thus in this context PCP signaling affects the organization of several cells. On the notum, the sensory bristles (consisting of 4 cells) are nicely polarized in the anterior-posterior axis (with the bristle itself pointing posteriorly). PCP establishment within the 4 cells of the bristle group is (at least partially) governed by the orientation of the axis of the mitotic spindle of the sensory organ precursor (SOP) cell. It has been shown that *fz*, *dsh* and *fmi* primarily control the orientation of the mitotic spindle axis of the first asymmetric cell division of the SOP.[53,54] Whereas in wild type the spindle is oriented in the anterior-posterior axis, in PCP mutants the orientation of the mitotic spindle is random. Although the spindle orientation has only been analyzed in detail for *fz*, *dsh* and *fmi*, the other "primary" PCP genes also affect this process as judged by the random orientation of the bristles on the notum (Table 1). Interestingly, in *C.elegans* the orientation of the mitotic spindle in some asymmetric divisions is also under the control of Fz family members, e.g., *mom-5* and *lin-17*,[55,56] suggesting that the employment of Fz mediated signaling in spindle orientation is a evolutionarily conserved feature.

### *c) Planar Polarity in the Eye*

In the *Drosophila* eye, PCP is reflected in the mirror image arrangement of ommatidia (or facets) across the dorsal-ventral (D/V) midline, the equator (Fig. 2 C). This pattern is generated when immature ommatidia rotate 90° towards the equator and adopt opposite chiral forms on either side (reviewed in refs. 12,57-59)(Fig. 6). Chirality is evident in mature ommatidia, as the R3 and R4 photoreceptors are asymmetrically positioned at the tip of the ommatidial trapezoids (Fig. 6). In mutants such as *fz* and *dsh*, both chirality and the direction and degree of rotation become random. In addition, in these mutants some ommatidia remain symmetrical, with either V or U-shaped clusters with non-chiral R3/R3 or R4/R4 photoreceptor pairs (Fig. 6, see also Fig. 3).

Genetic analysis of *fz* mutants has indicated that the R3/R4 pair, in particular the assignment of cell identity within the pair, is critical for establishing chirality and direction of ommatidial rotation.[16,60,61] Prior to rotation of the ommatidia the R3/R4 precursors are symmetrically arranged in the ommatidial precluster, with R3 being closer to the equator than R4. As it is assumed that the PCP signal, emanates from the equator, it is received first, or at higher levels, in the R3/R4 precursor that is closer to the equator. Fz signaling then specifies the R3 fate[16,61,62] and activates the expression of *Delta* (Dl) in this cell. Dl expression in R3 in turn leads to Notch activation in the neighboring cell, specifying it as R4.[61-63] This two-tiered mechanism determines the polarity/chirality of the ommatidium and thus the direction of rotation.

Accordingly, both indiscriminate activation or loss of Fz/PCP signaling in R3/R4 results in symmetric or random *Dl* expression in these cells: the chirality decision thus becomes stochastic and the direction and degree of rotation become random. Although the respective transcription factors downstream of Fz have not yet been fully identified, the AP-1 (Jun/Fos) transcription factor is a target of Fz/PCP signaling in R3.[43,45]

Interestingly, although *stbm* shows similar phenotypic features in the eye as fz and dsh,[23] its requirement in the eye is, at close inspection, opposite to that of *fz/dsh*: *stbm* is required in R4 and not in the R3 cell. Thus the role of Stbm might be to antagonize Fz signaling.

## General and Evolutionary Implications

Although the end result of epithelial cellular polarization is very distinct in different tissues and organisms, there is a growing body of evidence that the underlying signaling pathway(s) and mechanistic aspects of polarization are conserved throughout evolution. First, the specific aspects of Fz/PCP signaling are the same for planar cell polarization in flies and convergent extension in vertebrates. In particular, the domain requirements of Dsh are identical in these two contexts and distinct from canonical Wnt/Fz signaling. Whereas the PDZ and DEP domains are important for Fz/PCP signaling and Rho/Rac/JNK activation, the DIX and PDZ domain are critical for canonical Wnt signaling and β-catenin stabilization.[7,10,27,37,38] The recent addition of the PCP gene Strabismus (Stbm/Vang) to the convergent extension process[6,8] (see also the mouse homologue Looptail; ref. 64) strongly suggests that most (if not all) of the PCP gene interplay and resulting signaling is evolutionarily conserved. This is further supported by the observation that mutations in atypical cadherins Pcdh15 (Ref. 47) and CDH23 (Ref. 48) affect the polarization of the stereo-cilia within the vertebrate inner ear epithelium. Thus taken together, there is good evidence for the conserved function of Fz/PCP and the involvement of Stbm and Flamingo/atypical cadherins in planar cell polarization (flies, inner ear) and convergent extension, suggesting that many of the other PCP genes in Drosophila will have a related role in vertebrates. The analysis of direct molecular interactions between the Fz/PCP cassette and the other PCP genes should give insight how these interactions shape up at the mechanistic level.

## References

1. Adler PN. The genetic control of tissue polarity in Drosophila. BioEssays 1992; 14:735-741.
2. Eaton S. Planar polarity in Drosophila and vertebrate epithelia. Curr Opin Cell Biol 1997; 9:860-866.
3. Gubb D. Genes controlling cellular polarity in Drosophila. Development Suppl 1993:269-277.
4. Dietrich W. Die Facettenaugen der Dipteren. Z. Wiss Zool 1909; 92:465-539.
5. Eaton S, Simons K. Apical, basal and lateral cues for epithelial polarization. Cell 1995; 82:5-8.
6. Darken RS. The planar polarity gene strabismus regulates convergent extension movements in Xenopus. EMBO J 2002; 21:976-985.
7. Heisenberg CP et al. Silberblick/Wnt11 mediates convergent extension movements during zebrafish gastrulation. Nature 2000; 405:76-81.
8. Park M, Moon RT. The planar cell-polarity gene stbm regulates cell behaviour and cell fate in vertebrate embryos. Nature Cell Biol 2002; 4:20-25.
9. Tada M, Smith JC. Xwnt11 is a target of Xenopus Brachyury: regulation of gastrulation movements via Dishevelled, but not through the canonical Wnt pathway. Development 2000; 127:2227-2238.
10. Wallingford JB et al. Dishevelled controls cell polarity during Xenopus gastrulation. Nature 2000; 405:81-85.
11. Noselli S. JNK signaling and morphogenesis in Drosophila. Trends Genet 1998; 14:33-38.
12. Mlodzik M. Planar polarity in the Drosophila eye: a multifaceted view of signaling specificity and cross-talk. EMBO J 1999; 24:6873-6879.

13. Shulman JM, Perrimon N, Axelrod JD. Frizzled signaling and the developmental control of cell polarity. Trends Genet 1998; 14:452-458.
14. Strutt DI, Weber U, Mlodzik M. The role of RhoA in tissue polarity and Frizzled signalling. Nature 1997; 387:292-295.
15. Theisen H et al. dishevelled is required during wingless signalling to establish both cell polarity and cell identity. Development 1994; 120:347-360.
16. Zheng L, Zhang J, Carthew RW. frizzled regulates mirror-symmetric pattern formation in the Drosophila eye. Development 1999; 121:3045-3055.
17. Chae J et al. The Drosophila tissue polarity gene starry night encodes a member of the protocadherin family. Development 1999; 126:5421-5429.
18. Feiguin F, Hannus M, Mlodzik M et al. The Ankyrin repeat protein Diego mediates Frizzled-dependent planar polarization. Developmental Cell 2001; 1:93-101.
19. Gubb D et al. The balance between isoforms of the prickle LIM domain protein is critical for planar polarity in Drosophila imaginal discs. Genes Dev 1999; 13:2315-2327.
20. Taylor J, Abramova N, Charlton J et al. Van Gogh: a new Drosophila tissue polarity gene. Genetics 1998; 150:199-210.
21. Usui T et al. Flamingo, a seven-pass transmembrane cadherin, regulates planar cell polarity under the control of Frizzled. Cell 1999; 98:585-595.
22. Vinson CR, Adler PN. Directional non-cell autonomy and the transmission of polarity information by the frizzled gene of Drosophila. Nature 1987; 329:549-551.
23. Wolff T, Rubin GM. strabismus, a novel gene that regulates tissue polarity and cell fate decisions in Drosophila. Development 1998; 125:1149-1159.
24. Adler PN, Charlton J, Park WJ. The Drosophila tissue polarity gene inturned functions prior to wing hair morphogenesis in the regulation of hair polarity and number. Genetics 1994; 137:1-8.
25. Collier S, Gubb D. Drosophila tissue polarity requires the cell-autonomous activity of the fuzzy gene, which encodes a novel transmembrane protein. Development 1997; 124:4029-4037.
26. Choi KW, Benzer S. Rotation of photoreceptor clusters in the developing Drosophila eye requires the nemo gene. Cell 1994; 78:125-136.
27. Habas R, Kato Y, He X. Wnt/Frizzled activation of rho regulates vertebrate gastrulation and requires a novel formin homology protein Daam1. Cell 2001; 107:843-854.
28. Bhanot P et al. A new member of the frizzled family from Drosophila functions as a Wingless receptor. Nature 1996; 382:225-230.
29. Vinson CR, Conover S, Adler PN. A Drosophila tissue polarity locus encodes a protein containing seven potential transmembrane domains. Nature 1989; 338:263-264.
30. Bhat KM. Frizzled and Frizzled2 play a partially redundant role in Wingless signaling and have similar requirements to Wingless in neurogenesis. Cell 1998; 95:1027-1036.
31. Cadigan KM, Fish MP, Rulifson EJ et al. Wingless repression of Drosophila frizzled2 expression shapes the Wingless morphogen gradient in the wing. Cell 1998; 93:767-777.
32. Chen CN, Struhl G. Wingless transduction by the Frizzled and Frizzled2 proteins of Drosophila. Development 1999; 126:5441-5452.
33. Kennerdell JR, Carthew RW. Use of dsRNA-mediated genetic interference to demonstrate that frizzled and frizzled2 act in the wingless pathway. Cell 1998; 95:1017-1026.
34. Mueller H, Samanta R, Wieschaus E. Wingless signaling in the Drosophila embryo: zygotic requirements and the role of the frizzled genes. Development 1999; 126:577-586.
35. Boutros M, Mlodzik M. Dishevelled: at the crossroads of divergent intracellular signaling pathways. Mech. Dev.1999; 83:27-37.
36. Klingensmith J, Nusse R, Perrimon N. The Drosophila segment polarity gene dishevelled encodes a novel protein required for response to the wingless signal. Genes Dev 1994; 8:118-130.
37. Axelrod JD, Miller JR, Shulman JM, Moon RT, Perrimon N. Differential requirement of Dishevelled provides signaling specificity in the Wingless and planar cell polarity signaling pathways. Genes & Dev 1998; 12:2610-2622.
38. Boutros M, Paricio N, Strutt DI et al. Dishevelled activates JNK and discriminates between JNK pathways in planar polarity and wingless signaling. Cell 1998; 94:109-118.
39. Eaton S, Wepf R, Simons K. Roles for Rac1 and Cdc42 in planar polarization and hair outgrowth in the wing of Drosophila. J Cell Biol 1996; 135:1277-1289.

40. Fanto M, Weber U, Strutt DI et al. Nuclear signaling by Rac abd Rho GTPases is required in the establsihememt of epithelial planar polarity in the Drosophila eye. Current Biology 2000; 10:979-988.
41. Paricio N, Feiguin F, Boutros M et al. The Drosophila STE20-like kinase Misshapen is required downstream of the Frizzled receptor in planar polarity signaling. EMBO J 1999; 18:4669-4678.
42. Winter CG et al. Drosophila Rho-associated kinase (Drok) links Frizzled-mediated planar cell polarity signaling to the actin cytoskeleton. Cell 2001; 105:81-91.
43. Weber U, Paricio N, Mlodzik M. Jun mediates Frizzled induced R3/R4 cell fate distinction and planar polarity determination in the Drosophila eye. Development 2000; 127:3619-3629.
44. Li L et al. Dishevelled proteins lead to two signalling pathways: regulation of Lef-1 and c-Jun N-terminal kinase in mammalian cells. J Biol Chem 1999; 274:129-134.
45. Ciapponi L, Jackson DB, Mlodzik M et al. Drosophila Fos mediates ERK and JNK signals via distinct phosphorylation sites. Genes Dev 2001; 15:1540-1553.
46. Yang-Snyder J, Miller JR, Brown JD et al. A frizzled homolog functions in a vertebrate Wnt signaling pathway. Curr Biol 1996; 6:1302-1306.
47. Alagramam KN et al. The mouse Ames waltzer hearing-loss mutant is caused by mutation of Pcdh15, a novel protocadherin gene. Nat Genetics 2001; 27:99-102 .
48. Bolz H et al. Mutation of CDH23, encoding a new member of the cadherin gene family, causes Usher syndrome type 1D. Nat Genet 2001; 27:108-112.
49. Axelrod JD. Unipolar membrane association of Dishevelled mediates Frizzled planar cell polarity signaling. Genes Dev 2001; 15:1182-1187.
50. Shimada Y, Usui T, Yanagawa S et al. Asymmetric colocalization of Flamingo, a seven-pass transmembrane cadherin, and Dishevelled in planar cell polarization. Curr Biol 2001; 11:859-863.
51. Strutt DI. Asymmetric localization of frizzled and the establishment of cell polarity in the Drosophila wing. Molecular Cell 2001; 7:367-375.
52. Adler PN, Lee H. Frizzled signaling and cell-cell interactions in planar polarity. Curr Opin Cell Biol 2001; 13:635-640.
53. Gho M, Schweisguth F. Frizzled signaling controls orientation of asymetric sense organ precursor cell divisions in Drosophila. Nature 1998; 393:178-181.
54. Lu B, Usui T, Uemura T et al. Flamingo controls the planar polarity of sensory bristles and asymmetric division of sensory organ precursors in Drosophila. Curr Biol 1999; 9:1247-1250.
55. Sawa H, Lobel L, Horvitz HR. The C.elegans lin-17, which is required for certain asymmetric cell divisions,encodes a putative seven-transmembrane protein simsilar to the Drosophila Frizzled protein. Genes Dev 1996; 10:2189-2197.
56. Thorpe CJ, Schlesinger A, Carter JC, Bowerman B. Wnt signaling polarizes an early C. elegans blastomere to distinguish endoderm from mesoderm. Cell 1997; 90:695-705.
57. Reifegerste R, Moses K. The genetics of epithelial polarity and pattern in the Drosophila retina. BioEssays 1999; 21:275-285.
58. Strutt H, Strutt D. Polarity determination in the Drosophila eye. Curr Opin Genet Dev 1999; 9:442-446.
59. Wolff T, Ready DF. In: Martinez-Arias MBA, ed. The development of Drosophila melanogaster. Cold Spring Harbor: Cold Spring Harbor Press, 1993:1277-1326.
60. Fanto M, Mayes CA, Mlodzik M. Linking cell-fate specification to planar polarity: determinationof the R3/R4 photoreceptors is a prerequisite for the interpretation of the Frizzled mediated polarity signal. Mech Dev 1998; 74:51-58.
61. Tomlinson A, Struhl G. Decoding vectorial information from a gradient: sequential roles of the receptors Frizzled and Notch in establishing planar polarity in the Drosophila eye. Development 1999; 126:5725-5738.
62. Fanto M, Mlodzik M. Asymmetric Notch activation specifies photoreceptors R3 and R4 and planar polarity in the Drosophila eye. Nature 1999; 397:523-526.
63. Cooper MTD, Bray SJ. Frizzled regulation of Notch signalling polarizes cell fate in the Drosophila eye. Nature 1999; 397:526-529.
64. Kibar Z et al. Ltap, a mammalian homolog of Drosophila Strabismus/Van Gogh, is altered in the mouse neural tube mutant Loop-tail. Nat Genet 2001; 28:251-252.

# CHAPTER 7

# Patterning the Vertebrate Neural Plate by Wnt Signaling

**Jean-Pierre Saint-Jeannet**


## Summary

A central problem in the study of vertebrate development is defining the cellular and molecular mechanisms responsible for the patterning of embryonic tissues and the differentiation of specific cell types. A large number of studies indicate that the Wnt family of secreted glycoproteins plays an important role in the processes leading to the development of a broad range of cell types, tissues and organs. In this chapter, the contribution of Wnt signaling to the induction and patterning of the nervous system will be reviewed. Emphasis will be specifically placed on Wnt functions during neural induction, anteroposterior and dorsoventral patterning of the spinal cord and neural crest formation.


## Introduction

Intercellular communication mediated by secreted polypeptides represents one of the primary strategies used by the developing embryo to achieve proper body plan organization. These factors released in the extracellular space can promote cell growth, cell fate decisions, establish tissue boundaries and initiate differentiation programs. In recent years five major signaling pathways have emerged as key regulators of embryonic development: the transforming growth factor (TGF)β,[1] fibroblast growth factor (FGF),[2] platelet derived growth factor (PDGF),[3] Hedgehog (hh)[4] and Wnt[5] families. Among these, molecules of the Wnt family of secreted glycoproteins have been shown to mediate a broad range of activities during development of a variety of tissues and organs (reviewed in refs. 6, 7) including the nervous system (reviewed in refs. 8, 9).

Initial evidence for the function of Wnt signaling during early development of the nervous system came from the analysis of neural defects resulting from the disruption of the Wnt-1 gene in mouse embryos.[10,11] In all vertebrates Wnt-1 expression follows a restricted and unique expression pattern (Fig. 1A). Wnt-1 transcripts are detected in the most posterior domain of the midbrain and at the junction with the hindbrain, also known as the isthmus.[12] Later during embryogenesis, this region of the midbrain will contribute to the anterior half of the cerebellum. The posterior half of the cerebellum is derived from the anterior portion of the hindbrain,[13] which does not express Wnt-1.[12] Mouse embryos homozygous for a targeted disruption of the Wnt-1 gene exhibit a severe truncation of the midbrain and the cerebellum (Fig. 1B). In the most extreme cases, the phenotype of Wnt-1 mutant mice is characterized by the total absence of cerebellum.[10,11] Subsequent studies have established that at the isthmus, Wnt-1 is implicated in the maintenance of engrailed expression in the mesencephalon and in the anterior

*Wnt Signaling in Development*, edited by Michael Kühl.

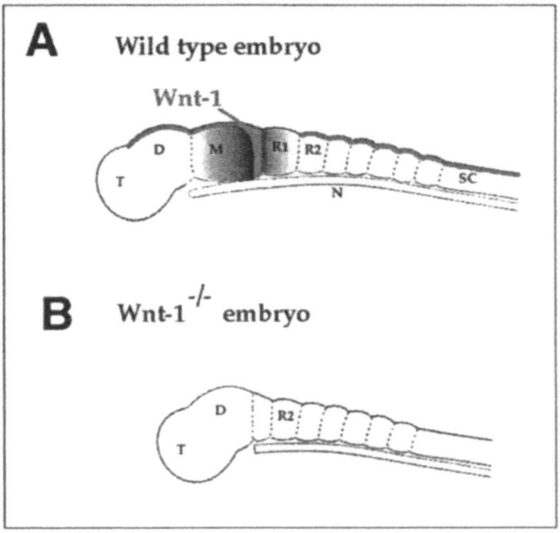


Figure 1. Wnt-1 signaling is required for midbrain development. Schematic representation of the central nervous system, lateral view, anterior is to the left. (A) In wild type embryo Wnt-1 is expressed in the midbrain, at the midbrain/hindbrain boundary, known as the isthmus, and in the dorsal spinal cord. The entire midbrain region is deleted in Wnt-1 deficient mice (B). Engrailed protein is expressed in two opposite gradients, anterior and posterior to the isthmus. Wnt-1 is required at the isthmus to maintain engrailed expression. T, telencephalon; D, diencephalon; M, mesencephalon; R, rhombomere; SC, spinal cord; N, notochord.

portion of the hindbrain.[14] Engrailed proteins are believed to be required for the proper outgrowth of the cerebellum.[14] Wnt-1 mutant embryos appear to have a normal spinal cord and hindbrain, even though Wnt-1 is also normally expressed in the posterior hindbrain and in the roof plate of the spinal cord (Fig. 1). The lack of a phenotype at other Wnt-1 expression domains is likely to be due to a functional compensation by other Wnt genes, including Wnt-3a, which is expressed in overlapping domains in the more caudal regions of the hindbrain and along the entire length of the spinal cord. [15] These results imply a prominent role of Wnt-1 in the development of specific regions of the brain, illustrating one of the numerous functions of Wnt-1 during embryogenesis.

Since the publication of this work in the early 90s,[10,11] the description of the sites of expression of several Wnt family members and their frizzled receptors have been reported in much detail in a number of species.[16-18] These descriptive studies indicate: (i) that the expression pattern of these molecules is very well conserved across species and (ii) that a majority of these molecules present very highly restricted expression domains throughout the nervous system. This implies that other Wnt proteins may exhibit patterning functions similar to those described for Wnt-1 at other time and locations.

Genetic studies in mouse and zebrafish have established that mutations in extracellular or intracellular components of the Wnt signaling pathway result in a broad range of defects affecting multiple aspects of neural development. As illustrated in Table 1, analysis of these mutant embryos indicate that during neurogenesis, Wnt signaling can regulate cell growth, cell determination and differentiation, tissue identity and patterning, as well as programmed cell death.

***Table 1. Mutations in components of the Wnt signaling pathway resulting in neural defects***

| Gene | Organism | Neural Phenotype | Reference |
|---|---|---|---|
| Dkk-1 | mouse | -Loss of forebrain. | 19 |
| Wnt-1 | mouse | -Loss of midbrain and cerebellum. | 10, 11 |
| Wnt-3A | mouse | -Duplicated neural tube. Reduced hippocampus. | 15, 20, 21 |
| Wnt-1/Wnt-3A | mouse | -Loss of neural crest progenitors. Loss of dorsal interneurons. | 22, 23 |
| Wnt-7A | mouse | -Immature glomerular rosettes in the cerebellum. | 24 |
| Fz-4 | mouse | -Cerebellar degeneration and ataxia. Deafness due to defect in the peripheral auditory system. | 25 |
| LRP5 | mouse | -Osteoporosis, pseudoglioma syndrome. | 26 |
| LRP6 | mouse | -Duplicated neural tube. Truncated cerebellum. | 27 |
| Dsh1 | mouse | -Abnormal social behavior. Sensorimotor gating deficits. | 28 |
| Axin | zebrafish mouse | -Reduced telencephalon. Expansion of the diencephalon. | 29, 30 |
| APC | mouse | -Apoptosis of cephalic and cardiac neural crest cells. | 31 |
| β-catenin | mouse | -Loss of midbrain and cerebellum. Loss of cranial neural crest cells. | 32 |
| LEF-1 | mouse | -Loss of dentate gyrus granule cells. | 33 |
| TCF-1/LEF-1 | mouse | -Duplicated neural tube. Loss of hippocampus. | 34 |
| TCF-3 | zebrafish | -Forebrain, midbrain and eyes defects. | 35 |

In this chapter, we will review recent results illustrating Wnt signaling functions in the control of neural induction and subsequent patterning decisions including regionalization of the neuraxis and neural crest formation.

## Induction of the Neural Plate

Neural induction constitutes the initial step in the generation of neural tissue from the ectoderm. This classical embryonic induction has been best understood in amphibians. During gastrulation, signals derived from the dorsal mesoderm (organizer) induce the neural plate, making it distinct from the adjacent non-neural ectoderm. These inductive signals, including noggin, chordin and follistatin, share the same molecular property, the ability to block bone morphogenetic protein (BMP) signaling in the ectoderm, thereby eliciting a neural differentiation program (reviewed in refs. 36-38). Thus, the model for neural induction in frogs suggests that neural fate is acquired in response to the activity of organizer-specific BMP antagonists. In zebrafish, a mutation in the chordin gene (known as *dino*, or *chordino*) causes embryos to develop a reduced neural plate[39], providing genetic evidence that BMP signaling inhibition in the ectoderm is necessary to promote neural development.

Studies in chick and mouse do not support a strict requirement for BMP antagonists activity to generate neural fate. For example, chordin/noggin double mutant mouse embryos form a neural plate. However, functional redundancy by other BMP inhibitors cannot be excluded

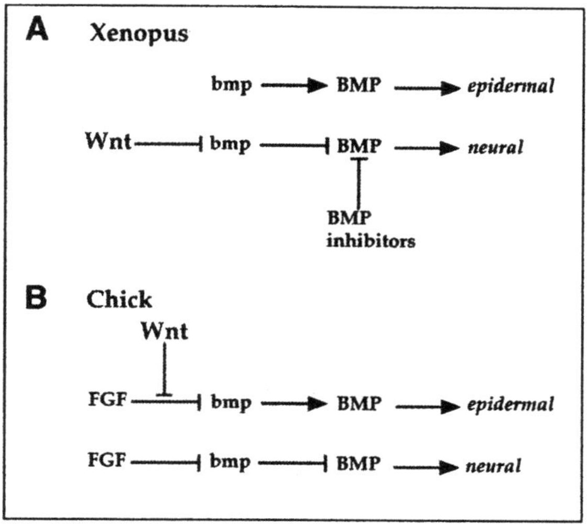


Figure 2. Regulation of neural versus epidermal fate by Wnt signaling. (A) In Xenopus Wnt blocks BMP transcription (bmp) in the ectoderm allowing in conjunction with BMP antagonists to generate neural fate. (B) In the chick epiblast Wnt signaling blocks FGF-mediated attenuation of BMP transcription (bmp) leading to epidermal fate.

in these studies.[40] In the chick the timing and pattern of expression of noggin, chordin and follistatin indicate that BMP antagonists are neither sufficient nor needed to generate neural fate.[41-43] Therefore, the requirement for BMPs inhibitors derived from the organizer may not be the entire basis of neural induction. In fact, BMP inhibition in the chick epiblast requires FGF signaling rather than BMP antagonists.[44,45] BMP transcripts are absent from epiblast explants specified as neural plate. Exposure of the epiblast to FGF receptor antagonists prevents down-regulation of BMP transcripts and convert the epiblast to an epidermal fate.[44]

Recently, it has been found that ectopic activation of Wnt signaling pathway in *Xenopus*, by overexpression of various Wnt ligands, or a dominant-negative GSK3 or a constitutively active β-catenin, is sufficient to activate neural markers in presumptive ectoderm.[46] In these explants Wnt neural inducing activity appears to be mediated by down-regulation of BMP mRNA expression, which in turn is sufficient to allow neural fate to develop. The mechanism by which Wnt signaling down-regulate BMP transcripts levels remains unclear. The function of Wnt signaling in neural induction in Xenopus[46] appears similar to the activity proposed for FGF signaling in the chick epiblast,[44,45] in that both lead to attenuation of BMP mRNA expression. Based on these observations, a revised model of neural induction in frog has been proposed.[47] While BMP antagonists block the activity of BMP proteins resident in the ectoderm, Wnt signaling, by an unknown mechanism, prevents further BMP transcription resulting in development of neural fate (Fig. 2A).

It is important to emphasize here that these findings contrast with several reports indicating that Wnt signaling is not sufficient to neuralize the frog ectoderm.[48-52] While it is difficult to reconcile these observations, differences in the amount of inducers used could explain some of these conflicting results. However, by suggesting that induction of neural tissues may occur through a Wnt-dependent pathway, this model clearly challenges our current view on neural induction in *Xenopus*.

In the chick, since BMP antagonists either alone or in combination with FGF signaling are not sufficient to induce neural fate in the prospective ectoderm,[44,45] additional signals might also be required for neuralization. It has been recently proposed that Wnt signaling may regulate epidermal and neural fate decisions in the chick epiblast.[53] During normal development medial and lateral epiblast will develop neural and epidermal fates respectively. Wnt-3a and Wnt-8C mRNA are expressed in the lateral but not in the medial epiblast cells at the blastula stage.[53] In a series of very elegant experiments Wilson et al[53] have shown that blockage of Wnt signaling in the lateral epiblast, by exposure to a truncated soluble fragment of the mouse Wnt receptor Frizzled 8 (MFz-8-CRD), prevents epidermal fate and allows FGF-mediated neuralization of lateral epiblast cells to occur. If both Wnt and FGF signaling pathways are simultaneously inhibited, lateral epiblast cells will develop their normal epidermal fate. Conversely, exposure of soluble Wnt to medial epiblast cells prevents FGF-mediated neuralization and promotes epidermal differentiation. Therefore, it has been proposed that endogenous Wnt signaling in the lateral epiblast blocks FGF-mediated attenuation of BMP transcription leading to the epidermal fate.[53] By contrast in the medial epiblast in which Wnt signaling is absent, FGF signaling function is to repress BMP mRNA expression required for progression to the neural fate (Fig. 2B).

There is an interesting parallel between Wnt function during neural induction in frog and chick embryos (Fig. 2). However, in each species Wnt activity leads to opposite effects on the prospective ectoderm. While in the chick Wnt signaling acts to promote epidermal fate by blocking BMP attenuation mediated by FGF, in *Xenopus* Wnt function is required in the ectoderm to promote neural fate by blocking accumulation of BMP transcripts. Therefore, Wnt proteins can act both as a positive (chick epiblast) and negative (frog ectoderm) regulator of BMP signaling. Neural induction in both species shares one common feature in that they both require inhibition of BMP transcription/signaling but differs in the type of molecular effectors used to achieve this function.

## Anteroposterior Patterning of the Neural Plate

Shortly after neural induction, the neuroectoderm becomes subdivided along the anteroposterior axis into forebrain, midbrain, hindbrain and spinal cord. Classical experiments in amphibian embryos have implicated a two-signal model of neural induction and anteroposterior patterning of the neuroectoderm.[54] According to this model, anterior neural tissue is induced by early-involuting mesendoderm during a step known as activation. Anterior neural tissue is subsequently modified to a more posterior character by a set of transforming signals emanating from the late-involuting mesoderm in a phase of transformation. During the latter step, a gradient of posteriorizing factor confers progressive posterior identity to the neural tissue.[54]

It is now well established that the activation step is largely mediated by BMP inhibitors secreted by the dorsal mesoderm, although FGFs and Wnts may also be implicated in this process. On the other hand, candidate-transforming molecules have remained elusive for quite some time. Proposed candidates include FGF and retinoic acid (RA) which can posteriorize anterior neural tissue with virtually no neural inducing activity on their own.[55-58]

The first evidence pointing in the direction of a Wnt signal involved in the posteriorization of the neural plate was established in *Xenopus* embryos. Expression of Xwnt-3a induces posterior and suppresses anterior neural markers in neuralized animal explants.[59] Conversely, in vivo blockage of Wnt ligands by expression of a dominant negative Xwnt-8 results in loss of posterior neural fates.[51,60]

Genetic evidence also indicates that Wnt signaling is implicated in the posteriorization of the neuroectoderm. In the zebrafish, *headless* mutant embryos carrying a mutation that abolishes the repressor function of TCF-3, a transcriptional modulator of Wnt signaling, have severe anterior defects.[35] Similarly, *masterblind* mutant embryos exhibit anterior brain defects as a result of a mutation of the scaffolding protein Axin1, a negative regulator of Wnt signaling.[29] Moreover, mouse embryos with mutation in the secreted Wnt antagonist Dkk-1 lack a forebrain,[19] whereas mice mutants for Wnt-3a[15] and the Wnt co-receptor LRP6[27] show posterior truncations. Altogether these results strongly argue that Wnt signaling is required for establishing an anteroposterior pattern in the neuraxis.

A recent study in *Xenopus* has provided evidence for the existence of an endogenous Wnt signaling gradient during gastrulation,[61] which display all the characteristics of the transforming agent as defined in Nieuwkoop's model of posteriorization.[54] When increasing doses of soluble Xwnt-8 are applied to neuralized animal explants, progressive posteriorization of the neural character of these explants can be generated, from forebrain to midbrain and hindbrain respectively. Importantly this activity is completely abolished when the treatment is performed in the presence of the Wnt antagonist MFz-8-CRD.[61] In vivo, at the late gastrula stage the neural plate presents an anteroposterior gradient of nuclear accumulation of β-catenin, with higher levels posteriorly and lower levels anteriorly.[61] This observation clearly supports the idea of a differential activation of the Wnt signal transduction pathway in the neuroectoderm at different axial levels. These results are consistent with previous work showing that overexpression of graded amounts of the Wnt effector dishevelled in animal explants induces neural markers with increasingly posterior identity.[62] Moreover, in zebrafish embryos increasing doses of Wnt-8 morpholino antisense lead to a progressive loss of posterior neural fates associated with expansion of neural tissues of anterior character.[63,64]

Based on these observations it has been proposed that in the more posterior regions of the frog embryo, graded amounts of Wnt signals, probably provided by multiple Wnts, are responsible for the development of a broad range of posterior structures. Whereas in the head region, a number of Wnt antagonists such as cereberus, Dkk-1, FrzBs and sFRPs,[65-68] produced by the anterior neuroectoderm and mesendoderm are believed to be responsible for defining a domain free of Wnt activity required for the development of forebrain structures. The combination of both activities may therefore contribute to the establishment of a repertoire of anteroposterior positional values along the neuraxis.[61]

These studies also suggest that a gradient of Wnt signaling in the neuroectoderm is not sufficient to account for the entire array of anteroposterior character found in the neuraxis and suggest the requirement for additional signals. These may include RA and FGF, which also have been implicated in posteriorization of the neural plate in several species (reviewed in ref. 36).

## Dorsoventral Patterning of the Neuroepithelium

During early stages in the development of the spinal cord, three major classes of cells are generated in the ventral neural tube, floor plate cells at the midline, motor neurons at a ventrolateral position and ventral interneurons more dorsally. Cells in the dorsal neural tube give rise initially to neural crest cells and subsequently to roof plate cells at the midline and to several classes of dorsal sensory interneurons. The inductive signals that control the identity and pattern of these cell types come initially from two distinct groups of non-neural cells, the notochord ventrally and the epidermal ectoderm dorsally (reviewed in ref. 69). The floor plate and notochord are important organizing centers of the spinal cord, and sources of Sonic hedgehog (Shh). A gradient of Shh has been shown to be involved in establishing neuronal diversity in the ventral spinal cord (reviewed in ref. 69).

Wnt gene expression in the spinal cord exhibits striking restrictions along the dorsovental axis prior to differentiation of neurons. In the ventral spinal cord Wnt-4 appear restricted to the floor plate, suggesting that it may have an important role in patterning the ventral portion of the spinal cord.[16,70,71] Wnt-4 deficient mice present severe kidney defects, consistent with Wnt-4 expression in the mesenchyme of the kidney and its derivatives.[72] Nevertheless, these mutant embryos develop a normal pattern of neurons in the ventral spinal cord. Redundancy of expression of other Wnt genes in the ventral spinal cord, such as Wnt-7a,[16] could explain the lack of an obvious phenotype in the spinal cord of Wnt-4 deficient mouse. Thus, the precise function of Wnt-4 in the floor plate remains to be elucidated.

Wnt-1 and Wnt-3a are both expressed in overlapping domains in the dorsal spinal cord, in a region extending caudal to the diencephalon.[12,15,73] Mouse embryos carrying a homozygous deletion of either one of these genes have a normal spinal cord,[10,11,15] due to functional redundancy. Nevertheless, compound mutant embryos lacking both Wnt-1 and Wnt-3a showed major deficiency in neural crest derivatives.[22] These include partial loss of dorsal root ganglia neurons and glia, reduction of pigment cells precursors, and loss or reduction of cranial skeletal structures.[22] It has been proposed that loss of Wnt-1 and Wnt-3a results in a broad deficiency in neural crest progenitors, supporting a role for Wnt-1 and Wnt-3a in expansion of neural crest population rather than in the control of specific cell fate.[22] This is consistent with transgenic analysis suggesting that Wnt signaling is involved in regulating cell proliferation in the dorsal spinal cord.[74,75] In mouse embryos, ectopic expression of Wnt-1 in the entire spinal cord under the control of the Hox-B4 enhancer has been shown to increase the number of dividing dorsal cells without affecting the overall polarity/fate of the spinal cord.[74] Altogether these observations indicate a major function Wnt-1 and Wnt-3a in regulating cell proliferation in the dorsal spinal cord.[22]

A more recent study suggests that Wnt-1 and Wnt-3a may also be required for the specification of individual population of dorsal interneurons. [23] Three subclasses of dorsal interneurons are arranged along the dorso-ventral axis of the spinal cord (Fig. 3A). They can be identified by the expression of a number of homeodomain-containing proteins. D1 interneurons express LH2, D2 interneurons express Islet1 and D3 interneurons express Pax-2. The progenitors from each population of dorsal interneurons are found in the ventricular zone of the spinal cord (Fig. 3 A) and are characterized by the expression of bHLH proteins such as Math1 (D1), Ngn1 (D2) and Mash1 (D3). In Wnt-1/Wnt-3a double mutant mouse embryos, only residual expression of LH2 and Islet 1 is detected on each side of the roof plate indicating that D1 and D2 neurons are virtually absent (Fig. 3B). On the other hand Pax-2 expressing-cells were significantly increased in the dorsal half of the spinal cord, suggesting that increased D3 interneurons compensated for the loss of D1 and D2 neurons (Fig. 3B). This loss of D1/D2 neurons correlated with a dramatic reduction of Math1 and Ngn1 progenitors in the ventricular zone, whereas Mash1 positive cells (D3 progenitors) expanded their boundary dorsally up to the roof plate (Fig. 3B). Thus, Wnt proteins are clearly required for proper development of D1 and D2 subclasses of interneurons. [23]

These observations were largely corroborated by in vitro analysis. Exposure of soluble Wnt-3a to chick explants of medial neural plate is sufficient to induce expression of LH2 and Islet1 and block Pax-2 expression, indicating that Wnt3a can directly induce D1 and D2 interneurons at the expense of D3 neurons.

Molecules of the BMP family are considered as the major source of signaling in the dorsal spinal cord for the development of dorsal interneurons. [76] Interestingly, the expression of a number of these molecules (including BMP-4, BMP-6, BMP-7 and Gdf7) was unaffected in Wnt-1/Wnt-3a deficient embryos, implying that Wnt signaling directly regulates the specification of D1 and D2 interneurons (Fig. 3A,B). Altogether these observations are of great importance as they establish the first evidence for a role of Wnt signaling in the specification of neuronal fate in the dorsal spinal cord, thereby contributing to the regionalization of the spinal cord along its dorso-ventral axis.

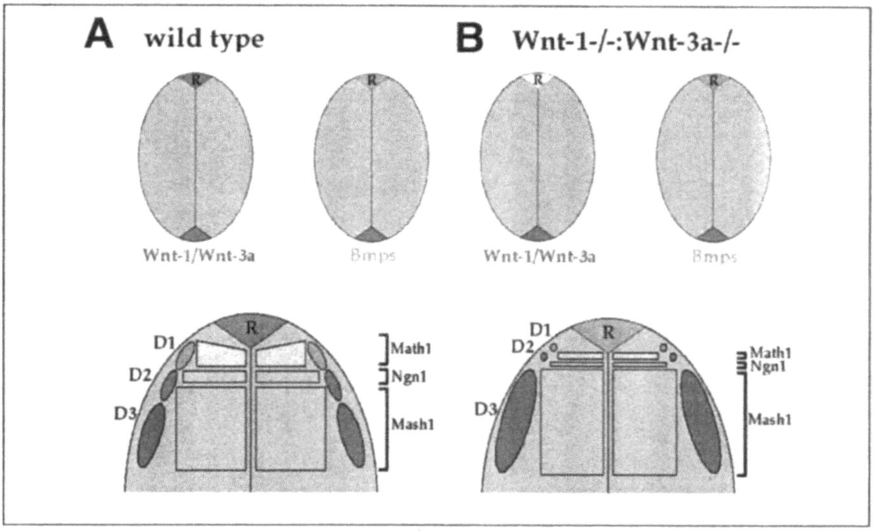


Figure 3. Patterning activity of Wnt-1 and Wnt-3a in the dorsal spinal cord. (A) Wnt-1, Wnt-3a and BMPs have overlapping domains of expression in the roof plate (R) of wild type embryos. (B) Bmps expression is not modified in the roof plate of Wnt-1-/Wnt-3a mutant embryos. As compared to a wild type embryos (A), the pattern of dorsal interneurons is affected in Wnt-1/Wnt-3a deficient mice (B). D1/D2 neurons are reduced while D3 neurons are greatly expanded. The loss of D1/D2 neurons correlates with a dramatic reduction of Math1 and Ngn1 progenitors in the ventricular zone, while Mash1 positive cells (D3 progenitors) expand their boundary dorsally.

## Neural Crest Induction and Diversification

The neural crest is a unique embryonic structure composed of a migratory population of multipotent cells arising at the lateral edges of the neural plate, at the junction between the neural and the non-neural ectoderm (reviewed in ref. 77). During neurulation, the neural crest becomes positioned at the dorsal most region of the neural tube. As the neuroepithelium closes, neural crest cells delaminate in a rostro-caudal wave and migrate throughout the embryo along specific and restricted routes. When they reach their final location, neural crest cells present the remarkable ability to differentiate into a large variety of cell types including peripheral and enteric neurons and glia, smooth muscle cells, craniofacial cartilage and bone, endocrine and pigment cells (reviewed in ref. 77).

The molecular signals controlling neural crest specification are not fully understood (reviewed in ref. 78). Signals emanating from the non-neural ectoderm, the paraxial mesoderm, or both are believed to be involved in the specification of the neural crest.[79-81] There is evidence that molecules belonging to the BMP family, expressed by the non-neural ectoderm and the dorsal neural tube, initiate the formation of the neural crest.[82,76] The neural inducers noggin and chordin act by neutralizing a BMP signal in the ectoderm[83,84] and it is believed that neural crest forms in a region where BMPs and BMP antagonists reach an appropriate balance.[85-87] Other classes of signaling molecules have also been proposed as potential inducers of neural crest, this is the case for members of the FGF family.[88,89,50]

A large number of studies in the mouse,[22,31,32,75] frog[48-52] and zebrafish[90,91] have implicated Wnt signaling in neural crest formation. In Xenopus, as illustrated in Figure 4, ectopic expression of some Wnt family members, as well as downstream components of the pathway,

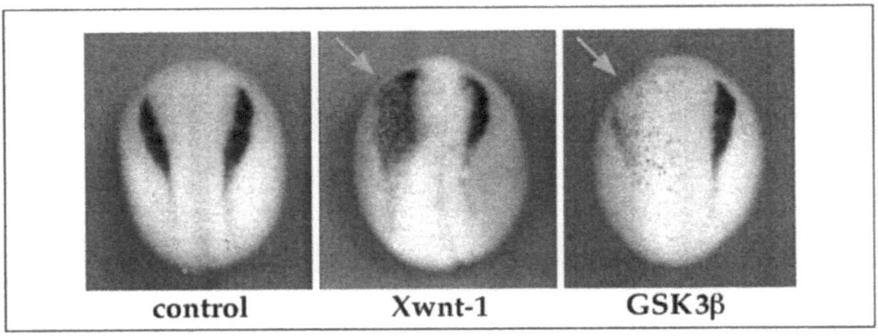


Figure 4. Slug expression in the neural crest forming region is regulated by Wnt signaling. The transcription factor Slug is one of the earliest genes activated in response to neural crest-inducing signals in *Xenopus* embryos. By in situ hybridization, Slug is detected at the lateral edge of the neural plate (control). Upon injection of Xwnt-1 plasmid, Slug domain of expression is expanded on the injected side (arrow, left side). Over-expression of GSK3beta, known to block Wnt signaling, prevents Slug expression (arrow, left side). In all panels, early neurula stage embryos are viewed from the dorsal side, anterior to top.

enhances production of neural crest progenitors,[48-52] whereas inhibition of Wnt signaling blocks neural crest formation.[48,50,52]

Using a functional assay in Xenopus it has been shown that signaling by XWnt-1, but not other Wnts, can be specifically enhanced by frizzled-3 (XFz-3). Consistent with this observation, XFz-3 and XWnt-1 are highly localized to dorsal neural tissues that give rise to neural crest,[52] suggesting that they may participate in neural crest formation. In the embryo, XFz-3 overexpression can induce ectopic neural crest formation, similar to XWnt-1. Whereas, loss of XFz-3 function, either by depletion with a XFz-3-directed morpholino antisense or by expression of an inhibitory form of XFz-3 (NFz-3) blocks XWnt-1-mediated neural crest induction and prevents neural crest formation in whole embryos.[52] These results, together with other studies,[48-51] clearly demonstrate the existence of a Wnt-dependent pathway required for neural crest formation.

Mouse embryos with targeted inactivation of the β-catenin gene, within Wnt-1 expression domain, have severe defects in the formation of cranial and dorsal root ganglia and in several craniofacial skeletal elements of neural crest origin.[32] In these embryos it is believed that signaling through β-catenin is required for survival and/or differentiation of cranial neural crest cells.[32] Loss of Wnt-1 and Wnt-3a results in a broad loss of neural crest derivatives and it has been proposed that Wnt signaling in the dorsal neural tube of these mutant embryos is primarily required for the expansion of neural crest progenitors.[22] In *Xenopus*, the increased in neural crest progenitors generated upon XWnt-3a overexpression has been shown to occur independently of cell proliferation,[48] suggesting that Wnt signaling may also be involved in the specification of neural crest fate.

The molecular process by which multipotent neural crest progenitors are assigned to specific lineages is believed to depend on environmental signals acting prior, during or after neural crest cell migration (reviewed in ref. 77). In vitro studies have identified several growth factors that can direct the differentiation of neural crest precursors along distinct lineages (reviewed in refs. 92, 93).

In zebrafish embryos activation of the Wnt signaling pathway by overexpression of β-catenin in individual cranial crest progenitors promotes pigment cell formation at the expense of neurons and glia.[90] Conversely, a dominant-negative Wnt favored neuronal fates at the expense of pigment cells.[90] These results suggest that Wnt signaling is required to specify the melanocyte lineage thereby contributing to neural crest cell diversity. Consistent with this activity, soluble Wnt-3a increases the number of pigment cells in quail neural crest cultures while decreasing the number of neurons and glial cells, without affecting cell proliferation.[94]

Further evidence that Wnt signaling directly regulates pigment cell fates came from analysis of the MITF promoter (microphthalmia-associated transcription factor), a basic helix-loop-helix transcription factor required for development of melanocytes.[95] The zebrafish homologue of mouse MITF, *nacre*, is expressed in the primordium of the retina and in premigratory and migrating neural crest cells fated to become melanocytes. nacre mutant embryos lack all neural crest-derived melanocytes throughout development.[96] MITF/*nacre* presents in its promoter region TCF/LEF binding sites, which can mediate Wnt responsiveness.[91] Binding assays and mutational analysis, have demonstrated that nacre is both necessary and sufficient for pigment cell formation in zebrafish.[91] Similarly, in vitro studies indicate that soluble Wnt-3a up-regulates MITF expression in a mouse melanocyte cell line and recruits β-catenin and LEF-1 to the LEF-1 binding site of MITF promoter.[97] These observations clearly indicate that Wnt signaling can not only induce and expand neural crest progenitors but also specify at least one neural crest-derived lineage by direct activation of MITF/*nacre* gene.

## Neural Crest Apoptosis

In the hindbrain, neural crest cells originate from segmental units, known as rhombomeres, and migrate into individual streams that populate the branchial arches. This migration into segregated streams is of great importance to allow uninterrupted transfer of patterning information from the neuraxis to the branchial arches (reviewed in ref. 98). In the chick and mouse embryos, rhombomere 3 (R3) and 5 (R5) are depleted of neural crest cells by programmed cell death, thereby defining two neural crest-free zones that contribute to the segregation of neural crest streams. It is now well established that apoptosis in R3 and R5 depends on signal(s) derived from adjacent even-numbered rhombomeres and requires Msx2 induction mediated by BMP-4 (reviewed in ref. 98). Consistent with this role, BMP-4 and Msx2 expression pattern in the hindbrain is concurrent with the pattern of neural crest apoptosis.[99]

Experiments in the chick embryo suggest that Wnt signaling may act as a regulator of neural crest apoptosis in the hindbrain.[100] The secreted frizzled related proteins (sFRP) act as Wnt antagonists. Over-expression of one member of this family (sFRP-2) prevents BMP-4 expression in R3 and R5 leading to ectopic production of neural crest cells from both rhombomeres. Conversely, loss of sFRP-2 or Wnt-1 overexpression results in elevated levels of Msx2 transcripts and ectopic apoptosis in R4. Interestingly, early on sFRP-2 is expressed throughout the hindbrain but is down-regulated in odd-numbered rhombomeres by the time cell death is initiated in R3 and R5.[100] These results indicate that while a number of Wnt molecules are broadly expressed throughout the hindbrain their activity is tightly regulated by sFRP-2 in even-numbered rhombomeres. Therefore, it has been proposed that Wnt signaling in R3 and R5 promote neural crest apoptosis by up-regulating BMP-4/Msx2 expression while in adjacent rhombomeres sFRP-2 blocks Wnt signaling allowing, by default, production of neural crest cells.

Additional evidence for a role of Wnt signaling in cranial neural crest apoptosis has also been reported in transgenic mouse embryos.[31] Adenomatous polyposis coli (APC) is a negative regulator of Wnt signaling through down-regulation of β-catenin. Loss of APC causes accumulation of β-catenin resulting in the activation of specific downstream target genes. The function of APC during neural crest development was analyzed through neural crest specific disruption of APC using the Cre-loxP recombination system. Mice carrying the APC gene flanked by loxP sites were mated with Protein 0 (P0)-Cre transgenic mice. P0-Cre transgenic animals expressed Cre in a wide range of cranial and trunk neural crest derivatives including, dorsal root ganglia, sympathetic and enteric nervous system, craniofacial mesenchyme and septum of the outflow tract of the heart.[101] P0-Cre/APC mutant mice display severe craniofacial and cardiac defects.[31] These defects are always associated with an early loss of neural crest progenitors by apoptosis. The sites of apoptosis in affected neural crest derivatives are found within regions of accumulation of β-catenin consistent with activation of Wnt signaling pathway.[31] Neural crest derivatives from the trunk region (peripheral nerves and melanocytes) appeared unaffected in these transgenic animals indicating that neural crest progenitors from different axial levels may differ in their ability to respond to Wnt signals. These studies indicate that activation of Wnt signaling in the neural crest-forming region promotes apoptosis in the cranial neural crest, and suggests that inhibition of Wnt signals may be required to promote cranial neural crest formation in the developing embryo.

These observations supporting an active role for Wnt signaling in programmed cell death in the rhombencephalic neural crest [100,31] contrast significantly with previous studies in mouse, frog and zebrafish that have established a positive role for Wnt signals in the generation of neural crest cells.[22,31,32,48-52,75,90,91] While it is yet difficult to reconcile both observations, it may indicate dose-dependence activities of Wnt proteins and/or differences in temporal-responsiveness to Wnt signals at the multiple stages of the neural crest development.

## Conclusions

During the last decade an impressive amount of novel information has been generated that furthered our understanding of the roles of Wnt proteins during vertebrate embryogenesis. In this chapter, only a portion of some of these activities has been reviewed in the context of the developing nervous system. As illustrated in Table 1, numerous other aspects of neural development are also regulated by Wnt signaling. These include, eye development,[26,102] axonal remodeling,[24,103] formation of the hippocampus[21,33,34,104] and social behavior.[28] One striking feature is that Wnt signaling as the ability within the same tissue and at different times to regulate activities as diverse as cell proliferation, cell fate specification and programmed cell death. This observation establishes the extremely versatile nature of this class of signaling molecules. Nevertheless, the molecular mechanisms underlying most of these activities are still poorly understood. A greater understanding of the role of the Wnt signaling during formation of the nervous system will certainly benefit from the more systematic development of temporal- and tissue-specific loss of function of various Wnt and frizzled family members.

## *Acknowledgements*

The author is extremely grateful to Trish Labosky for critical reading of the manuscript and to Natasha Saint-Germain for helping with bibliography. Research in the author's laboratory is supported by the National Institutes of Health, March of Dimes, Deafness Research Foundation, Johnson & Johnson and Whitehall Foundation.

## References

1. Massague J, Chen Y. Controlling TGF-β signaling. Genes Dev 2000; 14:627-644.
2. Szebenyi G, Fallon J. Fibroblast growth factors as multifunctional signaling factors. Int Rev Cytol 1999; 185:45-106.
3. Betsholtz C, Karlsson L, Lindhal P. Developmental roles of platelet-derived growth factors. BioEssays 2001; 23:494-507.
4. Ingham PW, McMahon AP. Hedgehog signaling in animal development: paradigms and principles. Genes Dev 2001; 15:3059-3087.
5. Nusse R, Varmus HE. Wnt genes. Cell 1992; 69:1073-1087.
6. Parr BA, McMahon AP. Wnt genes and Vertebrate development. Curr Biol 1994; 4:523-528.

7. Cadigan KM, Nusse R. Wnt signaling: a common theme in animal development. Genes & Dev 1997; 11:3286-3305.
8. Patapoutian A, Reichardt L. Roles of Wnt proteins in neural development and maintenance. Curr Opinion Neurobiol 2000; 10:392-399.
9. Wilson SW, Rubenstein JL. Induction and dorsoventral patterning of the telencephalon. Neuron 2000; 28:641-651.
10. McMahon AP, Bradley A. The Wnt-1 (int-1) proto-oncogene is required for development of a large region of the mouse brain. Cell 1990; 62:1073-1085.
11. Thomas KR, Capecchi MR. Targeted disruption of the murine int-1 proto-oncogene resulting in severe abnormalities in midbrain and cerebellar development. Nature 1990; 346:847-850.
12. Wilkinson DG, Bailes JA, McMahon AP. Expression of the proto-oncogene int-1 is restricted to specific neural cells in the developing mouse embryo. Cell 1987; 50:79-88.
13. Hollonet MER, Teillet MA, Le Douarin NM. A new approach to the development of the cerebellum provided by the quail-chick marker system. Development 1990; 108:19-31.
14. Joyner AL. Engrailed, Wnt and Pax genes regulate midbrain-hindbrain development. Trends Genet 1996; 12: 15-20.
15. Takada S, Stark KL, Shea MJ et al. Wnt-3a regulates somite and tailbud formation in the mouse embryo. Genes Dev 1997; 48:174-189.
16. Parr BA, Shea MJ, Vassileva G et al. Mouse Wnt genes exhibit discrete domains of expression in the early embryonic CNS and limb bud. Development 1993; 119:247-261.
17. Hollyday M, McMahon JA, McMahon AP. Wnt expression patterns in chick embryo nervous system. Mech Dev 1995; 52:9-25.
18. Gradl D, Kuhl M, Weldich D. Keeping a close eye on Wnt-1/wg signaling in Xenopus. Mech Dev 1999; 86:3-15.
19. Mukhopadhyay M, Shtrom S, Rodriguez-Esteban C et al. Dickkopfl is required for embryonic head induction and limb morphogenesis in the mouse. Dev Cell 2001; 1:423-434.
20. Yoshikawa Y, Fujimori T, McMahon AP et al. Evidence that absence of Wnt3a signaling promotes neuralization instead of paraxial mesoderm development in the mouse. Dev Biol 1997; 183:234-242
21. Lee SMK, Tole S, Grove E et al. A local Wnt-3a signal is required for development of mammalian hippocampus. Development 2000; 127:457-467.
22. Ikeya M, Lee SMK, Johnson JE et al. Wnt signaling is required for expansion of neural crest and CNS progenitors. Nature 1997; 389:966-970.
23. Muroyama Y, Fujihara M, Ikeya M et al. Wnt signaling plays an essential role in neuronal specification of the dorsal spinal cord. Genes Dev2002; 16:548-553.
24. Hall AC, Lucas FR, Salinas PC. Axonal remodeling and synaptic differentiation in the cerebellum is regulated by Wnt-7a signaling. Cell 2000; 100:525-535.
25. Wang Y, Huso D, Cahill H et al. Progressive cerebellar, auditory, and esophageal dysfunction caused by targeted disruption of the frizzled-4 gene. J. Neurosci. 2001; 21:4761-4771.
26. Gong Y, Slee RB, Fukai N et al. LDL receptor-related protein 5 (LRP5) affects bone accrual and eye development. Cell 2001; 107:513-523.
27. Pinson KI, Brennan J, Monkley S et al. An LDL-receptor-related protein mediates Wnt signalling in mice. Nature 2000; 407:535-538.
28. Lijam N, Paylor R, McDonald MP et al. Social interaction and sensorimotor gating abnormalities in mice lacking Dvl1. Cell 1997; 90:895-905.
29. Heisenberg CP, Houart C, Take-Uchi M et al. A mutation in the Gsk3-binding domain of zebrafish Masterblind/Axin1 leads to a fate transformation of telencephalon and eyes to diencephalon. Genes Dev 2001; 15:1427-1434.
30. Zeng I, Fagotto F, Zhang T et al. The mouse fused locus encodes Axin, an inhibitor of the Wnt signaling pathway that regulates embryonic axis formation. Cell 1997; 90:181-192.
31. Hasegawa S, Sato T, Akazawa H et al. Apoptosis in neural crest cells by functional loss of APC tumor suppressor gene. Proc Natl Acad Sci USA 2002; 99:297-302.
32. Brault V, Moore R, Kutsch S et al. Inactivation of the β-catenin gene by Wnt1-Cre-mediated deletion results in dramatic brain malformation and failure of cranifacial development. Development 2001; 128:1253-1264.

33. Galceran J, Miyashita-Lin EM, Devaney E et al. Hippocampus, development and generation of dentate gyrus granule cells is regulated by LEF-1. Development 2000; 127: 469-482.
34. Galceran J, Farinas I, Depew MJ et al. Wnt3a-/- -like phenotype and limb deficiency in LEF-1(-/-)TCF-1(-/-) mice. Genes Dev 1999; 13:709-717.
35. Kim CH, Oda T, Itoh M et al. Repressor activity of Headless/TCF-3 is essential for vertebrate head formation. Nature 2000; 407:913-916.
36. Sasai Y, De Robertis EM. Ectodermal patterning in vertebrate embryos. Dev Biol 1997; 182: 5-20.
37. Weinstein DC, Hemmati-Brivanlou A. Neural Induction. Ann Rev Cell Dev Biol 1999; 15: 411-433.
38. Wilson SI, Edlund T. Neural induction: toward a unifying mechanism. Nature (Neuro. Supp.) 2001; 4:1161-1168.
39. Hammerschmidt M, Pelegri F, Mullins MC et al. dino and mercedes, two genes regulating dorsal development in the zebrafish embryo. Development 1996; 123:95-102.
40. Bachiller D, Klingensmith J, Kemp C et al. The organizer factors, Chordin and Noggin are required for mouse forebrain development. Nature 2000; 403:658-661.
41. Streit A, Lee KJ, Woo I et al. Chordin regulates primitive streak development and the stability of induced neural cells, but is not sufficient for neural induction in the chick embryo. Development 1998; 125:507-519.
42. Storey KG, Crossley JM, De Robertis EM et al. Neural induction and regionalization in the chick embryo. Development 1992; 114:729-741.
43. Levin M. The roles of activin and follistatin signaling in chick gastrulation. Int J Dev Biol 1997; 42:553-559.
44. Wilson SI, Graziano E, Harland RM et al. An early requirement for FGF signalling in the acquisition of neural cell fate in the chick embryo. Curr Biol 2000; 10:421-429.
45. Streit A, Berliner AJ, Papanayotou C et al. Initiation of neural induction by FGF signalling before gastrulation. Nature 2000; 406:74-78.
46. Baker JC, Beddington RSP, Harland RM. Wnt signaling in Xenopus embryos inhibits Bmp4 expression and activates neural development. Genes Dev 1999; 13:3149-3159.
47. Harland RM. Neural induction. Curr Opin Genet Dev 2000; 10:357-362.
48. Saint-Jeannet J-P, He X, Varmus HE et al. Regulation of dorsal fate in the neuraxis by Wnt-1 and Wnt-3a. Proc Natl Acad Sci USA 1997; 94:13713-13718.
49. Chang C, Hemmati-Brivanlou A. Neural crest induction by Xwnt7B in Xenopus. Dev Biol 1998; 194:129-134.
50. LaBonne C, Bronner-Fraser M. Neural crest induction in Xenopus: evidence for a two-signals model. Development 1998; 125:2403-2414.
51. Bang AG, Papalopulu N, Goulding MD et al. Expression of Pax-3 in the lateral neural plate is dependent on a Wnt-mediated signal from the posterior non-axial mesoderm. Dev Biol 1999; 212:366-380.
52. Deardorff MA, Tan C, Saint-Jeannet J-P et al. A role for frizzled-3 in neural crest development. Development 2001; 128:3655-3663.
53. Wilson SI, Rydström A, Trimborn T et al. The status of Wnt signalling regulates neural and epidermal fates in the chick embryo. Nature 2001; 411:325-330.
54. Nieuwkoop PD. Activation and organization of the central nervous system in amphibians. J Exp Zool 1952; 120:1-108.
55. Doniach T. Basic FGF as an inducer of anteroposterior neural pattern. Cell 1995; 83: 1067-1070.
56. Holowacz T, Sokol S. FGF is required for posterior neural patterning but not for neural induction. Dev. Biol. 1999; 205:296-308.
57. Taira M, Saint-Jeannet J-P, Dawid I. Role of Xlim-1 and Xbra genes in anteroposterior patterning of the neural tissue by the head and trunk organizer. Proc Natl Acad Sci USA 1997; 94: 895-900
58. Kolm PJ, Apekin V, Sive H. Xenopus hindbrain patterning requires retinoic signaling. Dev Biol 1997; 192:1-16.
59. McGrew LL, Lai CC, Moon RT. Specification of the anteroposterior neural axis through synergistic interaction of the Wnt signaling cascade with noggin and follistatin. Dev Biol 1995; 172:337-342.
60. McGrew LL, Hoppler S, Moon RT. Wnt and FGF pathways cooperatively patterns anteroposterior neural ectoderm in Xenopus. Mech Dev 1997; 69:105-114.

61. Kiecker C, Niehrs C. A morphogen gradient of Wnt/beta-catenin signalling regulates anteroposterior neural patterning in Xenopus. Development 2001; 128:4189-4201.
62. Itoh K, Sokol SY. Graded amounts of Xenopus dishevelled specify discrete anteroposterior cell fates in prospective ectoderm. Mech Dev 1997; 61:113-125.
63. Erter CE, Wilm TP, Basler N et al. Wnt8 is required in lateral mesendodermal precursors for neural posteriorization in vivo. Development 2001; 128:3571-3583.
64. Lekven AC, Thorpe CJ, Waxman JS et al. Zebrafish wnt8 encodes two wnt8 proteins on a bicistronic transcript and is required for mesoderm and neurectoderm patterning. Dev Cell 2001; 1: 103-114.
65. Piccolo S, Agius E, Leyns L et al. The head uinducerr Cerberus is a multifunctional antagonist of nodal, BMP and Wnt signals. Nature 1999; 397:707-710.
66. Kazanskaya O, Glinka A, Niehrs C. The role of dickkopfl in prechordal plate specification and neural patterning. Development 2000; 27:4981-4992.
67. Bradley L, Sun B, Collins-Rice L et al. Different activities of the frizzled-related proteins frzb2 and sizzled2 during Xenopus anteroposterior patterning. Dev Biol 2000; 227:118-132.
68. Pera EM, De Robertis EM. A direct screen for secreted proteins in Xenopus embryos identifies distinct activities for the Wnt antagonist crescent and Frzb-1. Mech Dev 1999; 96:183-195.
69. Tanabe Y, Jessell TM. Diversity and pattern in the developing spinal cord. Science 1996; 274:1115-1123.
70. McGrew LL, Otte AP, Moon RT. Analysis of Xwnt-4 in embryos of Xenopus laevis: a Wnt family member expressed in the brain and the floor plate. Development 1992; 115:463-473.
71. Ungar AR, Kelly GM, Moon RT. Wnt4 affects morphogenesis when misexpressed in the zebrafish embryo. Mech Dev 1995; 52:153-164.
72. Stark K, Vainio S, Vassileva G et al. Epithelial transformation of metanephric mesenchyme in the developing kidney regulated by Wnt-4. Nature 1994; 372:679-683
73. Wolda SL, Moody CJ, Moon RT. Overlapping expression of Xwnt-3A and Xwnt-1 in neural tissue of Xenopus laevis embryos. Dev Biol 1993; 155:46- 57.
74. Dickinson M E, Krumlauf R, McMahon AP. Evidence for a mitogenic effect of Wnt-1 in the developing mammalian central nervous system. Development 1994; 120:1453-1471.
75. Dunn KJ, Williams BO, Li Y et al. Neural crest-directed gene transfer demonstrates Wnt1 role in melanocyte expansion and differentiation during mouse development. Proc Natl Acad Sci USA 2000; 97:10050-10055.
76. Liem KF,Tremml G, Jessell TM. A role for roof plate and its resident TGF-β-related proteins in neuronal patterning in the dorsal spinal cord. Cell 1997; 91:127-138.
77. LeDouarin NM, Kalcheim C. The neural crest. 1999; Second Edition, London: Cambridge University Press.
78. LaBonne C, Bronner-Fraser M. Molecular mechanisms of neural crest formation. Annu Rev Cell Dev Biol 1999; 15:81-112.
79. Selleck MA, Bronner-Fraser M. Origins of the avian neural crest: The role of neural plate-epidermal interactions. Development 1995; 121:525-538.
80. Mancilla A, Mayor R. Neural crest formation in Xenopus laevis: mechanism of Xslug induction. Dev Biol 1996; 177:580-589.
81. Bonstein L, Elias S, Frank D. Paraxial-fated mesoderm is required for neural crest induction in Xenopus embryos. Dev Biol 1998; 193:156-168.
82. Liem KF, Tremml G, Roelink H et al. Dorsal differentiation of neural plate cells induced by BMP-mediated signals from epidermal ectoderm. Cell 1995; 82:969-979.
83. Zimmerman LB, de Jesus-Escobar JM, Harland R. The Spemann organizer signal noggin binds and inactivates bone morphogenetic protein 4. Cell 1996; 86:599-606.
84. Piccolo S, Sasai Y, Lu B et al. Dorsoventral patterning in Xenopus: inhibition of ventral signals by direct binding of chordin to BMP-4. Cell 1996; 86:589-598.
85. Morgan R, Sargent MG. The role in neural patterning of translation initiation factor eIF4AII; induction of neural fold genes. Development 1997; 124:2751-2760.
86. Merchant L, Linker C, Ruiz P et al. The inductive properties of mesoderm suggest that the neural crest cells are specified by a BMP gradient. Dev Biol 1998; 198:319-329.

87. Nguyen V H, Schmid B, Trout J et al. Ventral and lateral regions of the zebrafish gastrula, including the neural crest progenitors, are established by a bmp2b/swirl pathway of genes. Dev Biol 1998; 199:93-110.
88. Kengaku M, Okamoto H. Basic fibroblast growth factor induces differentiation of neural tube and neural crest lineages of cultured ectoderm cells from Xenopus gastrula. Development 1993; 119:1067-1078.
89. Mayor R, Guerrero N, Martinez C. Role of FGF and noggin in neural crest induction. Dev Biol 1997; 189:1-12.
90. Dorsky RI, Moon RT, Raible DW. Control of neural crest cell fate by the Wnt signalling pathway. Nature 1998; 396:370-373.
91. Dorsky RI, Raible DW, Moon RT. Direct regulation of nacre, a zebrafish MITF homolog required for pigment cell formation, by the Wnt pathway. Genes Dev 2000; 14:158-162.
92. Anderson DJ. Cellular and molecular biology of neural crest cell lineage determination. Trends Genet 1997; 13:337-345.
93. Dorsky RI, Moon RT, Raible DW. Environmental signals and cell fate specification in premigratory neural crest. BioEssays 2000; 22:708-716.
94. Jin E-J, Erickson CA, Takada S et al. Wnt and BMP signaling govern lineage segregation of melanocytes in the avian embryo. Dev Biol 2001; 233:22-37.
95. Golding CR. Mitf from neural crest to melanoma: signal transduction and transcription in the melanocyte lineage. Genes Dev 2000; 14:1712-1728.
96. Lister JA, Robertson CP, Lepage T et al. Nacre ecodes a zebrafish microphthalmia-related protein that regulates neural-crest-derived pigment cell fate. Development 1999; 126:3757-3767.
97. Takeda K, Yasumoto K, Takada R et al. Induction of Melanocyte-specific Microphthalmia-associated transcription factor by Wnt-3a. J. Biol. Chem. 2000; 19:14013-14016.
98. Graham A, Koentges G, Lumsden A. Neural crest apoptosis and the establishment of craniofacial pattern: an honorable death. Mol Cell Neurosc 1996; 8; 6-83.
99. Graham A, Francis-West P, Brickell P et al. The signaling molecule BMP4 mediates apoptosis in the rhombencephalis neural crest. Nature 1994; 372:684-686.
100. Ellies DL, Church V, Francis-West P et al. The Wnt antagonist cSFRP-2 modulates programmed cell death in the developing hindbrain. Development 2000; 127:5285-5295.
101. Yamauchi Y, Abe K, Mantani A et al. A novel transgenic technique that allows specific marking of the neural crest cell lineage in mice. Dev Biol 1999; 212:191-203.
102. Rasmussen JT, Deardorff MA, TanC et al. Regulation of eye development by frizzled signaling in Xenopus. Proc Natl Acad Sci USA 2001; 98:3861-3866.
103. Salinas PC. Wnt factors in axonal remodelling and synaptogenesis. Biochem Soc Symp 1999; 65:101-109.
104. Roelink H. Hippocampus formation: an intriguing collaboration. Curr Biol 2000; 10:R279-R281.

# CHAPTER 8

# Wnts in Kidney and Genital Development

**Andreas Kispert and André W. Brändli**


## Abstract

The finding that *Wnt* genes play important roles in numerous developmental processes has prompted the search for *Wnt*-dependent functions during kidney and genital development. A major driving force behind these studies was the notion that Wnts may act as ureter-derived tubule-inducing signals in metanephric development. Although current thinking makes it unlikely that Wnts occupy such a role, detailed studies have succeeded in identifying specific expression of *Wnt* genes in kidney development and assigning functional importance to *Wnt-4*. Genetic analysis in the mouse has unambiguously established an essential role for *Wnt-4* as a mesenchymal inducer of tubulogenesis in the developing metanephric kidney. *Wnt*-deficient mouse strains have also revealed pivotal functions for several *Wnt* genes in genital development. These include *Wnt-4*, which is required for normal female sexual differentiation, *Wnt-5a*, which controls formation of the genital tubercle, and *Wnt-7a*, which regulates female sex duct differentiation. More recently, studies in lower vertebrates have begun to unravel an evolutionary conserved role for *Wnts* in vertebrate intermediate mesoderm differentiation.


## Introduction

Despite the fact that the excretory and the reproductive organs represent two distinct organ systems fulfilling quite different physiological purposes in the vertebrate body, they are collectively known as the urogenital system and usually end up as a single chapter in most textbooks of embryology and anatomy. There are several good reasons for doing this. First, kidneys and gonads both arise from the intermediate mesoderm. Second, development of both organ systems is spatially as well as functionally interconnected; and third, important structural components are shared as illustrated by the urethra, which serves both the removal of urine and the transport of sperm in the adult male.

Here, we will review the evidence for functions of *Wnt* genes during kidney as well as gonad development. While *Wnt-4* gene function is essential for the development of both organ systems, other *Wnt* genes appear to fulfill specific functions only in one or the other system. The developmental roles of *Wnt* genes will therefore be discussed separately for each organ system. We will first provide a synopsis of metanephric kidney development in the mouse, and subsequently discuss the function of specific Wnt family members as revealed by the analysis of Wnt deficient mouse strains. We will conclude the kidney section by raising the issue of evolutionary conservation of renal *Wnt* gene functions and reviewing the presently available evidence from studies of lower vertebrates, particularly amphibia and fish. In contrast, essentially all our knowledge about *Wnt* gene function during development of the vertebrate reproductive

*Wnt Signaling in Development*, edited by Michael Kühl.

system derives from studies in the mouse. We will therefore focus our analysis exclusively on describing the genital defects observed in *Wnt*-deficient mouse strains.

## Kidney Development in the Mouse

Development of the mouse excretory system is characterized by successive inductive interactions between two tissue compartments, which are both derived from the intermediate mesoderm.[1] The two players are represented by the epithelial nephric duct and the nephrogenic mesenchyme, which lies adjacent to the nephric duct. Because of these tissue interactions, three embryonic kidneys are laid down from anterior to posterior in time and space, the pronephros, the mesonephros, and the metanephros (Fig. 1). The initial excretory organ, the pronephros, is only a very transient structure established at 8 days post coitum (d p.c.) at the level of the future forelimb buds. The mesonephros extends by posterior elongation of the nephric duct (or Wolffian duct) and subsequent tubule induction in the adjacent mesonephrogenic mesenchyme between 9.5 and 11 d p.c. between the forelimb and hindlimb anlagen. Although forming elaborate tubules, it is still unclear whether the mesonephros of the rodent embryo becomes functional with respect to the production of urine.[1] However, in the male embryo, it will contribute to the ductal network of the rete testis. Metanephric development is initiated when a bud emerges from the Wolffian duct at the level of the hindlimbs around 10.5 d p.c. This bud subsequently invades the metanephric blastema, a mesenchymal condensation at the posterior end of the intermediate mesoderm. The bud branches several times, while growing peripherally, and eventually forms the collecting duct system of the mature kidney. Shortly after ureteric bud formation, mesenchymal cells condense around the tip of the ureter. Some of them aggregate into pretubular cell clusters and undergo a mesenchymal-epithelial transition to form simple renal vesicles. These elongate and fuse with the collecting duct system. Driven by further morphogenetic movements the vesicles change shape through the intermediate stages of the comma-shaped and S-shaped bodies to form renal tubules. Patterning along the proximal-distal axis of this tubule and further cell differentiation eventually creates the functional unit of the metanephric kidney, the nephron (Fig. 2). The proximal end of the nephron harbors the primary filtration apparatus, the glomerulus. Here, intimate contact is made between a capillary bowel of the vasculature system and specialized cells of the nephron, the podocytes, to achieve ultrafiltration of the blood and form the primary urine. The urine is then transported through a tubular system consisting of the proximal tubule, the loop of the Henle and the distal tubule to the collecting duct system. During that time the primary urine dramatically changes composition until almost all water and useful solvents have been taken up again by both active and passive transport processes across the tubular epithelial membranes. The final urine consisting primarily of waste products is discharged from the body via the ureter, the bladder, and the urethra. The latter two structures actually do not arise from the intermediate mesoderm but are derived from ectodermal infoldings.

Nephron formation ceases seven to ten days post partum most likely as the mesenchymal stem cells in the periphery of the kidney are exhausted. The complex architecture of the mature kidney (Fig. 3 a-d), consisting of some 30.000 nephrons connected to an elaborate collecting duct system[2], can only be achieved, if the morphogenetic and cellular differentiation programs of both the epithelial compartment, the nephric duct, and the nephrogenic mesenchymal derivatives are highly integrated. This makes it very likely that multiple signaling systems between and within the two compartments are operative (for recent reviews on kidney development see refs. 1, 3-9).

Classical kidney organ culture experiments have primarily focused on the signals exchanged between mesenchyme and ureter upon their initial contact. Separation and recombination experiments pioneered in the 1950s by Grobstein and Saxén have shown that isolated metanephric mesenchyme undergoes apoptosis unless provided with a permissive stimulus which

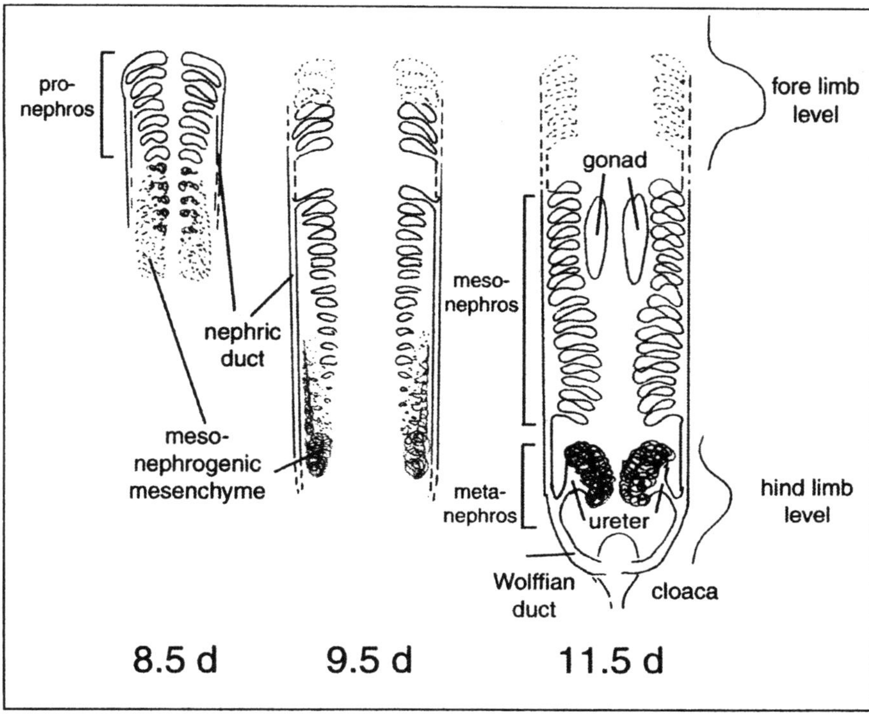


Figure 1. Schematic representation of the development of the murine excretory system. Successive establishment of the three embryonic kidneys is initiated in the intermediate mesoderm at 8 d p.c. By 8.5 d p.c., the pronephros has emerged in the intermediate mesoderm at the level of the future forelimbs. Posterior extension of the nephric duct and tubule induction in the adjacent mesonephrogenic mesenchyme results in the formation of the mesonephros by 9.5 p.c. By 11.5 d p.c., the pronephros has disappeared and the mesonephros has reached its maximum extension between forelimb and hindlimb anlagen. At the hindlimb level, the metanephros has started to develop. The ureter, a branch of the Wolffian duct, has invaded the adjacent metanephric blastema. The Wolffian duct has extended posteriorly and made contact with the cloaca. (Figure adapted from ref. 1).

leads to epithelial tubule formation.[1,10] Other embryonic tissues of various sources most notably the dorsal spinal cord can replace the ureter as potent inducers of tubulogenesis in these assays (Fig. 3 e-h). Similarly, ureteric tissue degenerates upon separation from metanephric mesenchyme, and undergoes limited or altered branching morphogenesis when recombined with heterologous mesenchymal tissues suggesting that metanephric mesenchyme secretes signals essential for ureteric survival and correct branching morphogenesis.[1,11,12]

In recent years, both genetic as well as biochemical approaches have identified a number of candidate molecules for ureteric and mesenchymal inducers. We now have some understanding about the complexity of these reciprocal signaling events. We know that a network of transcription factors, which include Pax-2, Eya1, and WT-1, sets up the metanephric blastema and controls the synthesis of glial derived neurotrophic factor, GDNF (for review see ref. 6). Mesenchymal activation of GDNF synthesis, which is under control of Eya1 and Pax-2[13,14], coordinates position and outgrowth of the ureteric bud via its receptors c-ret and GFRα1. In turn, sequential and combinatorial signaling seems necessary for ureter-mediated tubule

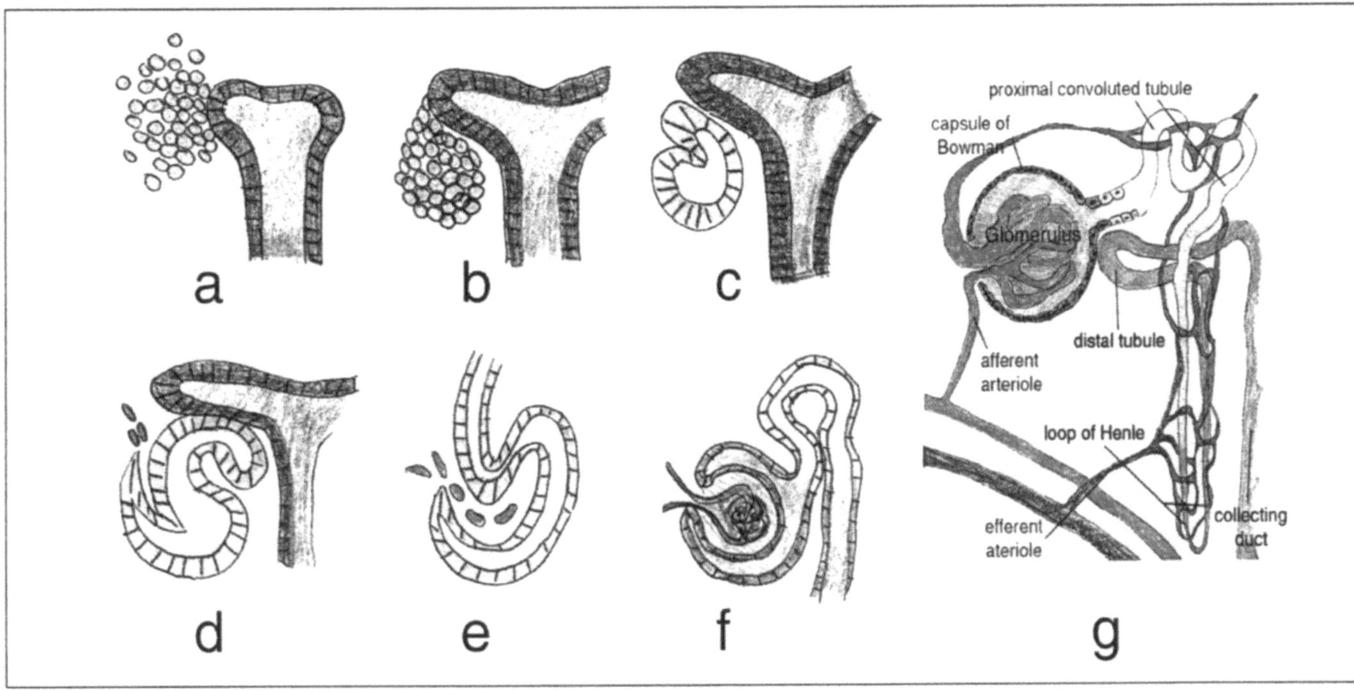


Figure 2. Schematic representation of nephron formation during metanephric kidney development. (a) Condensation of mesenchymal cells around the ureteric tip. (b) Formation of a pretubular cell cluster by aggregation of mesenchymal cells. (c) Mesenchymal-epithelial transition of the pretubular cluster generates a renal vesicle, which undergoes morphogenesis to form a comma-shaped body. (d) Subsequent morphogenesis results in the formation of an S-shaped body, which will undergo fusion at its distal end with the ureter. (e) Formation of a slit at the proximal end occurs during extension of the renal tubule. Endothelial precursor cells invade the slit to initiate glomerular development. (f) Regionalization and patterning of the tubule occurs along the proximal-distal axis. The glomerulus matures to the functional filtration unit. (g) Structure of a mature nephron, the functional unit of the adult metanephric kidney, with the associated vascular system.

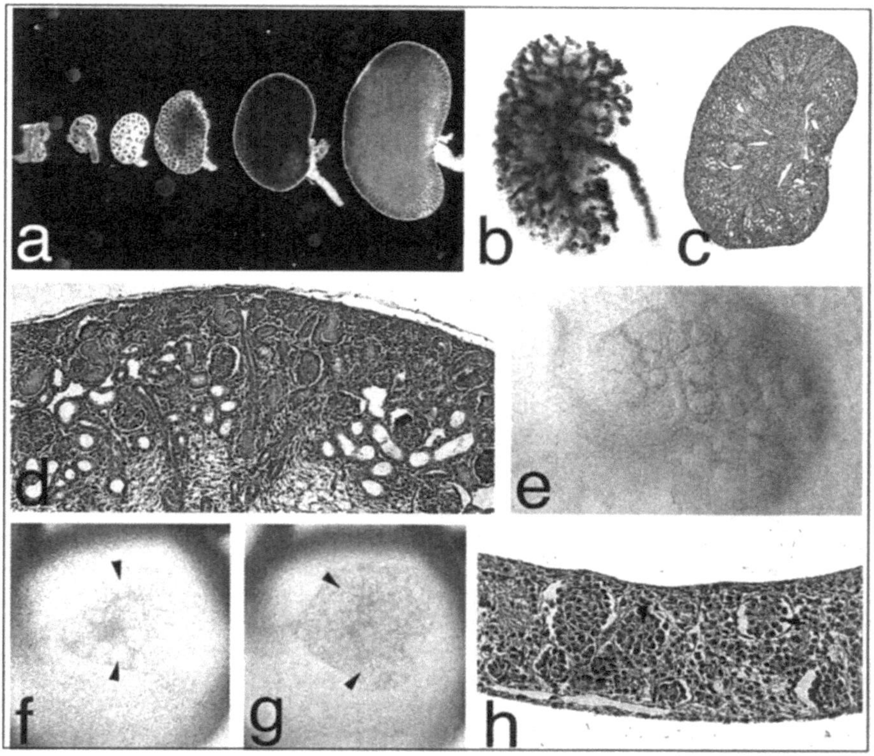


Figure 3. Murine metanephric kidney development in pictures. (a) Dramatic size increase during metanephric development. The metanephric kidneys shown were isolated from embryos at 11.5, 12.5, 13.5, 14.5, 16.5, and 18.5 d p.c., respectively. In situ hybridizations for expression of *Wnt-11* were performed to visualize ureteric tips (blue staining) at various stages of kidney development. (b) Branching morphogenesis of the ureter. In the 14.5 d p.c. kidney, an elaborate ureteric tree has developed by continuous branching morphogenesis. Ureter and collecting duct system are marked by a *lacZ* transgene under the control of a *Pax-2* enhancer element.[128] (c) Haematoxylin-eosin staining of an 18.5 d p.c. kidney section. (d) Enlargement of the peripheral region in (c) showing elements of nephrogenesis at different stages. (e) Brightfield photograph of a kidney rudiment isolated from an 11.5 d p.c. embryo and cultured for 2 days on to a nucleopore filter. (f-g) Induction of tubulogenesis in isolated metanephric mesenchyme by spinal cord tissue. Metanephric mesenchyme was isolated from an 11.5 d p.c. embryo and combined with pieces of dorsal spinal cord derived from embryos of the equivalent age. (f) Appearance of clusters of aggregating mesenchyme (arrowheads) in the metanephric mesenchyme after 2 days of culture. (g) Tubulogenesis in the mesenchyme (arrowheads) becomes obvious after 4 days. (h) Haematoxylin-eosin staining of sections of cocultures after 6 days. Glomeruli are readily seen (arrowheads).

induction. Leukemia inhibitory factor (LIF), transforming growth factor β (TGFβ), and basic fibroblast growth factor (bFGF) seem to be the important ingredients in this cocktail.[15,16] In addition to these classes of secreted factors, analysis of embryonic expression patterns has suggested prominent roles for members of the *Wnt* gene family in mouse metanephric kidney development.

## Role of *Wnt* Genes in Metanephric Kidney Development

The first *Wnt* shown to be critically involved in kidney development was *Wnt-4*[17]. *Wnt-4* is first expressed at 11.5 d p.c. in the aggregating metanephric mesenchyme shortly after the ureteric bud has invaded the metanephric blastema and has induced mesenchymal condensation (Fig. 4a). *Wnt-4* expression is maintained in simple renal vesicles but is gradually lost at the S-shaped body stage. Mice homozygous for a targeted allele of *Wnt-4* die within hours after birth due to hypoplastic non-functional kidneys. Histological analysis revealed the presence of collecting duct epithelium and interspersed interstitial mesenchyme. Further phenotypic analysis on the molecular level clearly showed that *Wnt-4* acts downstream of the ureteric inducer and controls the formation of pretubular aggregates. Without pretubular aggregates, tubule formation comes to a halt. Intriguingly, ureter development continues for some time, arguing that branching morphogenesis is independent of tubulogenesis.

As mentioned above classical kidney recombination experiments have shown that numerous embryonic tissues most notably embryonic spinal cord can elicit tubulogenesis in isolated metanephric mesenchyme (see refs. 1, 10, and references therein). This was interpreted that signals from these embryonic tissues mimic the endogenous ureteric inducer(s). The finding that cells engineered to produce Wnt-1 (which is not expressed in the kidney) induce tubulogenesis in isolated metanephric mesenchyme,[18] together with the fact that embryonic spinal cord is a rich source of various Wnt family members,[19] strongly argued for an endogenous ureter-derived tubule-inducing Wnt activity. However, things turned out differently.

Coculture experiments with isolated metanephric mesenchyme and cell lines engineered to produce those Wnt proteins, which were known to be expressed either in the spinal cord or in the kidney, revealed two classes of Wnts: those inducing tubulogenesis and those who did not.[20] Wnt-4, which is necessary for tubule formation as described above, was amongst the tubule-inducing class of Wnt proteins. Other Wnt proteins with tubule-inducing activity were either not expressed in the kidney or their renal expression patterns were incompatible with the role of a ureteric bud-derived tubule inducer. In the light of these findings thoughts emerged that spinal cord-derived signals may actually mimic the mesenchymal inducer Wnt-4. If this holds true, coculture of spinal cord tissue with mesenchyme from *Wnt-4* mutant embryos should rescue tubulogenesis. In contrast, if spinal cord derived signals were to mimic a yet unidentified ureteric Wnt protein *Wnt-4* mutant mesenchyme should degenerate. Surprisingly, the first scenario turned out to be correct. Spinal cord-derived signals efficiently induced tubulogenesis in *Wnt-4* deficient mesenchyme arguing that these signals act at the level or downstream of *Wnt-4*.[20] Thus, it is likely that one or several of the spinal cord-derived Wnts, including Wnt-4, may represent this activity. Subsequently, it was shown that renal tubules induced by exposure to Wnt-4 expressing cells were indistinguishable from those induced by spinal cord tissue (Fig. 4b-f).[20] Two important conclusions can be drawn from these findings. First, the classical tissue recombination experiments between spinal cord and isolated metanephric mesenchyme mimic a mesenchymal rather than a ureteric bud-derived activity, and second, *Wnt-4* is not only required but also sufficient for tubulogenesis. More recent studies have demonstrated that ureteric bud-derived tubule-inducing signals act both in combinatorial as well as in a synergistic fashion to mediate survival and condensation of metanephric mesenchymal cells. This will subsequently lead to activation of *Wnt-4* expression and formation of pretubular clusters. LIF, TGFβs, and FGFs are believed to be important components of this initial signal.[15,16]

A tight regulation of *Wnt-4* expression and activity during metanephrogenesis appears to be desirable given its powerful activity as mesenchymal inducer of renal tubulogenesis. Unfortunately, genetic analysis of the *Wnt-4* locus has not yet led to the identification of the key genomic control regions required for directing and restricting *Wnt-4* expression to the

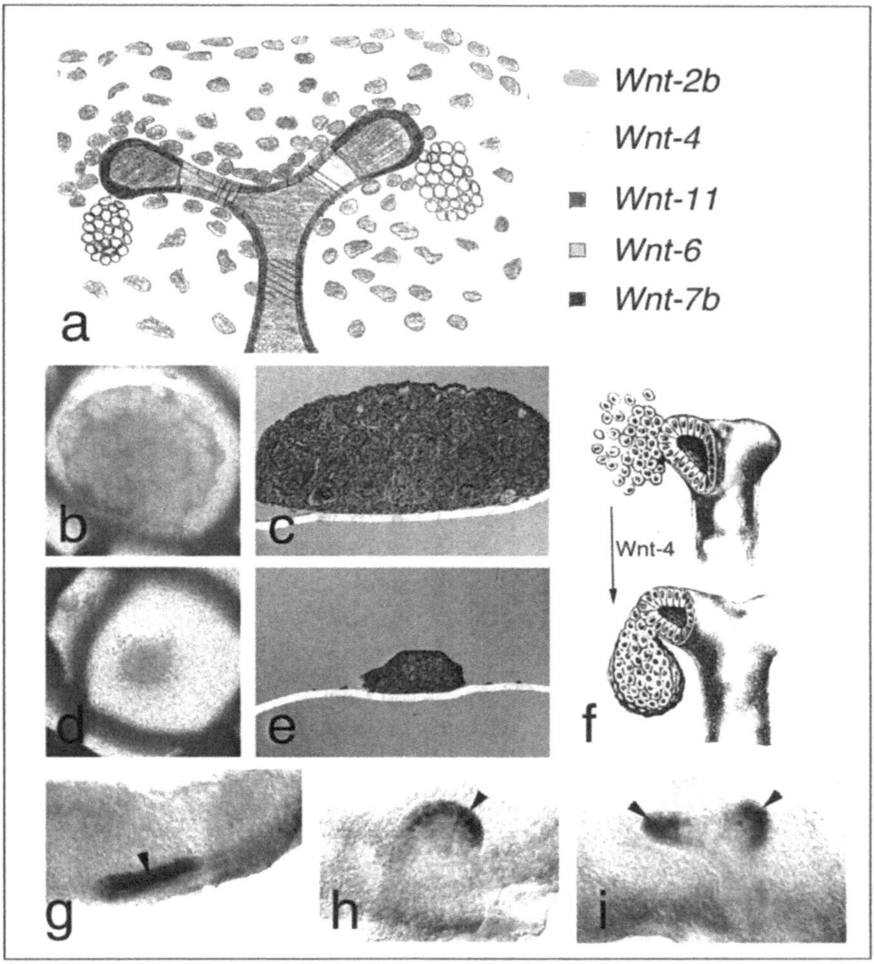


Figure 4. Wnt genes in metanephric kidney development. (a) Schematic drawing of a ureteric tip region during metanephric kidney development with the expression domains of various *Wnt* genes indicated. (b-e) Induction of tubulogenesis in isolated metanephric mesenchyme by *Wnt-4* expressing cells. Metanephric mesenchyme was isolated from an 11.5 d p.c. embryo and placed on to nucleopore filters overlying cells expressing *Wnt-4* and lacZ, respectively. (b) 2 day coculture with *Wnt-4* expressing cells. (c) Haematoxylin-eosin staining of sections of cocultures with *Wnt-4* cells after 6 days. Tubules are readily seen. (d) 2 day coculture with *lacZ* expressing cells. (c) Haematoxylin-eosin staining of sections of cocultures with *lacZ* control cells after 6 days. The metanephric mesenchyme has undergone degeneration. (f) Scheme of Wnt-4 function in nephrogenesis. Wnt-4 is required for induction and formation of pretubular cell aggregates. (g-h) Expression of *Wnt-11* in metanephric kidney development as detected by in situ hybridization analysis on whole kidneys. (g) Upregulation of *Wnt-11* expression (arrowhead) in the epithelium of the Wolffian duct contacting the metanephric blastema. (h) During invasion of the ureteric bud into the blastema, *Wnt-11* expression (arrowhead) is associated with the ureteric tip only. (i) *Wnt-11* expression is confined to the tips of the branching ureter (arrowheads).

aggregating mesenchyme. However, recent findings suggest that Wnt-4 signaling activity is also regulated at the protein level. *sFRP-2*, a member of the family of secreted frizzled related proteins (sFRPs) is expressed in an overlapping fashion with *Wnt-4*.[21,22] Coexpression is detected both in pretubular aggregates as well as in simple epithelial renal bodies. Furthermore, *sFRP-2* expression is induced by Wnt-4 signaling and requires *Wnt-4* gene function for sustained expression. Finally, biochemical experiments provide evidence that sFRP-2 can directly bind to Wnt-4 protein.[22] Taken together, Wnt-4 signaling appears to induce expression of its own protein inhibitor and thereby limits Wnt-4 activity in time and space. Surprisingly, however, administration of sFRP-2 protein to metanephric explant cultures does not seem to inhibit tubulogenesis.[23] Rather, sFRP-1, a second sFRP family member, which also binds avidly to Wnt-4, inhibits tubule formation in vitro. *sFRP-1* is expressed outside the condensing mesenchyme in the medullary and cortical stroma.[21,23] This suggests that sFRP-1 may be important in preventing inappropriate tubulogenesis outside differentiated areas. A possible role for sFRP-2 is indicated by the observation that addition of sFRP-2 to metanephric cultures can suppress the inhibitory effects of sFRP-1 on tubulogenesis.[23] sFRP2 appears therefore to counteract the inhibitory effects of sFRP-1 on Wnt-4 signaling. In other words, sFRP-2 may promote tubulogenesis by protecting Wnt-4 from the antagonistic effects of sFRP-1. Taken, together, a model emerges where sFRP-1 and sFRP-2 compete locally to regulate Wnt-4 signaling during metanephric kidney development.

The importance of tightly controlling the activity of Wnts in kidney epithelial development is further demonstrated by transgenic experiments. Mice expressing an oncogenic form of β-catenin develop severe polycystic lesions in tubules and collecting ducts, which are caused by increased proliferation and apoptosis of epithelial cells.[24] Unfortunately, there is no experimental data yet supporting a role for β-catenin as an intracellular mediator of the Wnt-4 signal. However, it is tempting to speculate that overexpression of the oncogenic form of β-catenin reflects a constitutively activated Wnt-4 signaling pathway. Yet another piece of evidence has emerged linking dysregulation of Wnt/β-catenin signaling to polycystic kidney diseases. The C-terminal cytoplasmic domain of polycystin, the gene product of the locus mutated in autosomal dominant polycystic kidney disease, activates Wnt/β-catenin signaling in vitro, suggesting that polycystin may modulate i.e., repress Wnt-4 signaling in renal development.[25]

Expression of several *Frizzled* genes, i.e., *Frizzled-2,Frizzled-3* and *Frizzled-7*, has been noted in the developing metanephric kidney.[26,27] At present, it is however unknown whether one or all three candidate genes function as receptors for Wnt-4 during tubulogenesis. Besides Frizzled receptors, proteoglycans have been implicated in the regulation of Wnt signaling and multiple evidences suggested roles for proteoglycan during early events of kidney development. Growth and branching of the ureteric bud in metanephric kidney cultures is disrupted in presence of inhibitors of glycosaminoglycan side chain sulphation.[28] Furthermore, a critical role of proteoglycans in Wnt-4 signaling could be demonstrated in coculture experiments. Administration of inhibitors of proteoglycan biosynthesis to cocultures of Wnt-4 expressing cells with metanephric mesenchyme completely abrogates tubule induction.[20] Finally, mice mutant for *heparan sulfate 2-sulfotransferase*, a gene coding for an enzyme of the biosynthetic pathway of heparan sulfate, exhibit a phenotype very similar to that of *Wnt-4* mice, since they lack formation of pretubular clusters.[29] This may suggest that proteoglycans act as co-receptors of the Wnt-4 signal in tubulogenesis.

While these findings shed some light on the mode of Wnt-4 signal transduction and regulation, our knowledge of the intracellular targets of this signaling pathway has remained surprisingly elusive. Although it seems likely that Wnt-4 signaling controls cell adhesion and/ or the mesenchymal-epithelial transition not a single gene has been identified as a direct molecular target to date.

A search for *Wnt* genes expressed during mouse metanephric development led to the identification of the previously uncharacterized *Wnt-11* gene (Kispert et al, 1996). A detailed analysis of embryonic expression revealed besides extranephric expression domains, highly restricted, strong and specific expression in the metanephric kidney (Fig. 4a, g-i). Metanephric expression is preceded by expression in the mesonephric duct. At 10.5 d p.c., when the Wolffian duct has just reached the metanephric blastema, *Wnt-11* expression is upregulated in the duct epithelium as it makes contact with the blastema. After ureteric bud formation, expression is confined to the tips of the ureter. During subsequent branching morphogenesis, *Wnt-11* expression is maintained at the ureteric tips but is lost in the intervening epithelium.[11]

The expression of *Wnt-11* at the ureteric tips and its nature as a secreted protein strongly argued for *Wnt-11* being the elusive ureteric bud-derived inducer of tubulogenesis in the adjacent metanephric mesenchyme. However, cell lines engineered to express *Wnt-11* failed to elicit tubulogenesis in metanephric mesenchyme in coculture experiments—while other Wnt proteins did.[20] Therefore, it seems unlikely that *Wnt-11* plays a role as a metanephric inducer. This notion is also supported by the fact that misexpression of *Wnt-11* in the section of the ureteric epithelium intervening the tips using a *Pax-2* ureteric bud enhancer resulted in abnormal ureteric branching rather than ectopic mesenchymal marker gene expression (A. Kispert, unpubl. observation).

Expression of *Wnt-11* during metanephric development seems to be regulated in at least two distinct steps. Upregulation of *Wnt-11* expression in the Wolffian duct epithelium contacting the metanephric blastema most likely is mediated by signals from the metanephric blastema, possibly GDNF, whereas maintenance of ureteric tip expression might be controlled by a combination of epithelial and mesenchymal factors.[11] Mesenchymally secreted GDNF may also be important in maintaining *Wnt-11* expression. Administration of GDNF to kidney cultures leads to upregulation of *Wnt-11* expression[30] whereas medium supplementation with dominant negative GDNF receptor c-ret results in downregulation of *Wnt-11*.[31] Proteoglycans may also participate in this regulation, as their removal quickly leads to loss of *Wnt-11* expression at the ureteric tips.[11] Taking together, expression analysis and functional data indicate a role of *Wnt-11* as a locally acting signaling molecule in branching morphogenesis in mouse metanephric development. Generation of *Wnt-11* deficient mice should provide further hints about the function of *Wnt-11* in metanephric kidney development.

Besides *Wnt-11*, a combined PCR-in situ hybridization screen for Wnt genes expressed in the developing mouse metanephros identified also *Wnt-7b*.[11] Expression of *Wnt-7b* can first be detected around 13.5 d p.c. in the ureteric epithelium (Fig. 4a). *Wnt-7b* expression persists in the presumptive collecting duct epithelium throughout kidney development. No expression could however be detected in the ureteric tip excluding a role for *Wnt-7b* in tubule induction.[11] Mouse embryos with a targeted mutation in the *Wnt-7b* gene were recently reported, but embryonic lethality prior to 11.5 d p.c. precludes analysis of *Wnt-7b* function in the developing metanephric kidney.[32]

Several groups have noticed a weak and transient expression of *Wnt-6* during metanephric development (see ref. 7 and A. Kispert, unpubl. observation). Expression was found in the ureteric epithelium overlapping with that of *Wnt-7b* (Fig. 4a). Targeted disruption of *Wnt-6* did not result in phenotypic consequences for kidney development (A. Kispert, unpubl. observation). *Wnt-6* may therefore only have a minor role in kidney development. Alternatively, *Wnt-6* and *Wnt-7b* may fulfill redundant functions in the developing metanephric kidney.

The latest addition to the collection of Wnt genes expressed during metanephric kidney development came with *Wnt-2b*.[33] *Wnt-2b* is expressed in perinephric mesenchymal cells in the region of the presumptive renal stroma (Fig. 4a). Functional experiments in kidney culture systems indicate that *Wnt-2b* is not capable of inducing tubulogenesis, but they provide evidence for a possible role of *Wnt-2b* in regulating ureteric bud growth and branching.[33] These findings are intriguing and raise the possibility of synergistic interactions between *Wnt-2b* and GDNF signaling. In vivo experiments are however now necessary to dissect the precise role of *Wnt-2b* in renal development.

## Wnt Signaling during Kidney Development—Perspectives from Studies in Alternative Vertebrate Models

Our knowledge about the role of Wnt signaling in kidney development is most advanced for the mammalian embryo. We have seen, that expression analyses led to the identification of *Wnt-4, Wnt-6, Wnt-7b, Wnt-11* and most recently *Wnt-2b* as those *Wnt* genes with specific expression during metanephric kidney development. To date, in vitro organ culture experiments and genetic studies have demonstrated a specific requirement for Wnt-4 in the transition of metanephrogenic mesenchyme to renal epithelia. However, a number of key questions about Wnt signaling during kidney development are still unanswered. The precise functions of the other renal *Wnt* genes, specifically *Wnt-2b, Wnt-6, Wnt-7b* and *Wnt-11*, remain to be determined. The identification of Wnt-4 as a mesenchyme-derived signaling molecule inducing tubulogenesis raises the issue about the nature of the downstream components, intracellular pathways employed, and the genetic program activated. While all these questions can be approached by genetic means in the murine metanephric kidney model, this methodology will undoubtedly be laborious and time consuming. Alternative experimental systems have recently emerged as attractive models for studies of kidney organogenesis. Let us take a closer look on the experimental advantages of these alternative kidney models before we discuss their contribution and relevance for studying Wnt function in metanephric development.

The chicken embryo with its advanced possibilities for microsurgery has been instrumental in defining the contributions of adjacent tissues, such as somites and ectoderm, to kidney development.[34,35] Lower vertebrates, such as the frog *Xenopus laevis* and the zebrafish, offer several distinct advantages over mammalian embryos for the molecular analysis of kidney organogenesis.[36,37,38] Females usually lay hundreds to thousands of eggs, which can be easily fertilized externally. The resulting embryos develop outside of the mother and are thus accessible to experimental manipulation. They become transparent such that internal organs can be observed without dissection. Finally, embryonic development occurs rapidly and excretory organs become fully functional within 2-3 days after fertilization. These features have been exploited in zebrafish for random mutagenesis screens to identify mutations affecting kidney development.[39] In *Xenopus*, in situ hybridization screens, the systematic cloning of cDNAs encoding orthologues of mammalian disease genes, and more recently EST database mining have been used successfully to define novel markers of the developing kidney.[36,40-44] Both models, zebrafish and *Xenopus*, are equally convenient to dissect gene function in vivo with gain- and loss-of-function experiments. Targeted misexpression of genes can be achieved by mRNA injection into selected blastomeres of the cleavage stage embryo. Conversely, injection of antisense morpholino oligonucleotides can be employed to specifically inhibit gene function in vivo.[45,46] In *Xenopus*, the repertoire of tools is further complemented by established renal cell lines, i.e., the epithelial A6 cells, and a simple in vitro organ culture model for renal tubulogenesis.[47] These tools will become increasingly useful where studies of renal gene functions ask for simplified assay systems. We will explore next whether the kidneys of lower vertebrates are valid models with relevance to mammalian kidneys and review the status of our knowledge on Wnt signaling during renal development in chicken, *Xenopus*, and zebrafish embryos.

## Pro- and Mesonephric Kidneys as Simplified Models of Nephrogenesis

While the experimental advantages of lower vertebrates are compelling, it is necessary to ask how relevant kidney development in zebrafish and *Xenopus* is to mammalian kidney organogenesis. Kidney development in all vertebrates is characterized by the successive formation of three separate excretory organs, the pronephros, mesonephros and the metanephros.[1] The three kidney forms arise from the intermediate mesoderm following a precise temporal and spatial sequence along the anterior-posterior body axis (Fig. 1). The pronephric kidney of mammalian embryos consists of a few tubule-like structures and has rudimentary, if any, secretory functions. Microsurgical experiments indicate that formation of the pronephric (Wolffian) duct is essential both for mesonephric and metanephric development.[48] In contrast, free-swimming fish and amphibian larvae develop fully functional pronephric kidneys that act as primary blood filtration and osmoregulatory organs. The pronephros is succeeded in all vertebrates by the mesonephros, a more complex excretory organ. It develops along the course of the nephric duct as result of mutual inductions between the duct and the mesonephrogenic mesenchyme. As a consequence, mesonephric tubules are generated in anterior-to-posterior sequence, with the anterior nephrons being formed first. The mesonephros is the permanent kidney in fish and amphibia and may function in embryos of higher vertebrates until the metanephros develops. As with the mesonephric kidney, reciprocal interactions between a derivative of the duct, the ureteric bud, and the adjacent mesenchyme govern development of the metanephric kidney. Importantly, however, the mesonephros develops from multiple short outgrowths from the nephric duct that do not branch, while the metanephros has a single bud that arborizes repeatedly.

All three kidney forms share a remarkably similar basic organization and differ mostly with respect to the number and organization of the nephrons, the functional unit of vertebrate kidneys. Pronephroi of fish and amphibia typically contain 1-3 nephrons, 10-50 nephrons are characteristic for mesonephroi, and the human metanephros may have up to 1 million nephrons. The generic organization of the vertebrate nephron with three modular components, the corpuscle, tubules, and duct, is conserved among all kidney forms. The basic design of the pronephroi in zebrafish and *Xenopus* is very similar.[36-39] The *Xenopus* pronephric kidney is shown here as an example (Fig. 5a, b). The pronephric kidney contains a single corpuscle, which is the site of blood filtration. It consists of the pronephric capsule with podocytes, the glomerular basement membrane, and a vascular component, the glomus. The pronephric capsule protrudes into the nephrocoel, which functions as the filtration chamber. Pronephric tubules are composed of polarized epithelia, which engage in selective reabsorption and secretion. As in the metanephros, pronephric tubules express important membrane transport proteins, which include ATP-driven pumps and sodium-coupled cotransporters.[39,41,42] Interestingly, expression of these genes occurs in a segment-specific manner along the proximal-distal axis of the nephron. Finally, the pronephric duct is involved in ion homeostasis and carries the modified urine to the exterior. The same basic structural units, corpuscles with glomeruli and tubules that drain into the nephric duct, are also found in the mesonephric kidney. Both kidney forms therefore contain a full range of specialized cell types typical of the complex kidneys of higher vertebrates.

The similarities between the three kidney forms extend down to the molecular level. Many genes known to be critical for metanephric kidney development are also expressed in the developing pro- and mesonephric kidneys. These include *Pax-2* in tubule and duct epithelia[34,43,49] and *WT-1* in podocytes.[39,40] This suggests that the genetic program controlling renal growth and development has been largely conserved between all kidney types. Pronephric and mesonephric kidneys represent therefore valid models of nephron development and epithelial cell differentiation.

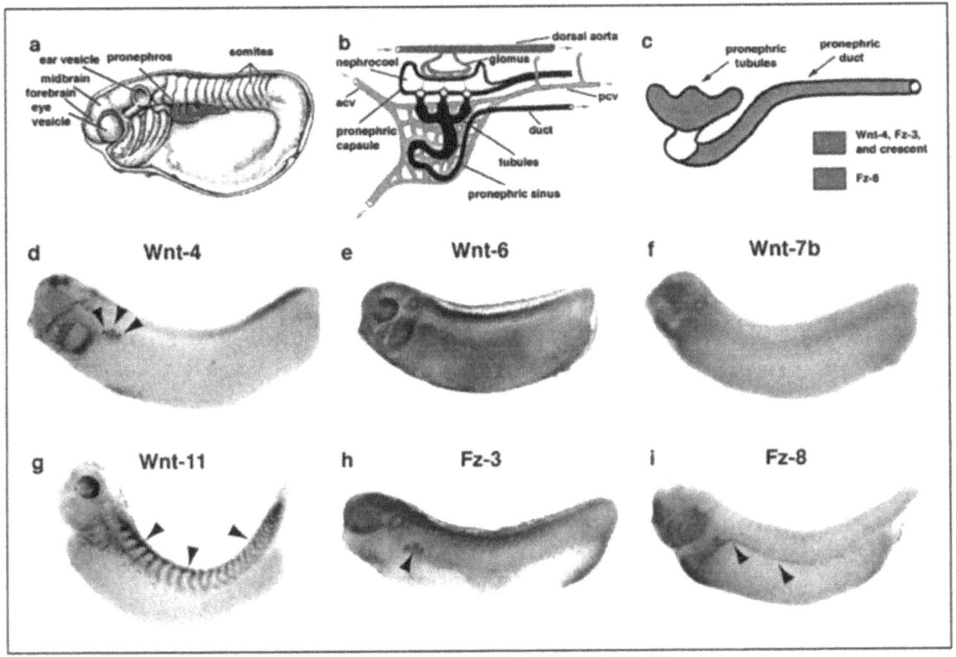


Figure 5. Wnt signaling in the developing *Xenopus* pronephric kidney. (a) A *Xenopus* embryo in an early tailbud stage with the skin removed illustrating the location of the developing pronephric kidney. (b) Schematic representation of the *Xenopus* pronephric kidney with its associated vascular system. See text for description. (c) Schematic composite summarizing expression patterns of *Wnts*, *Frizzleds*, and *sFRPs* in the pronephric kidney of a stage 30 *Xenopus* embryo. (d-i) Expression patterns of selected *Wnt* and *Frizzled* genes in tailbud stage *Xenopus* embryos. Gene expression was visualized by whole mount in situ hybridization. Embryos are shown in lateral view with anterior to the right. (d) *Wnt-4*, stage 31. Expression in the intermediate mesoderm is restricted to developing pronephric tubules (arrowheads). (e) *Wnt-6*, stage 29. Expression is found throughout the epidermis. (f) *Wnt-7b*, stage 31. No obvious expression is detectable in the pronephric region. (g) *Wnt-11*, stage 34. Expression is primarily associated with the developing somites. (h) *Fz-3*, stage32. Pronephric expression is limited to tubules (arrowhead). (i) *Fz-8*, stage 34. Pronephric expression is confined to the developing duct (arrowheads). Abbreviations: acv, anterior cardinal vein; common cardinal vein; posterior cardinal vein. (Panels a and b adapted from refs. 129 and 36, respectively).

## Wnt Signaling in the Pro- and Mesonephric Kidneys

The remarkable structural, cellular, and molecular conservation between the three vertebrate kidney forms suggests that studying the function of components of the Wnt signaling pathway during the pro- and mesonephric kidneys development may become useful in predicting functions of the Wnt signaling pathway in the metanephros. Vertebrate genomes express at least 19 different Wnt proteins, which play essential roles during early embryogenesis and organogenesis.[50-52] We have seen, that at least five of these genes are expressed in the developing mouse metanephric kidney: *Wnt-2b, Wnt-4, Wnt-6, Wnt-7b* and *Wnt-11* (Fig. 4a). Orthologues for several of these genes have been identified in chicken (*Wnt-2b, Wnt-4, Wnt-6, Wnt-7b, Wnt-11*), *Xenopus* (*Wnt-2b, Wnt-4, Wnt-6, Wnt-7b*, and *Wnt-11*), and zebrafish (*Wnt-4/4b, Wnt-11*). We will now describe the embryonic expression patterns of these genes and discuss studies, which have been initiated to analyze the functional importance of Wnt signaling pathways in pro- and mesonephric development. Table I provides a synopsis comparing the expression patterns of Wnt signaling components across several vertebrate species.

In the mouse, expression of *Wnt-2b* (previously known as *Wnt-13*) is associated with perinephric mesenchymal cells of the developing mesonephros and metanephros.[33] Chicken *Wnt-2b* is transiently expressed in the intermediate mesoderm and lateral plate mesoderm at the level of the presumptive forelimb.[53] While misexpression experiments demonstrate that *Wnt-2b* acts via β-catenin and has limb-inducing activity, its role during early kidney development is unknown. A partial *Xenopus* cDNA for *Wnt-2b* (previously known as *Wnt-2*) was isolated, but its precise embryonic expression pattern still remains to be established.[54] In summary, the *Wnt-2b* expression patterns in mouse and chicken embryos are consistent with a role of *Wnt-2b* in mesonephric and metanephric kidney development.

Genes encoding Wnt-4 proteins have been found in many vertebrate genomes, but also in invertebrates such as the cephalocordate amphioxus.[55-61] As we have seen earlier, Wnt-4 acts in the mouse embryo as a mesenchyme-derived signal promoting mesenchyme to epithelia transition during metanephric kidney development.[17,20] In the mouse embryo, *Wnt-4* expression is also detected in the mesonephric mesenchyme surrounding the nephric ducts and subsequently in mesonephric tubules.[17,19] Similarly, expression of *Wnt-4* occurs in mesonephric and metanephric tubules of the chicken embryo.[17,62] Expression of *Xenopus Wnt-4* is initially associated with the pronephric anlage and later with developing pronephric tubules, but never with the pronephric duct (Fig. 5d).[36,63] Onset of *Wnt-4* expression coincides with the epithelialization of the tubule anlage and ceases once tubule morphogenesis is completed. The findings in mouse, chicken and *Xenopus* embryos collectively suggest that the role of Wnt-4 may be conserved in the development of all kidney forms. Gain- and loss-of-function experiments to directly address this issue were performed in *Xenopus*.[64] Misexpression of *Wnt-4* in the intermediate and lateral plate mesoderm causes abnormal morphogenesis of pronephric tubules, but is not sufficient to initiate ectopic tubule formation. Disruption of *Wnt-4* gene function was achieved with a morpholino antisense oligonucleotide-based gene knock-down strategy. *Xenopus* embryos injected with antisense *Wnt-4* morpholinos fail to form pronephric tubules, while pronephric duct development precedes largely unaffected. These findings demonstrate unequivocally that, as in the metanephric kidney, *Wnt-4* function is critically required for tubulogenesis in the pronephric kidney.

Two closely related *Wnt-4* genes, *Wnt-4* and *Wnt-4b*, have been isolated in zebrafish.[55,57,60] Surprisingly, however, neither gene is expressed in the zebrafish pronephros.[55,57] This could indicate that another Wnt protein may function in place of *Wnt-4* during tubulogenesis in the zebrafish pronephros. A *Wnt-4* orthologue has also been recently described in amphioxus, which represents that closest living invertebrate relative of vertebrates.[59] The excretory system of amphioxus develops as a segmentally arranged series of nephridia without any contribution from condensing mesenchyme. Whereas the nephridium rudiment (also known as Hatchek's nephridium) expresses the *Pax-2* homologue *AmphiPax-258*, no expression of *Wnt-4* is detected.[59,65] This suggests that during vertebrate evolution renal Wnt-4 expression was acquired with the development of excretory organs composed of mesenchyme-derived tubular epithelia.

The murine ureteric epithelium expresses low levels of *Wnt-6* (Fig. 4a). In the chicken embryo, *Wnt-6* expression is confined to the epidermis. The intermediate mesoderm is devoid of detectable expression.[66] Similar results were obtained for embryonic expression of *Xenopus Wnt-6* (Fig. 5e; D. Saulnier and A. Brändli, unpublished observation). Furthermore, Northern blot analysis in *Xenopus* indicates that *Wnt-6* is not expressed in the adult kidney.[54] Thus, presently available data from chicken and *Xenopus* embryos do not support a role of *Wnt-6* in early kidney development.

Murine *Wnt-7b* is expressed in the posterior segment of the nephric duct and the ureteric bud of the metanephros, but is not detectable in the mesonephros.[11,67] Expression of chicken *Wnt-7b* in the embryonic nervous system was reported in the past[68], but no data is currently available on the renal expression of this *Wnt*. Similarly, in *Xenopus*, *Wnt-7b* expression is found in the dorsal neural tube and in the epidermis,[69] but expression could not be detected in the developing pronephric kidney (Fig. 5f; D. Saulnier and A. Brändli, unpublished observation).

During kidney development in the mouse, *Wnt-11* expression is upregulated in the posterior nephric duct and later becomes highly localized to the tips of the branching ureteric bud.[11] Similarly, expression of chicken *Wnt-11* is restricted the tips of the ureteric bud.[62] In more anterior regions, where pronephric and mesonephric development take place, *Wnt-11* expression is confined to the intermediate mesoderm, but is not detected in the renal ductal system.[66,70] *Wnt-11* genes have also been characterized in *Xenopus* and zebrafish, but again no expression could be detected in the developing nephric duct (Fig. 5g).[71,72] Collectively, this indicates that *Wnt-11* may primarily have a function in metanephric kidney development. The mesonephros contains multiple short outgrowths from the nephric duct that do not branch, while the metanephros has a single duct-derived bud that undergoes extensive branching. *Wnt-11* may therefore function in the regulation of ureteric bud morphogenesis. It should be noted that *Wnt-11* is also highly expressed in the somites (Fig. 5g), which lie adjacent to the developing pronephric anlage as shown in Figure 5a. It is thus conceivable that somitic Wnt-11 signals may influence pronephric development. Analysis of pronephric development in the zebrafish mutant *silberblick*, which is mutant for *Wnt-11*[73] may resolve this issue.

As several *Wnt* genes are differentially expressed in the kidney, the question of which Fz receptors could mediate Wnt signals in the developing kidney arises. Up to ten distinct *Fz* genes are present in vertebrate genomes,[74] but ligand-receptor specificities of Wnts and Fz proteins are poorly characterized. A screen of metanephric kidneys to identify potential receptors for Wnt-4 in the mouse has produced three candidate receptors (Fz-2, Fz-3, and Fz-7), but the renal cell types expressing these genes are unknown.[27] Separate studies indicate that the intermediate mesoderm strongly expresses *Fz-1* and the nephric duct is positive for both *Fz-6* and *Fz-9*.[75,76] Northern blot and RT-PCR analysis have further implicated *Fz-4* and *Fz-5* in metanephric kidney development.[74,77] A similar wide range of *Fz* gene expression was reported for the human kidney.[78] Taken together, overlapping *Fz* expression patterns may suggest redundancy of renal *Fz* gene function in metanephric development.

Comparative analyses of renal *Fz* gene expression across different species may help to identify the principal players, since their expression may have remained evolutionary conserved as demonstrated for the ligand Wnt-4. In the chicken embryo, only *Fz-4* (out of seven *Fz* genes analyzed) is unequivocally associated with kidney development.[62,79] Interestingly, *Fz-4* expression is restricted to condensing nephrogenic mesenchyme and developing tubules of all kidney forms. The renal expression pattern of *Fz-4* is therefore strikingly similar to that of Wnt-4 indicating that Wnt-4 may signal through Fz-4 during tubulogenesis. It should however be noted that kidney development is not affected in *Fz-4* deficient mice.[80]

In *Xenopus*, whole mount in situ hybridization studies have identified several candidate *Fz* genes with expression in the intermediate mesoderm and possibly the developing pronephros. These include *Fz-1*,[81] *Fz-3*,[82] *Fz-4*,[83] *Fz-7*[81,84-87] and *Fz-8*.[88] The intermediate mesoderm not only gives rise to the pronephric kidney, but also contributes to the hematopoetic system and the vasculature of the trunk. At present, only two *Frizzled* genes, *Fz-3* and *Fz-8*, are unquestionably expressed in the pronephric kidney (Fig. 5h, i; D. Saulnier and A. Brändli, unpublished observations). Interestingly, their expression is confined to two distinct compartments of the pronephric kidney. *Fz-3* is specifically associated with developing pronephric tubules, and thus is a candidate receptor for Wnt-4 (Fig. 5d, h). On the other hand, *Fz-8* expression is limited to the developing pronephric duct (Fig. 5i). It therefore appears that distinct Fz receptors function during development of tubular and duct epithelia in the pronephros, respectively. Functional studies specifically addressing this issue are currently in progress.

Zebrafish genes encoding orthologues of *Fz-2, Fz-7, Fz-8, Fz-9* and *Fz-10* are presently known.[89-93] Surprisingly, only *Fz-8a* appears to be expressed during pronephric kidney development. There, expression is confined to the early pronephric duct.[90] Therefore, expression of *Fz-8* during pronephric duct development has been conserved between fish and amphibia.

Secreted frizzled-related proteins (sFRPs) represent a major group of Wnt binding proteins, which have been implicated in the modulation of Wnt protein activity. Of the sFRPs (sFRP-1 to sFRP-5) studied so far, at least three have been implicated in kidney organogenesis.[21,22] They show characteristic, and frequently complementary expression patterns. *sFRP-1* expression was primarily in the stroma, while *sFRP-2* was expressed complementary in condensing mesenchyme and newly forming epithelia of the metanephros, similar to *Wnt-4*. Furthermore, *sFRP-4* is weakly expressed in the condensing and epithelializing mesenchyme.[21] Similar metanephric expression patterns were reported for *sFRP-1* and *sFRP-2* in the rat embryo.[23] Expression of *sFRP-2* is also detected in mesonephric mesenchyme and subsequently in mesonephric tubules, but not in the nephric duct.[21,22]

Genes encoding sFRPs have been characterized in chicken and *Xenopus*, but not yet in zebrafish. Four sFRP family members (*sFRP-1, sFRP-2, sFRP-3/FrzB, crescent*) have been isolated in chicken, but no systematic analysis of their expression during kidney development has been performed to date. A survey of the published literature on embryonic *sFRP* expression indicates that *sFRP-2* expression is localized to the developing mesonephros similar to its murine counterpart.[94] None of the other known *sFRP* genes appear to be present in the mesonephros. In *Xenopus*, orthologues for *sFRP-1* (*FrzA*), *sFRP-2, sFRP-3* (*FrzB*), *crescent* (*FrzB2*), and *sizzled* have been described to date. *Crescent* expression is detectable in the pronephric anlage and remains associated with developing pronephric tubules.[95-97] Expression could not be detected in the pronephric duct. In contrast, *sFRP-2* is broadly expressed in the intermediate and anterior lateral plate mesoderm.[96] The emerging picture indicates that expression of *sFRP-2* in nephrogenic mesenchyme is conserved from mouse to *Xenopus*, while renal expression of other *sFRPs* occurs rather in a species-specific manner.

## Wnts in Kidney Organogenesis: Conclusions from a Multi-Species Approach

The comparison of the embryonic expression patterns for Wnt signaling components has revealed both highly conserved features as well as distinct differences between mouse, chicken, *Xenopus*, and zebrafish (Table 1). At the level of Wnt proteins, expression and function of Wnt-4 during tubulogenesis appears to be conserved from amphibia to mammals. Hence, a common signaling pathway may control a key step in tubulogenesis in different vertebrate excretory organs: mesenchymal aggregation and epithelialization to renal vesicles. Apart from *Wnt-4*, there is little evidence for a role of other Wnts in the development of pronephric and

***Table 1. Species comparison of Wnt, Frizzled, and sFRP expression patterns during kidney development***

***Wnt* genes**

| Gene | Mouse | Chicken | *Xenopus* | Zebrafish |
|---|---|---|---|---|
| *Wnt-2b/13* | PnM | IM | ?[1] | ø |
| *Wnt-4* | MsT, MtT | MsT, MtT | PrT | - (Wnt-4)<br>- (Wnt-4b) |
| *Wnt-6* | UB | - | - | ø |
| *Wnt-7b* | ND, UB | ? | - | ø |
| *Wnt-11* | ND, UB | IM, UB | - | - |

[1] *Xenopus Wnt-2b*[34] was previously known as Wnt-2.

***Frizzled* genes**

| Gene | Mouse | Chicken | *Xenopus* | Zebrafish |
|---|---|---|---|---|
| *Fz-1* | IM | - | + | ø |
| *Fz-2* | + | - | - | - |
| *Fz-3* | + | ø | PrT | ø |
| *Fz-4* | + | MsT, MtT | + | ø |
| *Fz-5* | + | ø | - | ø |
| *Fz-6* | ND | ø | ø | ø |
| *Fz-7* | + | - | + | - |
| *Fz-8* | ? | - | PrD | PrD (Fz-8a)<br>- (Fz-8b) |
| *Fz-9* | ND | - | ø | - |
| *Fz-10* | ? | - | -[2] | - |

[2] *Xenopus Fz-10*[87] was previously known as *Fz-9*.

***sFRP* genes**

| Gene | Mouse | Chicken | *Xenopus* | Zebrafish |
|---|---|---|---|---|
| *sFRP-1 (FrzA)* | MtS | - | ? | ø |
| *sFRP-2* | MsT, MtT | MsM | IM | ø |
| *sFRP-3 (FrzB)* | - | - | - | ø |
| *sFRP-4* | MtM | ø | ø | ø |
| *sFRP-5* | - | ø | ø | ø |
| *crescent* | ø | - | PrT | ø |
| *sizzled* | ø | ø | - | ø |

Abbreviations: +, renal expression, but no further characterization available; -, no renal expression; ?, renal expression unknown; IM, intermediate mesoderm; MsM, mesonephric mesenchyme; MsT, mesonephric tubules; MtM, metanephric mesenchyme; MtS, metanephric stroma; MtT, metanephric tubules; ND, nephric (Wolffian or mesonephric) duct; PnM, perinephric mesenchyme; PrD, pronephric duct; PrT, pronephric tubules; UB, ureteric bud; ø, orthologue not described yet. See text for references.

mesonephric kidneys. Hence, expression of *Wnt-6*, *Wnt-7b* and *Wnt-11* in the metanephric kidney may relate to a ureteric outgrowth and differentiation program.

Expression of *Fz* genes in the mammalian metanephric kidney is complex and largely unresolved, suggesting that Fz receptors may have functionally redundant activities during metanephric kidney development. A significantly simplified picture emerges from expression analysis in the pronephros and mesonephros. Expression of *Fz-3* (in *Xenopus*) or *Fz-4* (in chicken) correlates with tubulogenesis, qualifying these Fzs as primary receptors for Wnt-4. In contrast, *Fz-8* expression is confined to the developing pronephric duct in *Xenopus* and zebrafish (see Fig. 5c, for a scheme). Separate Fz receptors possibly coupling to different intracellular Wnt signaling pathways may therefore differentially control development of tubules and duct in the pronephric kidney.

Only limited conclusions can be drawn from the analysis of *sFRP* expression patterns. Studies in the mouse demonstrate that members of the *sFRP* family are expressed in restricted patterns within the metanephric kidney. In *Xenopus*, only *crescent*, for which no mouse orthologue has been described yet, is clearly expressed in the developing pronephric kidney. Remarkably, *crescent* expression overlaps with *Wnt-4*, whereas in the mouse *sFRP-2* expression correlates with *Wnt-4*. Modulation of Wnt-4 activity by sFRPs during tubulogenesis may therefore be a further feature conserved across vertebrate species and different kidney forms. In summary, the comparison of *Wnt*, *Frizzled*, and *sFRP* expression patterns has been a valuable tool for the assessment of Wnt function, Wnt regulation and Wnt-Frizzled interactions during kidney development. Future studies will have to address the precise functions of the different candidate genes in relation to morphogenesis and differentiation of the kidney.

## Genital Development in the Mouse

Mammalian sex organ formation is a unique developmental process since it ultimately results in the establishment of two different organ systems from an initially identical set of precursor structures. Although sexual development takes place during embryogenesis, full sexual characteristics are only established late at postnatal stages (for recent reviews on mammalian sexual differentiation and its genetic control see refs. 98-101).

Sex development can be broken down into several distinct phases. First, indifferent or bipotential gonad anlagen arise from the urogenital ridge. Second, sex determination, i.e., the commitment of the indifferent gonad anlagen into testis or ovary, occurs according to a genetic program dependent on the chromosomal sex. Activation of the *Sry* gene within a critical time window triggers testis differentiation in the male. Subsequently, testicular hormones mediate the next phase of sex development, masculinization of internal and external genitalia, both of which arise from indifferent anlagen. Female differentiation occurs in the absence of *Sry*. Indifferent gonads differentiate into ovaries and the bipotential anlagen of internal and external genitalia differentiate into female reproductive organs. In a fourth phase, further sexual differentiation of genitalia, of physiological processes and of behavior leads to full sexual maturity. This final phase mainly occurs postnatally and is finished after puberty. Interestingly, primordial germ cells arise outside the gonads. They are set aside from the somatic cell lineages early in development, undergo extensive migration though the embryo to reach the developing gonad, where they differentiate into oogonia and spermatogonia, once somatic sex has been determined.[102] We will now have a closer look at the various levels of sex development to better understand where Wnt signaling is involved in.

The urogenital ridge arises as an outfolding of the intermediate mesoderm and its covering coelomic epithelium into the coelomic cavity. This process occurs along the posterior body wall between the anlagen of the limbs starting from around 9 d p.c. The urogenital ridge is tightly associated with the mesonephros, and cell lineages contributing to the adrenal cortex, the gonads and kidneys are tightly intermingled. Several genes have been identified which control the formation of the urogenital ridges. *Lim-1* seems to control formation of the intermediate mesoderm since targeted deletion of the gene results in mice characterized by complete absence of gonads and kidneys.[103,104] *WT-1*, identified as a tumor suppressor gene, is required for setting up the indifferent gonad anlagen as well as the metanephric blastema.[105] *SF-1* encoding a transcription factor of the nuclear orphan hormone receptor family plays a similar role as revealed by the loss of gonads (and adrenal glands) in knockout mice.[106] SF1 is probably also involved in other aspects of sexual development as it regulates enzymes mediating steps in steroidogenesis and transcription of the gene for Anti-Müllerian hormone.[107,108] Finally, several studies suggest that SF-1 and WT-1 cooperate on the transcriptional and/or protein level.[109]

The indifferent gonads of the mouse appear around 11 d p.c. At that time, the gonads look like slender sausages. They are composed of mesenchyme and covered by an epithelial layer of mesothelium. The epithelial layer proliferates, invades the underlying mesenchyme and forms primary sex cords, which will host the incoming primordial germ cells. Primordial germ cells can be detected at 8.0 d p.c. in the extraembryonic mesoderm. Leaving the allantoic region shortly later, they migrate through the hindgut mesentery and reach the urogenital ridge at 10.0 d p.c. From 11 d p.c. onwards, the primordial germ cells associate with the primary sex cords.

Progression from the bipotential gonad anlagen towards testicular differentiation is mediated by gonadal and autosomal genes led by the Y chromosome-encoded *Sry* gene (for reviews see refs. 101, 110-112). *Sry* becomes expressed at 11.5 d p.c. in cells of the indifferent gonad of the male embryo. Sry activated processes mediated by genes, such as *Sox9* and *DMRT1*, promote testis differentiation. The sex cords continue to proliferate forming an internal network of tubules. Cells of the cords differentiate into Sertoli cells whereas interstitial mesenchymal cells differentiate into Leydig cells. Both cell types produce hormones, which are instrumental in the third phase, the sexual differentiation of internal and external genitalia. In males, *Sry* represses *Dax1* expression during testicular development. In females on the other hand, which lack the Y chromosome and therefore do not carry an *Sry* gene, no repression of *Dax1* occurs and indifferent gonads will be able to differentiate to ovaries. Towards this goal, the initial sex cords degenerate and a cortical network of cords forms instead to enclose the primordial germ cells. Germ cells will proliferate reaching a maximum number at birth and differentiate into oocytes. The epithelial cells of the cords differentiate into granulosa cells, while interstitial cells mature to thecal cells. Both granulosa and thecal cells secrete female steroid hormones (estrogens and gestagens) and interact with the oocytes to form the primary follicles.

During early stages of development (from 11.5-14.5 d p.c. in the mouse), mammalian embryos have two pairs of sex ducts, irrespective of their karyotype. We have discussed above the first of these ducts, the Wolffian duct (also called mesonephric or nephric duct) in the context of kidney development. The Wolffian duct emerges from the pronephric region, extends posteriorly along the mesonephrogenic mesenchyme and generates a side branch, the ureter that invades the metanephric blastema. The Wolffian duct grows progressively towards the cloaca, to finally open into the urinary bladder. The second duct system arises by an infolding of the coelomic epithelium in the urogenital ridge region at around 11 d p.c. As the Wolffian duct, the Müllerian duct is established in both sexes and can be detected until 14.5 d p.c. Maintenance, differentiation, and regression of both ductal systems are under hormonal control (Fig. 6, 7 a-d). Pre-Sertoli cells differentiate into Sertoli cells of the testes subsequent to activation of the Sry gene.[113] Expression of the AMH (Anti-Müllerian hormone or MIS, Müllerian inhibiting substance) gene in Sertoli cells is under control of *Sry* and *SF1* genes.[109,114] Secreted AMH exerts its action by binding to a receptor present on mesenchymal cells

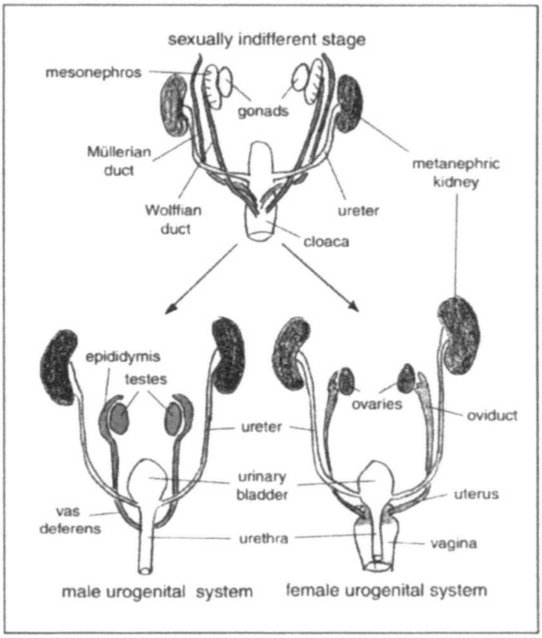


Figure 6. Schematic representation of genital system development in the mouse. Both the Wolffian duct and the Müllerian duct are present at the sexually indifferent stage. Gonads are bipotential. After completion of sexual determination and differentiation in the male, the Wolffian duct has differentiated and the Müllerian duct has regressed. In contrast, the Müllerian duct has differentiated into the oviduct, the uterus, and the upper part of the vagina, while the Wolffian duct has degenerated in the sexually mature female. Figure adapted from ref. 130.

surrounding the Müllerian duct.[115] This leads to regression of the Müllerian duct in the male. Additionally, interstitial cells of the testis, known as Leydig cells, secrete testosterone. In the presence of testosterone, Wolffian ducts differentiate into the epididymis and the vas deferens, and seminal vesicles bud off. In the female, neither AMH nor testosterone are present and as a consequence the Wolffian duct regresses and the Müllerian duct matures to form the oviduct, uterus and the upper part of the vagina. This maturation of the Müllerian duct largely occurs postnatally under the influence of female steroid hormones, estrogens, and gestagens.

Following the paradigm of gonad and genital duct development, external genitalia are initially laid down as sexually indifferent structures.[111] The genital tubercle is surrounded by the urethral fold and labioscrotal swellings in the 11.5 d p.c. mouse embryo. Differentiation of the indifferent anlage into the male external genitalia depends on circulating testosterone produced by Leydig cells of the testis. In the peripheral target tissues, testosterone is converted to dihydrotestosterone by 5$\alpha$-reductase. In response to dihydrotestosterone, the genital tubercle will extend and differentiate into the penis with its bulbous tip, the glans penis. The urethral folds will fuse to form the ventral part of the penis shaft, and the labioscrotal swellings will become the scrotum. In females, where testosterone is absent, the clitoris is formed from the genital tubercle, the labia minora develops from the unfused urethral folds and the labia majora is derived from the labiocrotal swellings. Final maturation of both reproductive systems is achieved during puberty and accompanied by establishment of gender-specific behavioral traits.

## Role of *Wnt* Genes in Genital Development

Wnt-4 fulfils a pivotal role in kidney development as a mesenchymal inducer of tubulogenesis. Careful phenotypic analysis of newborn mutant mice revealed that *Wnt-4* is also required at multiple points during genital development. Female mutants show partial sex reversal, whereas male mutant newborns show normal internal and external genitalia but fail to establish a Müllerian duct during development.[116] During genital development, *Wnt-4* expression is first detected in the mesonephric mesenchyme at 9.5 d p.c. By 11.5 d p.c., *Wnt-4* expression is found in the indifferent gonads of both sexes. Subsequently, expression is maintained in the female gonad, but is downregulated in the male gonad after sex determination has occurred.[116] Phenotypic analysis of *Wnt-4* mutants suggests that expression both in the early mesonephric mesenchyme as well as in the ovary is important for normal female differentiation.

*Wnt-4* expression is crucial for Müllerian duct formation as analysis of *Wnt-4* mutant animals has revealed.[116] Müllerian ducts fail to be established in *Wnt-4* mutant female embryos resulting in a complete loss of Müllerian duct derivatives (oviduct, uterus, upper part of the vagina). A detailed molecular analysis of the mutants has not yet been performed, but it is conceivable that Wnt-4 signaling from the mesonephric mesenchyme triggers the infolding of the coelomic epithelium and thus Müllerian duct establishment. Interestingly, *Wnt-4* is also expressed in the Müllerian duct derivatives in postnatal stages.[117] Shortly after birth, expression is found in the mesenchyme of the uterine horns and the oviduct, shortly later also in the sinus vagina epithelium. During the oestrus cycle in the adult mouse, *Wnt-4* expression is undetectable during diestrus, but found at low levels in the uterine epithelium during proestrus. During estrus expression in the uterine stroma and epithelium is upregulated, but maintained only in the stroma during metestrus. Further analysis of late *Wnt-4* functions in the adult female reproductive tract is however precluded by the early requirement for *Wnt-4* in Müllerian duct development.

In *Wnt-4* mutant females, the Wolffian duct is maintained and differentiates into epididymus and vas deferens (Fig. 7e,f). In contrast, no masculinization of external genitalia was observed in newborn female mutants. Thus, *Wnt-4* females are partially sex-reversed.[116] As *Wnt-4* is not expressed in the Wolffian duct at any time, a direct requirement for Wnt-4 signaling within the duct resulting in degeneration in the female can be excluded. Rather, Wnt-4 may normally act indirectly on Wolffian duct development by suppressing steroidogenesis, i.e., testosterone production, in the female. As described, differentiation of the male internal genitalia depends on testosterone, which is synthesized in interstitial cells of the testis, the Leydig cells. For differentiation of the bipotential anlagen of the external genitalia along the male pathway, 5α-dihydrotestosterone is required. 5α-DHT is derived from testosterone by a metabolic step occurring at the target site. In line with the above hypothesis, marker enzymes for testosterone biosynthesis, which are normally only present in the developing testis but not in the ovary, are detected in the *Wnt-4* mutant ovary. This may suggest that Wnt-4 signaling suppresses differentiation of the supporting cell lineages into Leydig cells and/or Sertoli cells in the ovary. Since there is no evidence for Sertoli cells in the ovary at embryonic stages, it is likely that loss of Wnt-4 signaling results in differentiation of Leydig cells in the mutant ovary. They secrete testosterone thereby preventing Wolffian duct regression. Lack of masculinization of external genitalia in the mutant may be due to lack of peripheral enzymes critical in metabolic activation of testosterone.

*Wnt-4* may also be required later for maturation of oocytes.[116] The number of oocytes is reduced to less than 10% of the normal number at birth in *Wnt-4* deficient females. At 14.5 d p.c., germ cell number is normal indicating that the migration and early proliferation is unaffected. Loss of oocytes probably causes a partial sex-reversal of the supporting cell lineages as has been reported in other cases.[118,119] In line with this, Sertoli cell markers become expressed

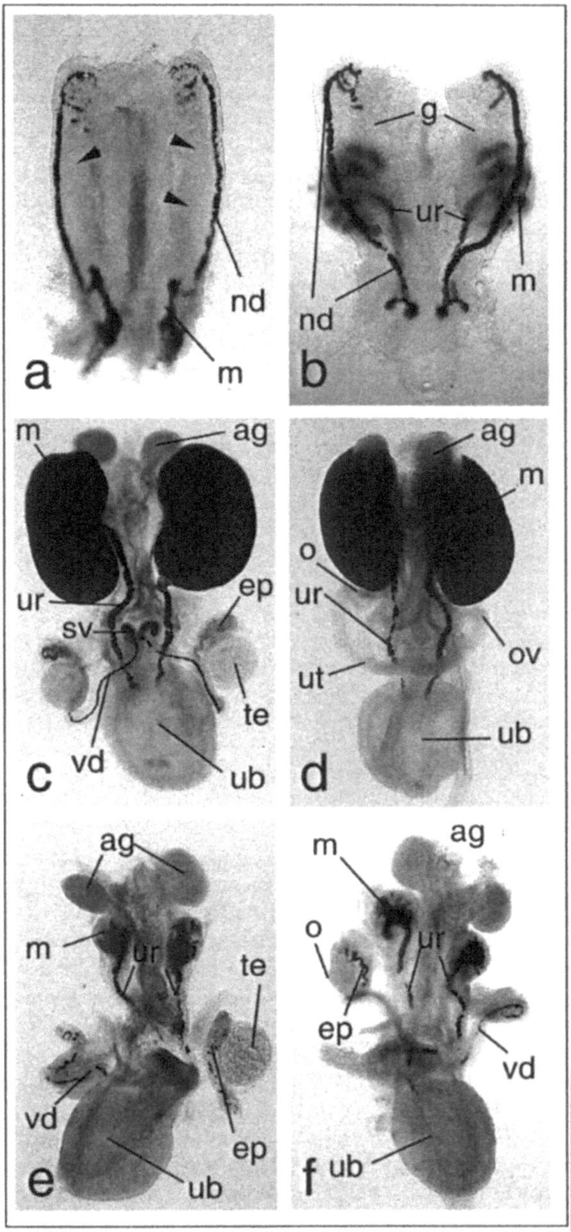


Figure 7. *Wnt-4* function in female reproductive tract development. Entire urogenital systems (UGS) were isolated, and the nephric ducts were visualized by β-galactosidase activity of a *lacZ* marker gene expressed under the control of a nephric duct-specific *Pax-2* enhancer.[128] (a) UGS isolated from an 11.5 d p.c. embryo. The ureter has budded at the caudal end of the nephric duct. It has started to invade the metanephric blastema and branched once. Gonads (arrowhead) are sexually indifferent, long sausage shaped structures adjacent to the nephric ducts. (b) UGS of a 12.5 d p.c. embryo (sexual indifferent stage). (c) UGS of an 18.5 d p.c. control male embryo. The nephric ducts have differentiated into epididymis and vas deferens. The testes have matured from the gonad anlagen and moved caudally. (d) UGS of an 18.5 d p.c. control female embryo. The nephric ducts have completely regressed and the Müllerian ducts have matured. (e) UGS of an 18.5 d p.c. *Wnt-4* mutant male. The genital system is normal, whereas the metanephric kidneys are hypoplastic. (f) UGS of an 18.5 d p.c. *Wnt-4* mutant female. The metanephric kidneys are hypoplastic. The genital system is characterized by absence of Müllerian duct derivatives and persistence of the nephric ducts. Abbreviations: ag, adrenal gland; ep, epididymis; g, gonad; m, metanephric kidney; nd, nephric duct; o, ovary; sv, seminal vesicle; te, testis, ub, urinary bladder; ur, ureter; ut, uterus; vd, vas deferens.

in Wnt-4 ovaries at birth but not at earlier stages. Finally, evidence from human genetics indicates that not only reduced levels of *Wnt-4* but also overexpression of *WNT-4* can cause partial sex reversal phenotypes. Overexpression of *WNT-4* in the male gonad may lead to activation of *DAX-1* resulting in feminization of XY males.[120]

Mice homozygous for a targeted mutation of *Wnt-7a*, and *postaxial hemimelia* mice that carry a mutation in the *Wnt-7a* gene, respectively, are viable but show defects in limb patterning.[121,122] However, no pregnancies were observed when homozygous males or females were mated suggesting male and female infertility. Detailed analysis of *Wnt-7a* expression and mutant analysis have clearly established an important role for Wnt-7a signaling in female reproductive tract development.[123,124]

*Wnt-7a* expression is associated with the development and function of the female reproductive tract during embryogenesis and adult life. In both sexes, expression is detected in the epithelium along the extension of the Müllerian duct from 12.5 d p.c. to 14.5 d p.c. *Wnt-7a* expression is lost in the male with regression of the Müllerian duct. In the female, epithelial-specific expression persists in all Müllerian duct derivatives well into early postnatal life. At around 10 days after birth, expression in the vaginal epithelium is lost. *Wnt-7a* expression however continues in the luminal epithelia of the uterus and the oviduct throughout life showing dependence on steroid hormone variations during the estrus cycle. Levels are highest during diestrus and metestrus and lower during proestrus and estrus suggesting regulation by circulating progesterone.[117,123,124]

Males homozygous for a targeted mutation of *Wnt-7a* establish Müllerian ducts in a process that is dependent on Wnt-4 signaling. Subsequently, the Müllerian ducts however fail to regress after 14.5 d p.c. and persist into adult life. The presence of these ectopic ducts appears to prevent the Wolffian ducts from establishing proper connections with the urogenital sinus. As a result, sperm passage becomes obstructed. Since regression of the Müllerian ducts in males is mediated by MIS signaling, it was tempting to speculate that persistence of the Müllerian ducts in mutant males might be caused by deregulation of MIS signaling. MIS signaling is mediated by MIS type II receptor, which is expressed in the mesenchyme surrounding the Müllerian ducts of both male and female embryos. Expression of *MIS type II receptor* is lost in the mesenchyme of the Müllerian ducts of *Wnt-7a* mutants. It appears therefore that Wnt-7a signaling from the epithelium is required for *MIS type II receptor* expression in periductal mesenchymal cells allowing Müllerian duct regression in the presence of MIS. The persisting Müllerian ducts in the male are thin and undifferentiated with no sign of regionalization along the anterior-posterior axis. Hence, continuous Wnt-7a signaling is required for morphogenesis of the Müllerian duct into a mature reproductive tract system.

As in *Wnt-7a* deficient males, the Müllerian ducts are initially established in *Wnt-7a* mutant females, which show rather normal embryonic development. However, postnatal differentiation of Müllerian ducts occurs abnormally resulting in a dysfunctional reproductive tract. Adult mutant females are characterized by reduced or absent oviducts. The uteri are smaller and thinner as in the wild type and their cytoarchitecture is reminiscent of the vagina. Molecular analysis has confirmed a fate shift from a uterine to a vaginal tissue type in the *Wnt-7a* mutant genital tract. The picture emerging from these studies suggests that Wnt-7a signaling is required for correct regionalization of the reproductive tract. In the mutant, the oviduct acquires uterine features, whereas the uterus displays vaginal characteristics suggesting a posteriorization along the proximal-distal axis. Wnt-7a may act through maintenance of *Hox* gene expression in uterine patterning. In the Wnt-7a mutant, expression of uterine specific *Hoxa-10* and *Hoxa-11* gene expression is initially established, but not maintained in postnatal stages. This in turn may also influence Wnt gene expression. In the mutant, expression of *Wnt-5a* and *Wnt-4* shifts into the uterine epithelium whereas stromal expression of *Wnt-5a* is not maintained. In contrast, *Wnt-7a* signaling is not required for vaginal patterning, which parallels early loss of *Wnt-7a* expression in the vaginal epithelium.

In addition to a role in controlling regionalization along the anterior-posterior axis, *Wnt-7a* may also be involved in radial patterning of the uterus. Endometrial glands, which derive from the luminal epithelium shortly after birth, are absent in the *Wnt-7a* mutant. Lack of these glands may explain the infertility of the mutant female. Newborn uterine mesenchyme differentiates into smooth muscle and stroma, a process, which depends on epithelial signaling. In the *Wnt-7a* mutant, the smooth muscle layer is disorganized. Smooth muscle cells are increased in number but it is unclear whether this may be due to a fate shift from stromal to smooth muscle cells. However, Wnt-7a signaling seems important in maintaining the stroma/smooth muscle boundary. The role of Wnt-7a in uterine radial patterning seems important with respect to effects of steroid hormones during embryogenesis and adulthood. Repression of *Wnt-7a* expression was repressed by fetal exposure to diethylstilbestrol.[125] Reduction of Wnt-7a signaling therefore may lead to myometrial disorganization as suggested by the mutant phenotype.[126]

High levels of *Wnt-5a* expression in the genital system are found in the genital tubercle starting from 10.5 d p.c. Expression is graded along the distal-proximal axis of this outgrowing structure, with highest levels at the distal end. It is maintained at least until 14.5 d p.c. Targeted disruption of the *Wnt-5a* gene clearly established a role for the gene in the formation of the external genitalia.[127] Homozygous mutants are born, but die shortly after birth. They are characterized by a complete lack of external genitalia. Closer analysis during development revealed that the genital tubercle is established but that distal extension of the structure is impaired. Wnt-5a not only plays an important role in genital tubercle outgrowth but seems to regulate tissue outgrowth in a more general sense. Both outgrowth of the primary body axis as well as that of secondary body extensions like limbs and branchial arches is severely impaired in the mutant. Thus, graded expression of *Wnt-5a* in all of these structures may be instrumental in establishing gradients of proliferative activities along different body axes.

Careful analysis of *Wnt* gene expression in female reproductive duct development also identified *Wnt-5a.*[117] At 17 d p.c., expression is restricted to the mesenchyme of the genital tract. In early postnatal days, *Wnt-5a* transcripts levels are maintained in the uterine stroma outside the smooth muscle forming region, decline form the vaginal stroma, and are found in the epithelium and mesenchyme of the oviduct. In the adult female reproductive system, *Wnt-5a* expression pattern changes during the estrous cycle suggesting that levels of steroid hormones regulate *Wnt-5a* expression. During diestrus, expression is detected both in the stroma and epithelium of the uterus. By proestrus, transcripts are confined to the stroma, but by estrus, epithelial expression reappears. Vaginal expression follows a different pattern. Here, expression in epithelium and stroma is low during diestrus, but becomes higher during proestrus. During estrus, expression is restricted and low in the stroma, but becomes undetectable in either compartment in metestrus. Early lethality of *Wnt-5a* homozygotes has precluded analysis of *Wnt-5a* requirement in female reproductive duct development.

## Sexy Wnts: Losing Them Makes a Difference

Loss-of-function studies in the mouse have revealed an unprecedented role for Wnt signaling in mammalian genital development. The requirement for *Wnt-4* gene function in genital development is three fold. First, Wnt-4 signaling controls Müllerian duct development in both sexes. Second, it suppresses Leydig cell differentiation and testosterone production in the female gonad, thus preventing maintenance of the Wolffian ducts in the female embryo. Third, *Wnt-4* is is essential for oocyte maturation. Similar to *Wnt-4, Wnt-7a* gene function is crucially required at several stages of genital development. Wnt-7a mediates regression of the Müllerian duct in males by inducing MIS receptor type II expression in the periductal mesenchyme. In addition, *Wnt-7a* is required for postnatal differentiation of the Müllerian duct. Here, Wnt-7a signaling is required for correct anterior-posterior regionalization as well as radial patterning of the uterus. Finally, *Wnt-5a* plays a more general, in regulating the outgrowth of the external genitalia, which is hampered in *Wnt-5a*-deficient animals most likely by a lack of cell proliferation.

In all three cases, little progress has been reported beyond the detailed description of the phenotypic alterations in the knockout mutants. Neither do we know anything about the Fz receptors mediating Wnt signaling in these situations, nor do we know about the intracellular signaling pathways involved or the downstream targets regulated. Additionally, early phenotypes have precluded the analysis of later requirements in the genital systems. A problem which may be tackled by tissue-specific gene targeting approaches. While in many organ systems including the kidneys as discussed above, studies in lower vertebrates have been complementary in functional characterization of *Wnt* genes, the genital system has been largely neglected in this respect.

## A Wnty but Promising Future

Expression analyses, functional studies including gene targeting approaches and gain-of-function studies in the mouse and in lower vertebrates have given us a deep appreciation of the importance of Wnt signaling in vertebrate kidney and genital development. Analysis of *Wnt* gene function has opened the doors to important biological problems in organogenesis. We are far from knowing what all is present behind these doors but the path is laid and the tools are now available to explore all the details hidden in the dark.

### *Acknowledgements*

A.W.B. would like to thank Hedyeh Ghanbari and Didier Saulnier for providing photographs for the Figure 5 and Doug Melton, Randy Moon, Edgar Pera, De-Li Shi, and Sergei Sokol for plasmids. Research in the laboratory of A.W.B. is supported by grants of the ETH Zürich and the Swiss National Science Foundation (No. 31-061766.00). Research in the laboratory of A. K. is supported by the Deutsche Forschungsgemeinschaft.

## References

1. Saxén L. Organogenesis of the kidney. Cambridge:, Cambridge University Press, 1987.
2. Kriz W, Koepsell H. The structural organization of the mouse kidney. Z. Anat Entwicklungsgesch 1974; 144:137-163.
3. Lechner MS, Dressler GR. The molecular basis of embryonic kidney development. Mech Dev 1997; 62:105-120.
4. Lipschutz JH. Molecular development of the kidney: a review of the results of gene disruption studies. Am J Kidney Dis 1998; 31:383-397.
5. Müller U, Brändli A. Cell adhesion molecules and extracellular-matrix constituents in kidney development and disease. J Cell Science 1999; 112:3855-3867.
6. Schedl A, Hastie ND. Cross-talk in kidney development. Curr Opin Genet Dev 2000; 10:543-549.
7. Carroll TJ, McMahon AP. Secreted molecules in metanephric induction. J Am Soc Nephrol 2000; 11:116-119.
8. Burrow CR. Regulatory molecules in kidney development. Pediatr Nephrol 2000; 14:240-253.
9. Bates CM. Kidney development: regulatory molecules crucial to both mice and men. Mol Genet Metab 2000; 71:391-396.
10. Grobstein C. Inductive epithelio-mesenchymal interactions in cultured organ rudiments of the mouse. Science 1953; 118:52-55.
11. Kispert A, Vainio S, Shen L et al. Proteoglycans are required for maintenance of Wnt-11 expression in the ureter tips. Development 1996; 122:3627–3637.
12. Sainio K, Suvanto P, Davies J et al. Glial-cell-line-derived neurotrophic factor is required for bud initiation from ureteric epithelium. Development 1997; 124:4077-4087.
13. Xu PX, Adams J, Peters H et al. Eya1-deficient mice lack ears and kidneys and show abnormal apoptosis of organ primordia. Nat Genet 1999; 23:113-117.
14. Brophy PD, Ostrom L, Lang KM, Dressler GR. Regulation of ureteric bud outgrowth by Pax2-dependent activation of the glial derived neurotrophic factor gene. Development 2001; 128:4747-4756.

15. Barasch J, Yang J, Ware CB et al. Mesenchymal to epithelial conversion in rat metanephros is induced by LIF. Cell 1999; 99:377-386.
16. Plisov SY, Yoshino K, Dove LF et al. TGF_2, LIF and FGF2 cooperate to induce nephrogenesis. Development 2001; 128:1045-1057.
17. Stark K, Vainio S, Vassileva G, McMahon AP. Epithelial transformation of metanephric mesenchyme in the developing kidney regulated by Wnt-4. Nature 1994; 372:679–683.
18. Herzlinger D, Qiao J, Cohen D et al. Induction of kidney epithelial morphogenesis by cells expressing Wnt-1. Dev Biol 1994; 166:815-818.
19. Parr BA, Shea MJ, Vassileva G, McMahon AP. Mouse Wnt genes exhibit discrete domains of expression in the early embryonic CNS and limb buds. Development 1993; 119:247-261.
20. Kispert A, Vainio S, McMahon AP. Wnt-4 is a mesenchymal signal for epithelial transformation of metanephric mesenchyme in the developing kidney. Development 1998; 125:4225-4234.
21. Leimeister C, Bach A, Gessler M. Developmental expression patterns of mouse sFRP genes encoding members of the secreted frizzled related protein family. Mech Dev 1998; 75:29-42.
22. Lescher B, Haenig B, Kispert A. sFRP-2 is a target of the Wnt-4 signaling pathway in the developing metanephric kidney. Dev Dyn 1998; 213:440-451.
23. Yoshino K, Rubin JS, Higinbotham KG et al. Secreted Frizzled-related proteins can regulate metanephric development. Mech Dev 2001; 102:45-55.
24. Saadi-Kheddouci S, Berrebi D, Romagnolo B et al. Early development of polycystic kidney disease in transgenic mice expressing an activated mutant of the beta-catenin gene. Oncogene 2001; 20:5972-5981.
25. Kim E, Arnould T, Sellin LK et al. The polycystic kidney disease 1 gene product modulates Wnt signaling. J Biol Chem 1999; 274:4947-4953.
26. Vainio SJ, Itäranta PV, Peräsaari JP, Uusitalo MS. Wnts as kidney tubule inducing factors. Int J Dev Biol 1999; 43:419-423.
27. Vainio SJ, Uusitalo MS. A road to kidney tubules via the Wnt pathway. Pediatr Nephrol 2000; 15:151-156.
28. Davies J, Lyon M, Gallagher J, Garrod D. Sulphated proteoglycan is required for collecting duct growth and branching morphogenesis but not nephron formation during kidney development. Development 1995; 125:1507-1517.
29. Bullock SL, Fletcher JM, Beddington RS, Wilson VA. Renal agenesis in mice homozygous for a gene trap mutation in the gene encoding heparan sulfate 2-sulfotransferase. Genes Dev 1998; 12:1894-1906.
30. Pepicelli CV, Kispert A, Rowitch DH, McMahon AP. GDNF induces branching and increased cell proliferation in the ureter of the mouse. Dev Biol 1997; 192:193-198.
31. Ehrenfels CW, Carmillo PJ, Orozco O et al. Perturbation of RET signaling in the embryonic kidney. Dev Genet 1999; 24:263-272.
32. Parr BA, Cornish VA, Cybulsky MI, McMahon AP. Wnt7b regulates placental development in the mice. Dev Biol 2001; 237:324-332.
33. Lin Y, Liu A, Zhang S et al. Induction of ureter branching as a response to Wnt-2b signaling during early kidney organogenesis. Dev Dyn 2001; 222:26-39.
34. Mauch TJ, Yang G, Wright M, Schoenwolf GC. Signals from trunk paraxial mesoderm induce intermediate mesoderm to express Pax-2. J Am Soc Nephrol 1999; 10:408A.
35. Obara-Ishihara T, Kuhlmann J, Niswander L, Herzlinger D. The surface ectoderm is essential for nephric duct formation in intermediate mesoderm. Development 1999; 126:1103-1108.
36. Brändli AW. Towards a molecular anatomy of the Xenopus pronephric kidney. Int J Dev Biol 1999; 43:381-395.
37. Carroll TJ, Wallingford J, Seufert D, Vize PD. Molecular regulation of pronephric development. Curr Top Dev Biol 1999; 44:67-100.
38. Drummond IA. The zebrafish pronephros: a genetic system for studies of kidney development. Pediat Nephrol 2000; 14:428-435.
39. Drummond IA, Majumdar A, Hentschel H et al. Early development of the zebrafish pronephros and analysis of mutations affecting pronephric funtion. Development 1998; 125:4655-4667.
40. Carroll TJ, Vize PD. Wilms' tumor suppressor gene is involved in the development of disparate kidney forms: evidence from expression in the Xenopus pronephros. Dev Dyn 1996; 206:131-138.

41. Eid SR, Brändli AW. Xenopus Na,K-ATPase: primary sequence of the b2 subunit and in situ localization of a1, b1, and g expression during pronephric kidney development. Differentiation 2001; 68:115-125.
42. Eid SR, Terrettaz A, Nagata K, Brändli AW. Embryonic expression of Xenopus SGLT-1L, a novel member of the solute carrier family 5 (SLC5), is confined to tubules of the pronephric kidney. Int J Dev Biol 2002; 46:177-184.
43. Heller N, Brändli AW. Xenopus Pax-2 displays multiple splice forms during embryogenesis and pronephric kidney development. Mech Dev 1997; 69:83-104.
44. Heller N, Brändli AW. Xenopus Pax-2/5/8 orthologues: novel insights into Pax gene evolution and identification of Pax-8 as the earliest marker for otic and pronephric cell lineages. Dev Genet 1999; 24:208-219.
45. Heasman J, Kofron M, Wylie C. β-catenin signaling activity dissected in the early Xenopus embryo: a novel antisense approach. Dev Biol 2000; 222:124-134.
46. Nasevicius A, Ekker SC. Effective targeted gene knockdown in zebrafish. Nat Genet 2000; 26:216-220.
47. Moriya N, Uchiyama H, Asashima M. Induction of pronephric tubules by activin and retinoic acid in presumptive ectoderm of Xenopus laevis. Develop Growth Differ 1993; 35:123-128.
48. Herzlinger D. Inductive interactions during kidney development. Semin Nephrol 1995; 15:255-262.
49. Krauss S, Johansen T, Korzh V, Fjose A. Expression of the zebrafish paired box gene pax[zf-b] during early neurogenesis. Development 1991; 113:1193-1206.
50. Cadigan KM, Nusse R. Wnt signaling: a common theme in animal development. Genes Dev 1997; 11:3286-3305.
51. Uusitalo M, Heikkilä M, Vainio S. Molecular genetic studies of Wnt signaling in the mouse. Exp Cell Res 1999; 253:336-348.
52. Wodarz A, Nusse R. Mechanisms of Wnt signaling in development. Annu Rev Cell Dev Biol 1998; 14:59-88.
53. Kawakami Y, Capdevila J, Büscher D et al. WNT signals control FGF-dependent limb initiation and AER induction in the chick embryo. Cell 2001; 104:891-900.
54. Wolda SL, Moon RT. Cloning and developmental expression in Xenopus laevis of seven additional members of the Wnt family. Oncogene 1992; 7:1941-1947.
55. Blader P, Strähle U, Ingham PW. Three Wnt genes expressed in a wide variety of tissues during development of the zebrafish, Danio rerio: developmental and evolutionary perspectives. Dev Genes Evol 1996; 206:3-13.
56. Gavin BJ, McMahon JA, McMahon AP. Expression of multiple novel Wnt-1/int-related genes during fetal and adult mouse development. Genes Dev 1990; 4:2319-2332.
57. Liu A, Majumdar A, Schauerte HE et al. Zebrafish wnt4b expression in the floor plate is altered in sonic hedgehog and gli-2 mutants. Mech Dev 2000; 91:409-413.
58. McGrew LL, Otte AP, Moon RT. Analysis of Xwnt-4 in embryos of Xenopus laevis: A Wnt family member expressed in the brain and floor plate. Development 1992; 115:463-473.
59. Schubert M, Holland LZ, Holland ND. Characterization of two Amphioxus Wnt genes (AmphiWnt4 and AmphiWnt7b) with early expression in the developing nervous system. Dev Dyn 2000; 217:205-215.
60. Ungar AR, Kelly GM, Moon RT. Wnt4 affects morphogenesis when misexpressed in the zebrafish embryo. Mech Dev 1995; 52:153-164.
61. Yoshioka H, Ohuchi H, Nohno T et al. Regional expression of the Cwnt-4 gene in developing chick central nervous system in relationship to the diencephalic neuromere D2 and a dorsal domain of the spinal cord. Bioch Biophys Res Commun 1994; 203:1581-1588.
62. Stark MR, Rao MS, Schoenwolf GC et al. Frizzled-4 expression during chick kidney development. Mech Dev 2000; 98:121-125.
63. Carroll TJ, Wallingford JB, Vize PD. Dynamic patterns of gene expression in the developing pronephros of Xenopus laevis. Dev Genet 1999; 24:199-207.
64. Saulnier DME, Ghanbari H, Brändli AW. Essential function of Wnt-4 for tubologenesis in the Xenopus pornephric kidney. Dev Biol 2002; 248:13-28.

65. Kozmik Z, Holland ND, Kalousova A, Paces J, Schubert M, Holland LZ. Characterization of an amphioxus paired box gene, AmphiPax2/5/8: Developmental expression patterns in optic support cells, nephridium, thyroid-like structures, and pharyngeal gill slits, but not in the midbrain-hindbrain region. Development 1999; 126:1295-1304.
66. Cauthen CA, Berdougo E, Sandler J, Burrus LW. Comparative analysis of the expression patterns of Wnts and Frizzleds during myogenesis in chick embryos. Mech Dev 2001; 104:133-138.
67. Patterson LT, Pembaur M, Potter SS. Hoxa11 and Hoxd11 regulate branching morphogenesis of the ureteric bud in the developing kidney. Development 2001; 128:2153-2161.
68. Hollyday M, McMahon JA, McMahon AP. Wnt expression patterns in chick embryo nervous system. Mech Dev 1995; 52:9-25.
69. Chang CB, Hemmati-Brivanlou A. Neural crest induction by Xwnt7B in Xenopus. Dev Biol 1998; 194:129-134.
70. Tanda N, Ohuchi H, Yoshioka H et al. A chicken WNT gene, WNT-11, is involved in dermal development. Biochem Biophys Res Commun 1995; 211:123-129.
71. Ku M, Melton DA. Xwnt-11: a maternally expressed Xenopus wnt gene. Development 1993; 119:1161-1173.
72. Makita R, Mizuno T, Koshida S et al. Zebrafish wnt11: pattern and regulation of the expression by the yolk cell and No tail activity. Mech Dev 1998; 71:165-176.
73. Heisenberg CP, Tada M, Rauch GJ et al. Silberblick/Wnt11 mediates convergent extension movements during zebrafish gastrulation. Nature 2000; 405:76-81.
74. Wang Y, Macke JP, Abella BS et al. A large family of putative transmembrane receptors homologous to the product of the Drosophila tissue polarity gene frizzled. J Biol Chem 1996; 271:4468-4476.
75. Borello U, Buffa V, Sonnino C et al. Differential expression of the Wnt putative receptors Frizzled during mouse somitogenesis. Mech Dev 1999; 89:173-177.
76. Wang Y, Spörle R, Paperna T et al. Characterization and expression pattern of the frizzled gene Fzd9, the mouse homolog of FZD9 which is deleted in Williams-Beuren syndrome. Genomics 1999; 57:235-248.
77. Ishikawa T, Tamai Y, Zorn AM et al. Mouse Wnt receptor gene Fzd5 is essential for yolk sac and placental angiogenesis. Development 2001; 128:25-33.
78. Sala CF, Formenti E, Terstappen GC, Caricasole A. Identification, gene structure, and expression of human frizzled-3 (FZD3). Biochem Biophys Res Comm 2000; 273:27-34.
79. Stark MR, Biggs JJ, Schoenwolf GC, Rao MS. Characterization of avian frizzled genes in cranial placode development. Mech Dev 2000; 93:195-200.
80. Wang Y, Huso D, Cahill H et al. Progressive cerebellar, auditory, and esophageal dysfunction caused by targeted disruption of the frizzled-4 gene. J Neurosci 2001; 21:4761-4771.
81. Brown JD, Hallagan SE, McGrew LL et al. The maternal Xenopus β-catenin signaling pathway, activated by frizzled homologs, induces goosecoid in a cell non-autonomous manner. Develop Growth Differ 2000; 42:347-357.
82. Shi DL, Goisset C, Boucaut JC. Expression of Xfz3, a Xenopus frizzled family member, is restricted to the early nervous system. Mech Dev 1998; 70:35-47.
83. Shi DL, Boucaut JC. Xenopus frizzled 4 is a maternal mRNA and its zygotic expression is localized to the neuroectoderm and trunk lateral plate mesoderm. Mech Dev 2000; 94:243-245.
84. Djiane A, Riou J, Umbhauer M et al. Role of frizzled 7 in the regulation of convergence extension movements during gastrulation in Xenopus laevis. Development 2000; 127:3091-3100.
85. Medina A, Reintsch W, Steinbeisser H. Xenopus frizzled 7 can act in canonical and non-canonical Wnt signaling pathways: implications on early patterning and morphogenesis. Mech Dev 2000; 92:227-237.
86. Sumanas S, Strege P, Heasman J, Ekker SC. The putative Wnt receptor Xenopus frizzled-7 functions upstream of β-catenin in vertebrate dorsoventral mesoderm patterning. Development 2000; 127:1981-1990.
87. Wheeler GN, Hoppler S. Two novel Xenopus frizzled genes expressed in developing heart and brain. Mech Dev 1999; 86:203-207.
88. Deardorff MA, Tan C, Conrad LJ, Klein PS. Frizzled-8 is expressed in the Spemann organizer and plays a role in early morphogenesis. Development 1998; 125:2687-2700.

89. El-Messaoudi S, Renucci A. Expression pattern of the frizzled 7 gene during zebrafish embryonic development. Mech Dev 2001; 102:231-234.
90. Kim SH, Park HC, Yeo SY et al. Characterization of two frizzled8 homologues in embryonic shield and prechordal plate of zebrafish embryos. Mech Dev 1998; 78:193-198.
91. Nasevicius A, Hyatt TM, Hermanson SB, Ekker SC. Sequence, expression, and location of Zebrafish frizzled 10. Mech Dev 2000; 92:311-314.
92. Sumanas S, Kim HJ, Hermanson S, Ekker SC. Zebrafish frizzled-2 morphant displays defects in body axis elongation. Genesis 2001; 30:114-118.
93. Van Raay TJ, Wang Y, Stark MR et al. Frizzled 9 is expressed in neural precursor cells in the developing neural tube. Dev Genes Evol 2001; 211:453-457.
94. Terry K, Magan H, Baranski M, Burrus LW. Sfrp-1 and sfrp-2 are expressed in overlapping and distinct domains during chick development. Mech Dev 2000; 97:177-182.
95. Bradley L, Sun B, Collins-Racie L et al. Different activities of the Frizzled-related proteins frzb2 and sizzled2 during Xenopus anteroposterior patterning. Dev Biol 2000; 227:118-132.
96. Pera EM, De Robertis EM. A direct screen for secreted proteins in Xenopus embryos identifies distinct activities for Wnt antagonists Crecent and Frzb-1. Mech Dev 2000; 96:183-195.
97. Shibata M, Ono H, Hikasa H et al. Xenopus crescent encoding a Frizzled-like domain is expressed in the Spemann organizer and pronephros. Mech Dev 2000; 96:243-246.
98. Hiort O, Holterhus PM. The molecular basis of male sexual differentiation. Eur J Endocrinol 2000; 142:101-110.
99. Hughes IA. Minireview: Sex differentiation. Endocrinology 2001; 142:3281-3287.
100. Parker KL, Schimmer B, Schedl A. Genes essential for early events in gonadal development. In: Scherer G, Schmid M, eds. Genes and mechanisms in vertebrate sex determination. Basel: Birkhäuser Verlag, 2001:11-24.
101. Veitia RA, Salas-Cortés L, Ottolenghi C et al. Testis determination in mammals: more questions than answers. Mol Cell Endocrinol 2001; 179:3-16.
102. Starz-Gaiano M, Lehmann R. Moving towards the next generation. Mech Dev 2001; 105:5-18.
103. Shawlot W, Behringer RR. Requirement for Lim1 in head-organizer function. Nature 1995; 374:425-430.
104. Tsang TE, Shawlot W, Kinder SJ et al. Lim1 activity is required for intermediate mesoderm differentiation in the mouse embryo. Dev Biol 2000; 223:77-90.
105. Kreidberg JA, Sariola H, Loring JM et al. WT-1 is required for early kidney development. Cell 1993; 74:679-691.
106. Luo X, Ikeda Y, Parker KL. A cell-specific nuclear receptor is essential for adrenal and gonadal development and sexual differentiation. Cell 1994; 77:481-490.
107. Shen WH, Moore CCD, Ikeda Y et al. Nuclear receptor steroidogenic factor 1regulates MIS gene expression: a link to the sex determination cascade. Cell 1994; 77:651-661.
108. Bakke M, Zhao L, Hanley NA, Parker KL. SF-1: a critical mediator of steroidogenesis. Mol Cell Endocrinol 2001; 171:5-7.
109. Nachtigal MW, Hirokawa Y, Enjeart-VanHouten DL et al. Wilms' tumor 1 and Dax-1 modulate the orphan receptor SF-1in sex-specific gene expression. Cell 1998; 93:445-454.
110. Koopman P, Bullejos M, Bowles J. Regulation of male sexual development by Sry and Sox9. J Exp Zool 2001; 290:463-474.
111. Ostrer H. Sexual differentiation. Semin Reprod Med 2000; 18:41-49.
112. Vilain E. Genetics of sexual development. Annu. Rev Sex. Res 2000; 11:1-25.
113. Albrecht KH, Eicher EM. Evidence that Sry is expressed in pre-Sertoli cells and Sertoli and granulosa cells have a common precursor. Dev Biol 2001; 240:92-107.
114. Haqq CM, King CY, Ukiyama E et al. Molecular basis of mammalian sexual determination: activation of Mullerian inhibiting substance gene expression by SRY. Science 1994; 266:1494-500.
115. Josso N, di Clemente N, Gouedard L. Anti-Mullerian hormone and its receptors. Mol Cell Endocrinol 2001; 179:25-32.
116. Vainio S, Heikkilä M, Kispert A et al. Female development in mammals is regulated by Wnt-4 signaling. Nature 1999; 397:405-409.
117. Miller C, Pavlova A, Sassoon DA. Differential expression patterns of Wnt genes in the murine female reproductive tract during development and the estrous cycle. Mech Dev 1998; 76:91-99.

118. Behringer RR, Cate RL, Froelick GJ et al. Abnormal sexual development in transgenic mice chronically expressing Müllerian inhibiting substance. Nature 1990; 345:167-170.
119. Lyet L, Louis F, Forest MG et al. Ontogeny of reproductive abnormalities induced by deregulation of anti-mullerian hormone expression in transgenic mice. Biol Reprod 1995; 52:444-454.
120. Jordan BK, Mohammed M, Saunders TC et al. Up-regulation of WNT-4 signaling and dosage-sensitive sex reversal in humans. Am J Hum Genet 2001; 68:1102-1109.
121. Parr BA, McMahon AP. Dorsalizing signal Wnt-7a required for normal polarity of D-V and A-P axes of the mouse limb. Nature 1995; 374:350-353.
122. Parr BA, Avery EJ, Cygan JA, McMahon AP. The classical mouse mutant postaxial hemimelia results from a mutation in the Wnt-7a gene. Dev Biol 1998; 202:228-234.
123. Parr BA, McMahon AP. Sexually dimorphic development of the mammalian reproductive tract requires Wnt-7a. Nature 1998; 395:707-710.
124. Miller C, Sassoon DA. Wnt-7a maintains appropiate uterine patterning during the development of the mouse female reproductive tract. Development 1998; 125:3201-3211.
125. Miller C, Degenhardt K, Sassoon DA. Fetal exposure to DES results in de-regulation of Wnt7a during uterine morphogenesis. Nat Genet 1998; 20:228-230.
126. Li S, Chiang TC, Davis GR et al. Decreased expression of Wnt7a mRNA is inversely associated with the expression of estrogen receptor-α in human uterine leiomyoma. J Clin Endocr Metabolism 2001; 86:454-457.
127. Yamaguchi TP, Bradley A, McMahon AP, Jones S. A Wnt5a pathway underlies outgrowth of multiple structures in the vertebrate embryo. Development 1999; 126:1211-1223.
128. Kuschert S, Rowitch DH, Haenig B et al. Pax-2 regulatory sequences that direct transgene expression in the Wolffian duct and its derivatives. Dev Biol 2001; 229:128-140.
129. Balinsky BI. An Introduction to Embryology. 5th ed. Philadelphia: Saunders College Publishing, 1981.
130. Gilbert SF. Developmental Biology. Sunderland: Sinauer Ass., 1997.

# Chapter 9

# Wnts in Muscle Development

**Bodo Christ and Ruijin Huang**

## Abstract

Wnt proteins control different steps in myogenesis in combination with other signaling molecules such as Sonic hedgehog (Shh) and Bone Morphogenetic Proteins (BMPs). Wnts are involved in the epithelialization of the paraxial mesoderm, the dorsoventral and mediolateral compartmentalization of somites, the mediolateral patterning of the dermomyotomes, and the specification of skeletal muscle cells. Wnt-1 and Wnt-3a which are expressed in the neural tube medialize the dermomyotome and induce epaxial myogenesis in concert with Shh. Wnts produced by the surface ectoderm such as Wnt-6 are capable to specify hypaxial muscle cells in combination with BMP-4 that lateralizes the dermomyotome.

## Introduction

Skeletal muscle derives from the somites that bud off as segmental spheres from the paraxial mesoderm in a cranial to caudal progression. The newly formed somite consists of an epithelial wall and a mesenchymal core and is surrounded by extracellular matrix (ECM) that is condensed at the external surface of the somite to form a basement membrane. The ECM connects the somite with adjacent structures: neural tube, notochord, surface ectoderm, endoderm, aorta, and Wolffian duct. Only laterally, there is a continuous cellular bridge that links the somite with the intermediate mesoderm. The somite is patterned along both its initial dorsoventral and mediolateral axes in response to local signals from surrounding tissues.[1] These signals have to pass through the ECM which may serve to modulate the local concentration of growth factors.[2-3]

The first molecular evidence of dorsoventral patterning is the downregulation of *Pax-3* in the ventral part of the somite including the mesenchymal core and the upregulation of *Pax-1* in this area.[4-7] Subsequently, the ventral half of the somite undergoes an epithelio-mesenchymal transition to form the sclerotome that gives rise to the vertebral column and ribs. The dorsal half of the somite retains its epithelial structure and continues to express *Pax-3*. This compartment, the dermomyotome, has the potential to form muscle, dermis, blood vessels, and the scapular blade.[1,8] Along its mediolateral extension the dermomyotome is patterned into a medial and a lateral part. In the chicken embryo, cells of the medial part start to express the *myogenic determination factors (MDFs) Myf-5* and *MyoD* and form the epaxial myotome from which the back musculature develops.[9] The hypaxial muscle, on the other hand, originates from the lateral part of the dermomyotome in which *Pax-3* and *Myf-5* is found to be expressed.[9-12] At limb levels and in the occipital region replicating muscle precursors detach from the lateral dermomyotome and invade the limb buds and the tongue anlage, respectively, where these cells continue to proliferate and later differentiate.[13] At interlimb level, lateral

*Wnt Signaling in Development*, edited by Michael Kühl.

dermomyotome cells form the hypaxial myotome. The cells of both the early epaxial and hypaxial myotome are postmitotic. They express *MDFs* and muscle-specific proteins such as desmin and can be regarded as terminally differentiated myocytes (Fig. 1).[11]

The cells of a newly formed somite retain plasticity with respect to their developmental fate. If the somites are rotated, or the dorsal half replaced by a ventral half, or the medial and lateral halves exchanged, the somite cells differentiate according to their new environment.[14-16] These results indicate that the differentiation of somite cells requires signals from cells and tissues that surround the somite.

Summarizing these observations, the early steps of myogenesis can be characterized by an epithelialization of the paraxial mesoderm, somitogenesis, compartmentalization of the somite into dermomyotome and sclerotome, proliferation of *Pax-3* expressing precursor cells, patterning of the dermomyotome along its mediolateral axis, migration of muscle precursors, and differentiation of myoblasts. It is the aim of this review to discuss the involvement of Wnt proteins in these processes.

## Epithelialization of Paraxial Mesoderm and Somite Formation

The paraxial mesoderm at trunk level consists of the somites and the segmental plate that extends from Hensen's node, or tail bud, to the most recently formed somite. The caudal part of the segmental plate is made up of a loosely arranged mesenchyme expressing *Fibroblast growth factor 8 (Fgf8)*.[17-19] In the cranial half of the segmental plate, cells become polarized with their long axes perpendicular to the surface ectoderm and begin to express *Paraxis*, a basic helix-loop-helix transcription factor. Knockout experiments have demonstrated that *Paraxis* is essential for epithelialization of the paraxial mesoderm and somite formation.[20] It has been shown that the initiation of *Paraxis* expression is dependent on inductive signals from the surface ectoderm.[21] This induction might be mediated by Wnts because *Paraxis* expression can be dramatically amplified after implantation of Wnt-1, Wnt-3a, and Wnt-4 secreting cells.[22] Additionally, indirect evidence has implicated the Wnt pathway in the process of segmental plate maturation. *Frizzled-1, Frizzled-2,* and *Frizzled-7* have been shown to be expressed in the segmental plate and the somites.[23] The Wnt antagonist, Wif-1, is expressed in the presomitic mesoderm, and its overexpression in *Xenopus* embryos interferes with proper somite formation.[24] Bellairs et al[25] have shown that cell adhesion increases during segmental plate maturation. Since the canonical Wnt signal transduction pathway leads to an accumulation of b-catenin[26] that links the cadherins to the cytoskeleton, the Wnts have been suggested to promote the process of segmental plate epithelialization.

The Eph family of receptor tyrosin kinases and their ligands, the ephrins, are assumed to be also involved in the regulation of cytoskeletal organization. *EphA4* is expressed in the segmental plate in a region where cells are undergoing changes in cell shape as a prelude of somite formation and patterning of somite boundaries.[27-28] Signals from the surface ectoderm are required for the induction of *EphA4* that is suggested to act downstream of *Paraxis.*[28]

*Wnt-6* is a candidate mediator for a contact-dependent inductive activity of the surface ectoderm on the underlying somite epithelium whose cells extend filamentous processes through the ECM to contact the ectodermal cells[29,30,23] (W. Denetclaw, personal communication).

## Formation, Maintenance, and Patterning of the Dermomyotome

The formation of the dermomyotome is a result of the dorsoventral patterning of the somite which is controlled by an antagonistic action of ventralizing and dorsalizing signals originating from adjacent tissues and traversing the ECM that surrounds the somite (Fig. 2). The notochord and the floor plate of the neural tube provide ventralizing signals, including Sonic hedgehog (Shh), that control sclerotome development.[31-32] Sclerotome formation is characterized by expression of *Pax-1* and an epithelio-mesenchymal transition of somitic cells.[33-35]

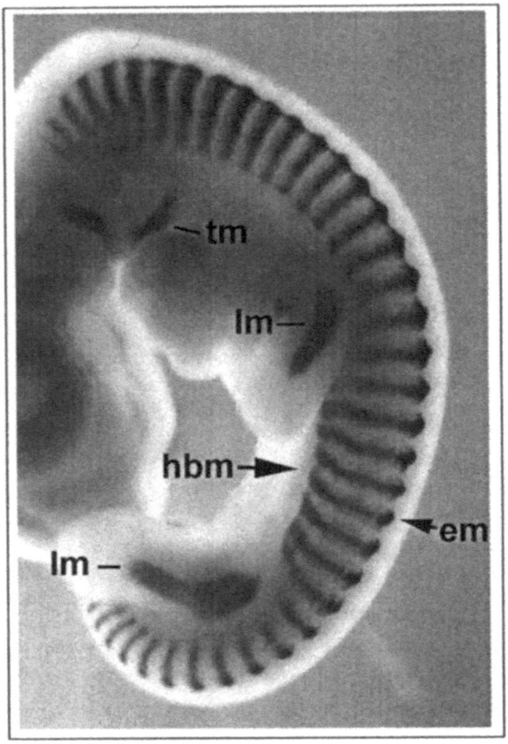


Figure 1. Epaxial and hypaxial muscle domains of a 4-day old chick embryo expressing *MyoD*. Hypaxial muscle: tongue muscle (tm), limb muscle (lm),hypaxial body wall muscle (hbm). em: epaxial muscle.

Dorsalizing signals that induce and maintain the dermomyotome originate from the dorsal part of the neural tube and the surface ectoderm.[29,36] In the absence of these structures, the epithelial organization of the dorsal somite compartment is lost and *Pax-3* expression is downregulated. Fan et al[30] have shown that cells expressing *Wnt-1, Wnt-3a, Wnt-4*, and *Wnt-6* can induce and maintain the expression of the dermomyotome markers *Pax-3, Pax-7,* and *Sim1* in tissue culture.

Ectopically expressed Wnt-1 in the presomitic mesoderm of chick embryos *in ovo* is able to completely repress sclerotomal markers and to enhance and expand the expression of dermomyotomal markers.[37] Delivery of an activated form of b-catenin to somitic mesoderm mimics the effects of Wnt-1. Wagner et al[22] have shown that ectopically implanted Wnt-1, Wnt-3a, and Wnt-4 secreting cells alter the process of dorsoventral somite compartmentalization *in vivo*. Implantation of Wnt-secreting cells ventrally between neural tube/notochord and epithelial somites of chicken embryos, leads to an enlarged dorsal compartment at the expense of the ventral compartment, the sclerotome. This shift of the border between the muscle and the skeleton forming areas was identified by an increase of the *Pax-3* expression domain and a complete loss of *Pax-1* expression. In mouse embryos lacking both *Wnt-1* and *Wnt-3a* the medial part of the dermomyotome is not formed and the expression of the lateral dermomyotome marker gene, *Sim1*, is expanded medially indicating that different members of the Wnt family are required to control the mediolateral pattern of the dermomyotome.[38] It has recently been

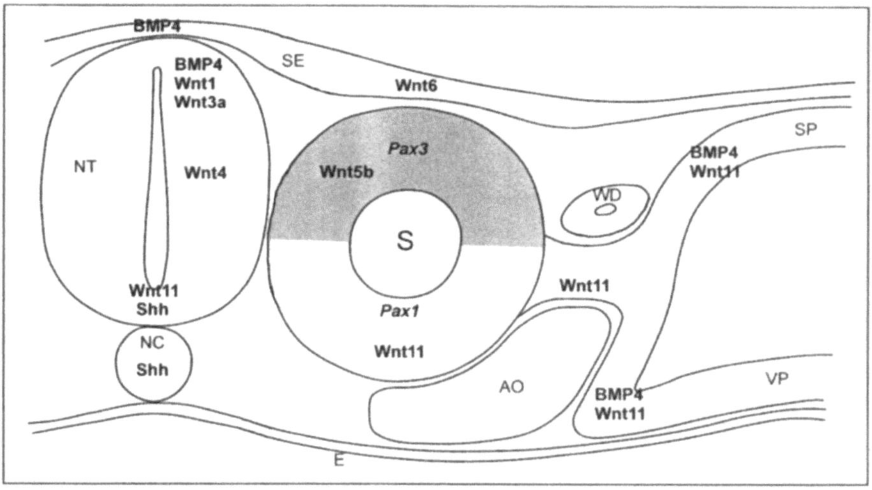


Figure 2. Expression domains of Wnts and other signaling molecules at epithelial somite stage. AO: aorta, E: endoderm, IM: intermediate mesoderm, NC: notochord, NT: neural tube, S: epithelial somite, SE: surface ectoderm, SP: somatopleura, VP: visceral pleura, WD: Wolffian duct.

shown that the development of the dorsal dermis which is a derivative of the medial dermomyotome is dependent on Wnt-1 signaling.[39] Taken together, the results show that Wnt-1, Wnt-3a, Wnt-4, and Wnt-6 induce and maintain dermomyotome development. Moreover, Wnt-1 and Wnt-3a signals are required to medialize the dermomyotome.

## Epaxial Myogenesis

Precursors of the epaxial myotome reside within the medial half of the dermomyotomal epithelium.[16,10,12] Ordahl et al[40] have shown that the dorsomedial lip of the dermomyotome drives growth and morphogenesis of both the epaxial myotome and the overlying dermomyotome. The early myotome consists of postmitotic, differentiated, *MyoD* and desmin expressing myocytes.

The process of epaxial muscle cell differentiation requires signals from axial structures (Figs. 2, 3). After ablation of both the neural tube and notochord, epaxial muscle does not develop.[41-42] Numerous in vitro and in vivo studies have given evidence that a combination of dorsal and ventral signals are necessary to initiate epaxial myogenesis. Notochord and dorsal neural tube are required to turn on and maintain *MyoD* expression in the epaxial myotome of avian embryos in vivo[43-44] and in vitro.[45-47] The notochord signal has been identified to be Sonic hedgehog (Shh).[48]

Two days after ablation of either the entire or the dorsal neural tube, *MyoD* expression is not longer maintained in the epaxial somite domain indicating that at least a second signal is required for the maintenance of epaxial muscle development[15,49,29,44] (R. Huang, unpublished data).

It has been shown that *Wnt-1, Wnt-3a* and *Wnt-4* are expressed in the dorsal half of the neural tube at the time when epaxial myogenesis begins to take place.[50,23] *Wnt-5b* is expressed in cells of the dorsomedial part of the somite and *Wnt-11* in cells of the dorsomedial dermomyotome lip from which the early epaxial myotomal cells arise.[23,51] Expression of *Wnt-1* and *Wnt-3a* in the neural tube are positively regulated by BMP-4 and BMP-7. *Wnt-1* and

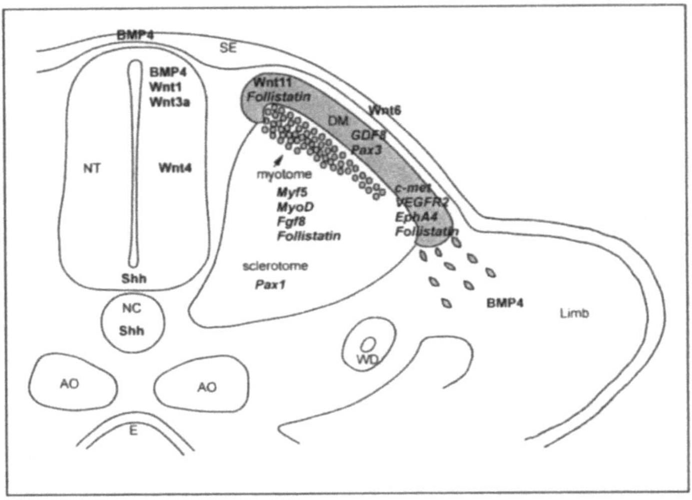


Figure 3. Expression domains of Wnts and other signaling molecules at the stage of a compartmentalized somite. AO: aorta, DM: dermomyotome, E: endoderm, NC: notochord, NT: neural tube, SE: surface ectoderm, WD: Wolffian duct.

*Wnt-3a,* in turn, initiate the expression of *Wnt-11* in the dorsomedial dermomyotome lip.[51,38,52] *Wnt-11* has recently been shown to be additionally expressed in the ventral somite colocalized with *Pax-1* and in the intermediate mesoderm.[23]

In explants of presomitic mesoderm from mouse embryos, Shh acts in concert with Wnt-1 or Wnt-7a to induce myogenesis as assayed by myosin light chain 3F,[53] which is consistent with the induction of *MyoD* expression in paraxial mesoderm explants of avian embryos.[54] The requirement of Wnt proteins for muscle development has been confirmed by transplacental delivery of the Wnt antagonist FrzB1 in the mouse.[55] Overexpression of FrzB1 causes a strong reduction in myogenesis in the embryo in which several genes downstream of the Wnt signaling pathway, such as *En1, Noggin,* and *Myf-5,* are downregulated. In the chick, the Wnt antagonist sFRP-2 has been shown to be expressed in the dorsomedial and ventrolateral lips of the dermomyotome where it may promote proliferation and prevent precocious differentiation of myogenic dermomyotome cells.[56] Lee et al[52] have shown that Shh not only upregulates its own receptor, *Ptc1,* but also *sFRP-2* which reduces the dermomyotome inducing activity of Wnt-1 and Wnt-4 but not that of Wnt-3a.

Epaxial myogenesis requires BMP-4 antagonizing signaling.[57] BMP-4 is expressed in the lateral plate mesoderm and tends to lateralize the dermomyotome (Fig. 3). The lateralization of the medial part of the dermomyotome is prevented by *Noggin* that is expressed in the dorsomedial dermomyotome lip in response to Wnt-1.[57,51,58] In the *Wnt-1/Wnt-3a* double knock out mouse, *Noggin* and *Wnt-11* expression is lost and the medial dermomyotome is lacking.

Wnt proteins have an additional effect on epaxial myogenesis by modulating the Shh signal transduction through the regulation of *Gli2* and *Gli3,* genes encoding zinc finger transcription factors that are effectors of Shh signaling in responding cells.[59] An overexpression of Wnt-1, 3a and 4 in the avian embryo at the level of the epithelial somite results in a considerable enlargement of the epaxial muscle domain at the expense of the sclerotome.[22] The shift of the border between the muscle-forming and skeleton-forming compartments is accompanied by an up-regulation of *Paraxis, Pax-3, MyoD*, and desmin in the premuscular compartment and a complete down-regulation of the sclerotome marker *Pax-1*.

## Hypaxial Myogenesis

Hypaxial muscle originates from the lateral part of the dermomyotome and gives rise to the muscles of the limbs, the tongue, the diaphragm, and the ventrolateral body wall.[1,60] At interlimb levels, the lateral myotome develops a "muscle bud" in conjunction with the adjacent dermomyotome. These muscle buds grow as epithelial structures into the somatopleure and give rise to thoracic and abdominal muscles. The attribute "hypaxial" is also applied to the limb muscle. At limb levels, the lateral dermomyotome disintegrates to release individually migrating muscle precursor cells that invade the limb buds where they continue to proliferate and later on differentiate.[13] Unlike epaxial myogenesis, early steps of hypaxial myogenesis do not require Shh signaling pathways.[42]

The mechanisms of hypaxial muscle cell specification are not well understood to date. It has been shown that the surface ectoderm is required to form the hypaxial dermomyotome.[28] BMP-4 expressed in the lateral plate mesoderm, is required to induce the expression of the lateral dermomyotome marker *Sim1* and also to induce lateral characteristics[61-62] including the expression of *Lbx1* and *c-Met*.[63-65] Ectodermal signals that are suggested to be involved in the specification of lateral dermomyotomal cells include Wnt-3, Wnt-4, Wnt-6, Wnt-5a, Wnt-7a, and Wnt-7b (Roelink, 1996; Dietrich et al, 1998). It has been shown in explants of mouse paraxial mesoderm that Wnt-7a and to a lesser extend Wnt-4 and Wnt-5a can activate hypaxial myogenesis.[53] Dietrich et al[64] have concluded that the hypaxial musculature is induced by lateralizing signals including BMP-4 and dorsalizing signals including Wnt proteins. Both types of signals are suggested to act in concert.

Cauthen et al[23] have recently shown for the chick embryo that Wnt-6 is uniformly expressed in the surface ectoderm immediately adjacent to the *Pax-3* expression domain in the dorsal somite. It has not been definitely determined if the specification of hypaxial muscle precursor cells does actually take place within the dermomyotome. Lateral dermomyotome cells additionally express specific cellular markers like VEGF receptor *Quek1*,[67] the homeobox gene *Meis2*,[68] and *EphA4*.[28] These cells give rise to endothelial cells of blood vessels and lymphatics[69-71] and even to cartilage cells forming the scapular blade.[8] This indicates that cells originating from the same part of the lateral dermomyotome can differentiate into different cell types. Single cell labeling experiments are required to study whether the lateral dermomyotomal cells are already committed.

In addition to the effect on muscle cell specification, Wnt proteins have been suggested to increase the number of muscle precursor cells in the dermomyotome. The question, however, whether Wnt signaling promotes proliferation or regulates cell survival of dermomyotome cells is still open to discussion. Münsterberg et al[54] have found an increase in cell proliferation of segmental plate mesoderm cultures after administration of Wnt-1 or Wnt-3. Removal of surface ectoderm has led to a downregulation of both *Pax-3* expression and proliferation of the dermomyotome. In both the dermomyotomes and the limbs, the effects of ectoderm ablation could be rescued by the application of BMP-4.[72-73] Interestingly, the application of BMP-4 beads caused excessive muscle growth in the presence of surface ectoderm. BMP-4 may activate members of the Wnt family that are expressed in the surface ectoderm.[51] Wnts, for their part, have been shown to upregulate *Pax-3* and to stimulate proliferation of muscle precursors,[45,46,30,74] or to activate *MyoD* and *Myf-5* expression.[53] On the other hand, *Wnt-1* and *Wnt-3a* double knock out mouse embryos do not show an obvious increase of cell death or a decrease of proliferation in the dorsal somite compartment although the medial part of the dermomyotome is lacking in these mice.[38]

## Concluding Remarks

Several lines of evidence indicate that Wnt signaling is required for dorsoventral somite compartmentalization, patterning of the dermomyotome, and myogenesis in both the epaxial and hypaxial domains (Figs. 2, 3). It remains challanging to determine the complex signaling network that integrates proliferation, migration and differentiation of muscle precursors. Exact analyses of the expression domains of *Wnts*, carefully performed cell lineage investigations, as well as gain of function and loss of function approaches are required to further increase our understanding of the functions of Wnt proteins on skeletal muscle development.

### *Acknowledgements*

We are grateful to Drs. B. Brand-Saberi, C. Schmidt, and J. Wilting for helpful comments and to Mrs. U. Uhl for typing of the manuscript. Work on this subject has been continuously supported by the Deutsche Forschungsgemeinschaft.

## References

1. Christ B, Ordahl CP. Early stages of chick somite development. Anat Embryol 1995; 191:381-396.
2. Wodarz A, Nusse R. Mechanisms of Wnt signaling in development. Annu Rev Cell Dev Biol 1998; 14:59-88.
3. Dhoot GK, Gustafsson MK, Ai X et al. Regulation of Wnt signaling and embryo patterning by an extracellular sulfatase. Science 2001; 293:1663-1666.
4. Goulding MD, Lumsden A, Gruss P. Signals from the notochord and floor plate regulate the region-specific expression of two Pax genes in the developing spinal cord. Development 1993; 117:1001-1016.
5. Goulding MD, Lumsden A, Paquette AJ. Regulation of Pax-3 expression in the dermomyotome and its role in muscle development. Development 1994; 120:957-971.
6. Ebensperger C, Brand-Saberi B, Wilting J et al. The ventralizing effect of the notochord on somite differentiation. Verh Anat Ges 1994; 89:464-465.
7. Williams BA, Ordahl CP. Pax-3 expression in segmental mesoderm marks early stages in myogenic cell specification. Development 1994; 120:785-796.
8. Huang R, Zhi Q, Patel K et al. Dual origin and segmental organisation of the avian scapula. Development 2000; 127:3789-3794.
9. Kiefer JC, Hauschka SD. Myf-5 is transiently expressed in nonmuscle mesoderm and exhibits dynamic regional changes within the presegmented mesoderm and somites I-IV. Dev Biol 2001; 232:77-90.
10. Denetclaw WF, Christ B, Ordahl CP. Location and growth of epaxial myotome precursor cells. Development 1997; 124:1601-1610.
11. Denetclaw WF, Ordahl CP. The growth of the dermomyotome and formation of early myotome lineages in thoracolmbar somtes of chicken embryos. Development 2000; 127:893-905.
12. Huang R, Christ B. Origin of the epaxial and hypaxial myotome in avian embryos. Anat Embryol 2000; 202:369-374.
13. Christ B, Jacob HJ, Jacob M. Experimental analysis of the origin of the wing musculature in avian embryos. Anat Embryol 1977; 150:171-186.
14. Aoyama H, Asamoto K. Determination of somite cells: Independence of cell differentiation and morphogenesis. Development 1988; 104:15-28.
15. Christ B, Brand-Saberi B, Grim M. Local signalling in dermomyotomal cell type specification. Anat Embryol 1992; 186:505-510.
16. Ordahl CP, Le Douarin NM. Two myogenic lineages within the developing somite. Development 1992; 114:339-353.
17. Crossley PH, Martin GR. The mouse Fgf8 gene encodes a family of polypeptides and is expressed in regions hat direct outgrowth and patterning in he developing embryo. Development 1995; 121:439-451.
18. Dubrulle J, McGrew MJ, Pourquié O. FGF signaling controls somite boundary position and regulates segmentation clock control of spatiotemporal Hox gene activation. Cell 2001; 106:219-232.

19. Stolte D, Huang R, Christ B. Spatial and temporal pattern of Fgf-8 expression during chicken development. Submitted 2002.
20. Burgess R, Rawls A, Brown D et al. Requirement of the paraxis gene for somite formation and musculoskeletal patterning. Nature 1996; 384:570-573.
21. Sosic D, Brand-Saberi B, Schmidt C et al. Regulation of paraxis expression and somite formation by ectoderm- and neural tube-derived signals. Dev Biol 1997; 185:229-243.
22. Wagner J, Schmidt C, Nikovits W et al. Compartmentalization of the somite and myogenesis in chick embryos are influenced by Wnt expression. Dev Biol 2000; 228:86-94.
23. Cauthen CA, Berdougo E, Sandler J et al. Comparative analysis of the expression patterns of Wnts and Frizzleds during early myogenesis in chick embryos. Mech Dev 2001; 104:133-138.
24. Hsieh JC, Kodjabachian L, Rebbert ML et al. A new secreted protein that binds to Wnt proteins and inhibits their activities. Nature 1999; 398:431-436.
25. Bellairs R, Curtis ASG, Sanders EJ. Cell adhesiveness and embryonic differentiation. J Embryol. exp Morph 1978; 46:207-213.
26. Hinck L, Nelson W, Pappkoff J. Wnt-1 modulates cell-cell adhesion in mammalian cells by stabilizing b-catenin binding to the cell adhesion protein cadherin. J Cell Biol 1994; 124:729-741.
27. Bergemann AD, Cheng HJ, Brambilla R et al. ELF-2, a new member of the Eph ligand family, is segmentally expressed in mouse embryos in the region of the hindbrain and newly forming somites. Mol Cell Biol 1995; 15:4921-4929.
28. Schmidt C, Christ B, Maden M et al. Regulation of EphA4 expression in paraxial and lateral plate mesoderm by ectoderm derived signals. Dev Dyn 2001; 220:377-386.
29. Spence MS, Yip J, Erickson CA. The dorsal neural tube organizes he dermamyotome and induces axial myocytes in he avian embryo. Development 1996; 122:231-241.
30. Fan C, Lee C, Tessier-Lavigne M. A role for Wnt proteins in induction of dermomyotome. Dev Biol 1997; 191:160-165.
31. Fan C-M, Tessier-Lavigne M. Patterning of mammalian somites by surface ectoderm and notochord: Evidence for sclerotome induction by a hedgehog homolog. Cell 1994; 79:1175-1186.
32. Johnson RL, Laufer E, Riddle RD. et al. Ectopic expression of Sonic hedgehog alters dorsal-ventral patterning of somites. Cell 1994; 79:1165-1173.
33. Deutsch U, Dressler GR, Gruss P. Pax-1, a member of a paired box homologous murine gene family, is expressed in segment structures during development. Cell 1988; 53:617-625.
34. Goulding MD, Chalepakis G, Deutsch U et al. Pax-3, a novel murine DNA binding protein expressed during early neurogenesis. EMBO J 1991; 10:1135-1147.
35. Brand-Saberi B, Ebensperger C, Wilting J et al. The ventralizing effect of the notochord on somite differentiation in chick embryos. Anat Embryol 1993; 188:239-245.
36. Schmidt C, Christ B, Patel K et al. Experimental induction of BMP-4 expression leads to apoptosis in the paraxial and lateral plate mesoderm. Dev Biol 1998; 202:253-263.
37. Capdevila J, Tabin C, Johnson RL. Control of dorsoventral somite patterning by Wnt-1 and b-catenin. Dev Biol 1998; 193:182-194.
38. Ikeya M, Takada S. Wnt signaling from the dorsal neural tube is required for the formation of the medial dermomyotome. Development 1998; 125:4969-4976.
39. Olivera-Martinez I, Thelu J, Teillet MA et al. Dorsal dermis development depends on a signal from the dorsal neural tube which can be substituted by Wnt-1. Mech Dev 2001; 100:233-244.
40. Ordahl CP, Berdougo E, Venters SJ et al. The dermomyotome dorsomedial lip drives growth and morphogenesis of both primary myotome and dermomyotome epithelium. Development 2001; 128:1731-1744.
41. Christ B. Experimente zur Lageentwicklung der Somiten. Anat Anz Erg-H Bd 1970; 126:555-564.
42. Rong PM, Teillet MA, Ziller C et al. The neural-tube notochord complex is necessary for vertebral but not limb and body wall striated muscle differentiation. Development 1992; 115:657-672.
43. Pownall M, Strunk K, Emerson C. Notochord signals control the transcriptional cascade of myogenic bHLH genes in somites of quail embryos. Development 1996; 122:1475-1488.
44. Dietrich S, Schubert FR, Lumsden A. Control of dorsoventral pattern in the chick paraxial mesoderm. Development 1997; 124:3895-3908.
45. Münsterberg AE, Lassar AB. Combinatorial signals from the neural tube, floor plate and notochord induce myogenic bHLH gene expression in the somite. Development 1995; 121:651-660.

46. Stern H, Brown A, Hauschka S. Myogenesis in paraxial mesoderm: preferential induction by dorsal neural tube and by cells expressing Wnt-1. Development 1995; 121:3675-3686.
47. Buffinger N, Stockdale FE. Myogenic specificationin somites: induction by axial structures: Development 1994; 120:1443-1452.
48. Fan CM, Porter JA, Chiang C et al. Long-range sclerotome induction by sonic hedgehog: Direct role of the amno-terminal cleavage product and modulation by the cyclic AMP signaling pathway. Cell 1995; 81:457-465.
49. Bober E, Brand-Saberi B, Ebensperger C et al. Initial steps of myogenesis in somites are independent of influence from axial structures. Development 1994; 120:3073-3082.
50. Hollyday M, McMahon J, McMahon A. Wnt expression patterns in chick embryos nervous system. Mech Dev 1995; 52:9-25.
51. Marcelle C, Stark M, Bronner-Fraser M. Coordinate actions of BMPs, Wnts, Shh and Noggin mediate patterning of the dorsal somite. Development 1997; 124:3955-3963.
52. Lee C, Buttitta L, May N et al. SHH-N upregulates sFRP-2 to mediate its competitive interaction with WNT1 and WNT4 in the somitic mesoderm. Development 2000; 127:109-118.
53. Tajbakhsh S, Borello U, Vivarelli E et al. Differential activation of Myf5 and MyoD by different Wnts in explants of mouse paraxial mesoderm and the later activation of myogenesis in the absence of Myf5. Development 1998; 125:4155-4162.
54. Münsterberg AE, Kitagewski J, Bumcrot DA et al. Combinatorial signaling by Sonic hedgehog and Wnt family members induces myogenic bHLH gene expression in the somite. Genes Dev 1995b; 9:2911-2922.
55. Borello U, Coletta M, Tajbakhsh S et al. Transplacental delivery of the Wnt antagonist Frzbl inhibits development of caudal paraxial mesoderm and skeletal myogenesis in mouse embryos. Development 1999; 126:4247-4255.
56. Ladher RK, Church VL, Allen S et al. Cloning and expression of the Wnt antagonists Sfrp-2 and Frzb during chich development. Dev Biol 2000; 218:183-198.
57. Hirsinger E, Duprez D, Jouve C et al. Noggin acts downstream of Wnt and Sonic Hedgehog to antagonize BMP4 in avian somite patterning. Development 1997; 124:4605-4614.
58. Reshef R, Maroto M, Lassar A. Regulation of dorsal somitic cell fates: BMPs and Noggin control the timing and pattern of myogenic regulator expression. Genes Dev 1998; 12:290-303.
59. Borycki AG, Brown AMC, Emerson CP. Shh and Wnt signaling pathways converge to control Gli gene activation in avian somites. Development 2000; 127:2075-2087.
60. Brand-Saberi B, Christ B. Genetic and epigenetic control of muscle development in vertebrates. Review article. Cell Tissue Res 1999; 296:199-212.
61. Pourquié O, Coltey M, Breant C et al. Control of somite patterning by signals from the lateral plate. Proc Natl Acad Sci USA 1995; 92:3219-3223.
62. Pourquié O, Fan CM, Coltey M et al. Lateral and axial signals involved in avian somite patterning: a role for BMP-4. Cell 1996; 84:461-471.
63. Jagla K, Dolle P, Mattei MG et al. Mouse Lbx1 and human LBX1 define a novel mammalian homeobox gene family related to the Drosophila lady bird genes. Mech Dev 1995; 53:345-356.
64. Dietrich S, Schubert FR, Healy C et al. Specification of the hypaxial musculature. Development 1998; 125:2235-2249.
65. Bladt F, Riethmacher D, Isenmann S et al. Essential role for the c-met receptor in the migration of myogenic precursor cells into the limb bud. Nature 1995; 376:768-771
66. Roelink H. Tripartite signalling of pattern: interactions between Hedgehogs, BMPs and Wnts in the control of vertebrate development. Curr Opin Neurobiol 1996; 6:33-40.
67. Eichmann A, Marcelle C, Brèant C et al. Two molecules related to the VEGF receptor are expressed in early endothelial cells during avian embryonic development. Mech Dev 1993; 42:33-48.
68. Cecconi F, Proetzel G, Alvarez-Bolado G et al. Expression of Meis2, a knotted-related murine homeoboxgene, indicates a role in the differentiation of the forebrain and the somitic mesoderm. Dev Dyn 1997; 210:184-190.
69. Wilting J, Brand-Saberi B, Huang R et al. Angiogenic potential of the avian somite. Dev Dyn 1995; 202:165-171.

70. Wilting J, Eichmann A, Christ B. Expression of the avian VEGF receptor homologues Quek1 and Quek2 in blood-vascular and lymphatic endothelial and non-endothelial cells during quail embryonic development. Cell Tiss Res 1997; 288:207-223.
71. Wilting J, Papoutsi M, Schneider M et al. The lymphatic endothelium of the avian wing is of somitic origin. Dev Dyn 2000; 217:271-278.
72. Amthor H, Christ B, Weil M et al. The importance of timing differentiation during limb muscle development. Curr Biol 1998; 8:642-652.
73. Amthor H, Christ B, Patel K. A molecular mechanism enabling continuous embryonic muscle growth—a balance between proliferation and differentiation. Development 1999; 126:1041-1053.
74. Maroto M, Reshef R, Münsterberg AE et al. Ectopic Pax-3 activates MyoD and Myf-5 expression in embryonic mesoderm and neural tissue. Cell 1997; 89:139-148

## CHAPTER 10

# Wnt Signaling in Limb Development

**Christine Hartmann**


## Abstract

The development of the vertebrate limbs is a complex multi-step process which involves specification of the early limb field, directed outgrowth, and patterning to establish the final three dimensional structure of the limb. All of these processes have to be temporally and spatially coordinated and thus require communication between cells and cell layers, a function which is mediated by specific extracellular signals. A number of Wnt molecules, in particular, have been implicated in playing specific roles during vertebrate limb development, including induction of the early limb bud, formation and maintenance of a specific ectodermal structure known as the apical ectodermal ridge (AER), outgrowth of the limb, and patterning of the limb bud axes. Components of this pathway have also been implicated in controlling differentiation, maturation and segmentation of skeletal elements, and bone homeostasis, and are associated with diseases of the joints such as rheumatoid arthritis.


## Introduction

The focus of this chapter is on the role of Wnt molecules during different stages of vertebrate limb bud development, ranging from the initiation of limb bud formation to skeletogenesis. The vertebrate limbs originate at discrete positions along the antero-posterior body axis from cellular contributions of two sources, the lateral plate and the somitic mesoderm.[1,2] Differential proliferation of lateral plate mesenchymal cells in the presumptive limb field leads to the initial formation of the bud. Shortly thereafter, cells from lateral edges of the adjacent somites begin migrating into the limb. These cells will go on and form the limb musculature. Like the limb muscles, nerves, pigment cells, and vasculature also originate from extra-limb regions, while all other limb tissues, including the skeletogenic mesenchyme, forming cartilage, bone and tendons, derive from the lateral plate mesoderm. An overlying epidermal sheet, the ectoderm, envelops the bulge of mesenchymal cells. It will eventually give rise to the limb epidermis as well as specialized structures such as feathers, hair, claws, and nails.

During early limb bud development, a specialized epithelial structure known as the apical ectodermal ridge (AER) forms in most tetrapods at the distal tip of the ectodermal pocket. The AER runs along the antero-posterior axis of the limb bud at the boundary of dorsal and ventral ectoderm and is necessary for proximo-distal outgrowth of the limb.[3-5] Soon after the formation of the AER, mesenchymal cells begin to aggregate in a proximo-distal fashion, forming first the anlagen of the proximal cartilage elements (humerus/femur). Concomitantly with the outgrowth of the limb, cells continue to condense in a proximal to distal wave and eventually branch into the future radius/ulna and tibia/fibula cartilage elements. This process continues in the distal regions, while proximally, the originally branched condensations of the humerus/

*Wnt Signaling in Development*, edited by Michael Kühl.

ulna/radius and femur/fibula/tibia become segmented into individual elements. Most of the future limb skeleton is thought to arise by this process of branching and segmentation of pre-existing prechondrogenic elements,[6,7] resulting in the formation of individual cartilaginous templates which prefigure the future skeletal limb components.[8]

Altogether, the final outcome of these developmental processes is the establishment of a prototypically patterned, fully functional tetrapod limb. In order to form this complex organ, three distinct axes, antero-posterior (AP), dorso-ventral (DV) and proximo-distal (PD), must first be established in the early limb bud. Signals originating from these three axes are then integrated to set up spatial coordinates to pre-pattern the limb mesenchyme, and finally the various cell types have to be specified. As described below, Wnt molecules have been implicated in a variety of these processes and thus serve a major role in regulating the formation of a functional tetrapod limb.

## A Role for Wnt Signaling in Limb Bud Initiation

Manipulations of chick embryos, including barrier experiments and surgical extirpations, originally hinted at the intermediate mesoderm (IM) as a possible source for signals required for limb bud initiation.[9-11] However, it should be noted here that this idea is somewhat controversial since more recent experiments by Fernandez-Teran and colleagues[12] have yielded conflicting results. If indeed, a factor(s) from the IM is required for limb bud initiation, *FGF-8*, which is expressed in the IM adjacent to the presumptive limb-forming region in the lateral plate (LP) at the time of limb bud initiation induction (Fig. 1), is a prime candidate to be the limb bud-inducing factor.[13] Especially, as several groups have shown that beads soaked in FGF proteins (including FGF-8) are able to induce the outgrowth of an ectopic limb when implanted into the flank region.[13-15] Interestingly, the limb-inducing activity of FGF-8 seems to be mediated via another *FGF* family member, FGF-10, which is expressed in the lateral plate mesoderm (LPM) and whose expression becomes restricted to the presumptive limb region around the time of limb bud induction (Fig. 1). Moreover, the observation that FGF-10 expression can be induced by FGF-8 in chick,[16] together with the *FGF-10* loss-of-function phenotype in mouse,[17,18] strongly suggests that FGF-10 activity is required for limb bud initiation. In turn, FGF-10 expression in the limb mesoderm is involved in the induction of *FGF-8* expression in the overlying ectoderm.[16]

Recently, work from Kawakami and colleagues also suggested a role for Wnt signaling in the process of limb bud initiation in the chick induction (Fig. 1).[19] Two Wnt-genes in particular *Wnt-2b* and *Wnt-8c* seem to be involved in mediating the FGF-8/FGF-10 loop that directs limb bud initiation. At the level of the presumptive forelimb (future wing), *Wnt-2b* is expressed in both the IM and the LPM. This expression pattern is strikingly similar to that of FGF-8 and FGF-10 and was a first hint for Kawakami and his colleagues that *Wnt-2b* could be involved in the FGF-8/FGF-10 loop.[19] In support of this notion, they found that local application of cells expressing *Wnt-2b* in the flank region induces FGF-10 expression as an early response, eventually resulting in ectopic limb outgrowth. Furthermore, ectopic application of FGF-8 was found to induce *Wnt-2b* expression prior to the induction of *FGF-10*, allowing them to place Wnt-2b downstream of FGF-8 and upstream of FGF-10 in the FGF-8/FGF-10 loop at the forelimb level.[19]

Interestingly, while a similar mechanism appears to be operating in the hindlimb, a different Wnt, Wnt-8c, has been implicated in mediating the FGF-8/FGF-10 loop in this region. In particular, *Wnt-8c* expression is restricted to the caudal LPM and precedes the confinement of FGF-10 expression to the LPM of the prospective hindlimb region. Furthermore, like Wnt-2b, Wnt-8c can induce FGF-10 expression. This, along with the spatio-temporal pattern of both *Wnt-8c* and *FGF-10*, makes *Wnt-8c* a likely candidate for the Wnt gene involved in mediating limb bud initiation in the hindlimb region.

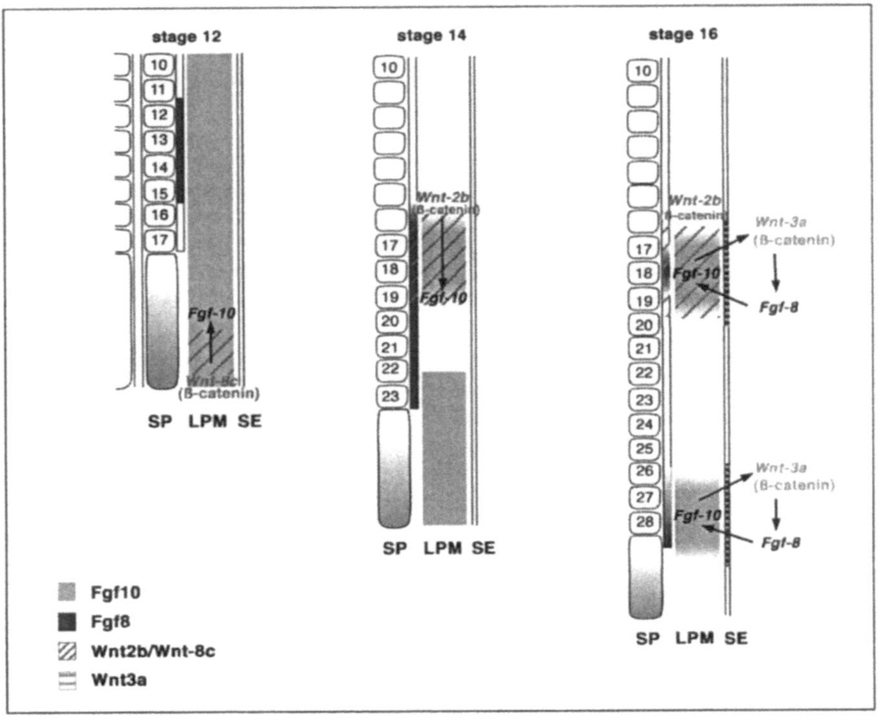


Figure 1. Model for the role of Wnt's during limb bud initiation in the chick embryo. At stage 12, prior to limb initiation, *FGF8* (dark blue/gray) is expressed in the intermediate mesoderm (IM) adjacent to the presumptive forelimb field. *FGF10* (light blue/gray) is expressed throughout the lateral plate mesoderm (LPM) and in the segmental plate (SP). Note *Wnt8c* (red, diagonal stripes) and *FGF-10* are co-expressed in the caudal region of the LPM. At stage 14, during limb initiation, *FGF-8* (dark blue/gray) and *Wnt-2b* (red, diagonal stripes) are co-expressed in the IM at the level of the forelimb. *Wnt-2b* is also expressed in the LPM at forelimb level and signals through β-catenin to regulate *FGF-10* expression in the LPM of the presumptive forelimb area. However, *FGF-10* continues to be expressed diffusely throughout the caudal region of the LPM. At stage 16, expression of *Wnt-2b* and *FGF-10* is confined to the forelimb area in the LPM. In addition, *FGF-10* is now confined to the presumptive hindlimb field. This restricted expression of *FGF-10* in the hindlimb area is likely to be controlled by *Wnt8c*. In both limb fields, fore- and hindlimb, FGF-10 is inducing the expression ofWnt-3a (orange, horizontal stripes) in the overlying surface ectoderm (SE). *Wnt-3a* signals through β-catenin inducing *FGF-8* (dark blue/gray) expression in the SE. Note for simplicity reasons, only the expression patterns in the relevant regions are shown, while the expression in somites and other areas is neglected (modified after Kawakami et al. Cell 2001; 104(6): 891-900).

Within the context of the FGF-8/FGF-10 feedback loop, there is evidence that both Wnt-2b and Wnt-8c are acting through the β-catenin/LEF pathway, also known as the canonical Wnt pathway.[19] In particular, implanting cell pellets that express a stabilized form of β-catenin, which acts as a constitutively active molecule, into the flank region can ectopically induce FGF-10 expression and limb bud outgrowth. In contrast, misexpression of Axin (a negative regulator of the β-catenin/LEF pathway) at early stages downregulates FGF-10 expression and negatively affects limb outgrowth.[19] Thus, while a common Wnt signaling pathway appears to be involved in the initiation of limb bud formation in both forelimbs and hindlimbs, distinct Wnt molecules appear to be activating this pathway in the two different limb fields. In this regard, it should be mentioned that similar expression patterns for *Wnt-2b* and *Wnt-8c* have not been reported in the mouse; however, it remains possible that other Wnt molecules may have been recruited to carry out this function in different species.

## Wnt Signaling and Limb Bud Outgrowth

As mentioned above, subsequent to the establishment of the limb-forming region in the lateral plate (as reflected by the restricted expression of *FGF-10* in the LPM), FGF-10 induces FGF-8 in the overlying limb ectoderm. Interestingly, yet another Wnt gene, *Wnt-3a*, has been proposed to act downstream of *FGF-10* in mediating this induction of *FGF-8*.[20] Wnt-3a is expressed prior to *FGF-8* in the ectoderm, and like *FGF-8* its expression becomes restricted to the AER during subsequent developmental stages. Significantly, previous studies had suggested that signaling by *Wnt-3a* is involved in the formation and maintenance of the AER.[20] In particular, *Wnt-3a* is capable of inducing ectopic FGF-8 expression as well as ectopic AER-like structures. Ectopic application of *FGF-8*, by contrast, does not induce ectopic *Wnt-3a* expression, at least not in the limb bud.[20] Again, it is interesting to note that the Wnt-3a signal, in this context, also appears to transduced through the canonical Wnt pathway.[19,20] Thus, the β-catenin/LEF pathway plays a major role both in the initiation of limb bud outgrowth and in the establishment of the AER, which is itself necessary for continuous, proximal-to-distal growth of the limb (Fig. 1).

Strikingly, like the Wnt genes implicated in mediating limb bud induction in the chick, the expression of *Wnt-3a* in the AER is not conserved between chick and mouse. Furthermore, limb development is normal in *Wnt-3a* mutant mice.[21,22] Nevertheless, there is evidence that a β-catenin/LEF-mediated Wnt signal does play a role in establishing or maintaining the AER in the mouse as well. In particular, mice lacking TCF-1 and LEF-1, two of the four members of the TCF/LEF-family of transcription factors, develop significantly smaller limb buds and lack a functional AER. The presence of a protrusion from the lateral plate at the forelimb level in these embryos suggests that limb bud outgrowth is initiated but stalls due to the absence of an AER.[23] These observations suggest that another member of the Wnt gene family, signaling through the β-catenin/LEF pathway, is involved in the formation and maintenance of the AER in the mouse.

Further outgrowth of the limb requires not only the presence of a functional AER but also continuous proliferation of the limb bud mesenchyme. Another member of the Wnt gene family, *Wnt-5a*, is expressed in the distal mesenchyme underneath the AER in both chick and mouse.[24,25] The analysis of a *Wnt-5a* loss-of-function mutation in the mouse revealed that *Wnt-5a* acts predominantly as a mitogen in the distal regions of multiple outgrowing structures of the embryo, including the limb bud.[26] In the limb, loss of *Wnt-5a* activity results in a progressive shortening of the appendicular skeleton in a proximal-to-distal fashion and a total loss of the most distal structures such as the digits.[26] At present, it is still not known through which intracellular signaling pathway Wnt-5a exerts its mitogenic activity. However, in other systems, Wnt-5a has been demonstrated to utilize either the canonical pathway[27] or the non-canonical $Ca^{2+}$ pathway[28,29] depending on which Frizzled receptors are present in the responding cells.

Interestingly, the various roles for Wnt molecules in the early vertebrate limb bud are reminiscent of the different requirements for *wingless* (*wg*) in the wing disc of *Drosophila*. For instance, localized expression of Wg protein in the early wing disc is necessary for the specification of the wing versus body wall structures.[30] This is somewhat similar to the situation in the chick, where Wnt-2b and Wnt-8c are involved in restricting *FGF-10* expression to the prospective fore- and hindlimb fields, respectively, thereby specifying limb versus body wall.[19] Subsequently, during a later phase of wing development, *Wg* expression becomes restricted to a stripe along the dorso-ventral boundary. Likewise, in the chick, *Wnt-3a* is expressed in the

AER, which runs along the DV boundary of the vertebrate limb bud. Furthermore, wg also acts as a local mitogen in the hinge region subsequent to wing specification[31], a situation reminiscent of the requirement for Wnt-5a for limb outgrowth.[19] Interestingly, this mitogenic activity of wg is executed independently from its role in patterning in the wing pouch, raising the possibility that such region-specific responses are elicited through the stimulation of different intracellular pathways.[31] Similarly, in vertebrates, region-specific differences in mitogenic versus cell-fate specification responses to a Wnt signal have also been reported for *Wnt-1* in the central nervous system of the mouse.[32] It is therefore tempting to speculate that the multiple roles served by a single Wnt molecule, wg, during *Drosophila* appendage development have been subdivided in the vertebrate limb and are now executed by different members of the Wnt family, while the relevant pathways for the different functions may have been broadly conserved in both the arthropod and vertebrate lineages.

## Limb Bud Patterning—A Role for Wnt Signaling in Establishing the Dorso-Ventral Axis and Beyond

The vertebrate limb develops along three primary axes: antero-posterior (thumb versus pinky), dorso-ventral (knuckles versus palm), and proximo-distal (shoulder versus fingers). These axes are set up very early during limb development, and the recent identification of molecular signals involved in patterning of the different axes has revealed a certain degree of interdependence between them. Considering first the AP and PD axes, *Sonic hedgehog* (*Shh*), a member of the hedgehog-family encoding secreted signaling molecules, is both necessary and sufficient for patterning of the AP axis, while members of the FGF superfamily are required for PD axis formation. Patterning of the DV axis, in turn, is controlled by a member of the Wnt-family, Wnt-7a.

In 1974, MacCabe and co-workers demonstrated by rotating the limb bud ectoderm 180°, that the DV polarity of the distal limb is determined by the ectoderm.[33] In this regard, Wnt-7a, which is expressed exclusively in the dorsal compartment of the ectodermal pocket in both chick[24] and mouse,[25,34] was an ideal candidate for the DV patterning molecule. Strikingly, loss-of-*Wnt-7a* function in the mouse resulted in embryos with ventralized paws having sole pads on both surfaces, showing that Wnt-7a was necessary for specifying dorsal cell fates in the distal limb.[35] Complementary gain-of-function experiments in the chick led to the identification of a Wnt-7a-specific target gene, the transcription factor *Lmx-1*, which is expressed in the dorsal mesenchyme, and further demonstrated that Wnt-7a acting through *Lmx-1*, is sufficient for dorsal cell fate specification (Fig. 2).[36,37] Concomitantly, the expression of *Lmx-1b* (the mouse homolog of *Lmx-1*) is lost in the dorsal mesenchyme of the distal limb of *Wnt-7a* deficient mice.[38] This observation, together with the chick data, established that *Wnt-7a* is both necessary and sufficient for *Lmx-1* expression.

The identification of *Lmx-1* (*Lmx-1b*) as a *Wnt-7a* target gene made it possible to ask in turn whether *Wnt-7a* regulates *Lmx-1* through the β-catenin/*LEF* pathway. To address this, Kengaku and colleagues conducted misexpression experiments in chick using viruses expressing either a stabilized form of β-catenin (resulting in constitutively active Wnt signaling through the canonical pathway) or a truncated form of the transcription factor *LEF-1* (acting as a dominant-negative molecule which blocks signaling through the canonical pathway), and analyzed the effects on Lmx-1 expression.[20] In contrast to the robust induction of Lmx-1 expression in the ventral mesenchyme in response to Wnt-7a misexpression, activation or disruption of the β-catenin/LEF pathway had almost no effect on *Lmx-1* expression, suggesting that *Wnt-7a* does not signal through the canonical Wnt pathway.[20]

To set up a proper DV axis restriction of *Wnt-7a* expression to the dorsal ectoderm requires the activity of the transcription factor *Engrailed-1* (*En-1*) in the ventral ectoderm.[39,40] In mice deficient for *En-1* function, *Wnt-7a* expands into the ventral ectoderm of the limb bud, and as

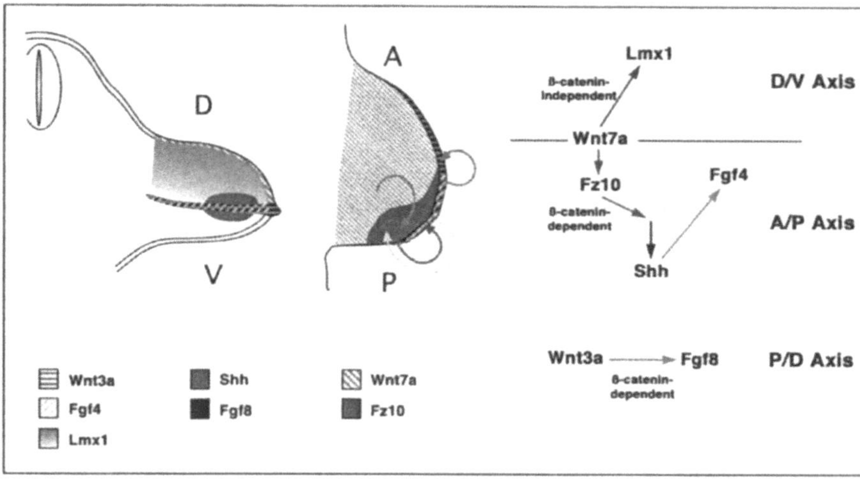


Figure 2. Expression of Wnt's and their roles during limb bud patterning. View from the posterior (on the left; dorsal (D) is up, ventral (V) is down) and from dorsal (middle; anterior (A) is up, posterior (P) is down) on to a chick limb bud at stage 21. *Wnt-7a* (green, reverse diagonal stripes) is expressed in the dorsal ectoderm of the limb bud. Wnt-7a signals through a β-catenin independent pathway activating *Lmx1* (purple/gray gradient) in the dorsal mesenchyme of the limb bud thereby determining the dorso-ventral axis. Further Wnt-7a is maintaining the expression of *Shh* (red/grey), expressed in a posterior domain. This role of *Wnt-7a* in AP patterning could be mediated through a specific interaction of *Wnt-7a* with the receptor *Fz-10* (dark green/gray) involving the β-catenin pathway. *Fz-10* is expressed in the posterior mesenchyme of the limb and has been shown to utilize the β-catenin pathway in other model systems. The maintenance of Shh involves a combinatorial action of *Wnt-7a* and *FGF-4* (yellow, diagonal stripes), which is expressed in the posterior half of the AER. Wnt-3a (orange, horizontal stripes) is co-expressed with *FGF-8* (dark/gray blue) in the AER and is maintaining *FGF-8* expression through a β-catenin-dependent pathway, thereby promoting proximal-distal axis formation.

a consequence, the limbs display aspects of double-dorsal paws.[39] Conversely, ectopic expression of *En-1* throughout the limb ectoderm in chick results in suppression of *Wnt-7a* in the dorsal ectoderm.[40] Interestingly, *En-1* deficient mice also display expansion of the ventral AER as well as the formation of an ectopic AER,[38,39,41] and significantly, the analysis of *En-1/Wnt-7a* double mutant embryos revealed that both the expansion of the AER and the formation of the ectopic AER-like structure on the ventral side of the limb bud are dependent on Wnt-7a activity.[38,41] However, while these observations suggest that Wnt-7a plays a role in the formation of secondary AER structures in *En-1* mutants, there is no evidence that Wnt-7a is required for normal AER formation within the context of the wild-type limb.

Besides its well-established role in DV patterning, *Wnt-7a* also serves an additional function during limb development, as Wnt-7a deficient mice also show a notable loss of posterior skeletal elements. In particular, both the ulna (in the zeugopod) and the most posterior digit 5 (in the autopod region) are frequently lost or abnormal in *Wnt-7a* mutant limbs.[35] This loss of posterior skeletal elements in the mutant limbs suggested that *Wnt-7a* signaling may be necessary for normal function of the zone of polarizing activity (ZPA). Consistent with this proposal, the expression of *Shh*, which mediates ZPA activity, is progressively lost in *Wnt-7a* deficient limbs.[35] The loss of *Shh* expression further suggests that *Shh* may be a direct target of Wnt-7a signaling in the posterior mesenchyme. This notion is further supported by experiments in the chick, where Yang and Niswander, through ablation experiments, observed that

*Shh* expression is dependent on signals from the dorsal ectoderm.[42] Significantly, following such ablations of the dorsal ectoderm, both *Shh* expression and posterior skeletal development can be rescued by applying Wnt-7a in combination with FGF-4, demonstrating that maintenance of Shh expression is under the combinatorial influence of Wnt and FGF signaling. Therefore, Wnt-7a not only plays a role in DV patterning and establishment of the AER, but is also contributes to AP patterning in the limb (Fig. 2).

The fact that the ectodermal signal Wnt-7a induces distinct responses in the underlying mesenchyme, such as induction of *Lmx-1* expression throughout the dorsal mesenchyme versus posteriorly restricted maintenance of *Shh* expression, raises the question of how can the same signal lead to such different responses. Recently, Kawakami and co-workers proposed that Frizzled-10 (Fz-10), which is expressed in a posterior-dorsal domain overlapping with the *Shh* domain in the chick limb bud mesenchyme, is the receptor through which Wnt-7a signals to maintain expression of *Shh*.[43] Interestingly, they observed that the expression of *Fz-10* is under the cooperative influence of Wnt-7a and Shh, suggesting that both signals are involved in the induction of *Fz-10* expression. Furthermore, the synergistic effect on the expression of additional target genes of the canonical Wnt pathway following co-injection of Wnt-7a and Fz-10 in *Xenopus* animal cap assays[43] suggests that Wnt-7a, through its interaction with Fz-10, might maintain *Shh* expression via the canonical Wnt pathway (Fig. 2). This hypothesis remains to be tested. Nevertheless, tissue-specific expression of different Frizzled receptors, which preferentially activate a particular signal transduction pathway (and which may also exhibit a degree of specificity for different Wnt molecules), provides an intriguing explanation for how the same or similar signal(s) might induce different responses in different tissues.

Alternatively, the recent identification of co-receptors such as LRP6 (a member of the low-density lipoprotein (LDL)-receptor-related protein family), which seems to direct Wnt-Frizzled signaling towards the β-catenin pathway,[44,45] provides another possibility to explain this phenomenon. Significantly, while the phenotypic observations in *Drosophila* mutants and the gain-of-function experiments in *Xenopus*[44,45] do support such a model, some aspects of the phenotypes described in *LRP6* mutant mice are less easy to reconcile with this proposal. On the one hand *LRP6* mutants exhibit developmental defects which are a striking composite of those caused by mutations in individual Wnt genes[46], some of which are thought to signal through the β-catenin pathway. Conversely, the limb phenotype which displays antero-posterior as well as dorso-ventral defects, is reminiscent of the one observed in *Wnt-7a* mutant mice, a Wnt gene which is thought to signal through a alternative pathway. In LRP6 mutant mice the posterior digit is most frequently lost in and this is presaged at earlier stages by a gradual loss of Shh expression in the ZPA.[46] However, one striking difference compared to the skeletal phenotype of *Wnt-7a* mutants is that *LRP6* mutant limbs show a loss of the radius instead of the ulna. While it is still unclear which signaling pathway Wnt-7a utilizes to maintain *Shh* expression (see above), it has been proposed that the role of Wnt-7a in DV patterning is independent of the canonical Wnt pathway.[20] Therefore, this observation makes a simple model, whereby signaling specificity is rendered through interaction with different co-receptor somewhat difficult to envision.

## Roles of Wnt Signaling in Appendicular Skeletogenesis

Long bones of the appendicular skeleton are formed from a cartilage template in a process referred to as endochondral ossification. Skeletal development in the limb is initiated in the early proximal limb when cells in the limb mesenchyme start to aggregate and condense to form the precartilaginous elements of the future bones. These develop into cartilaginous elements accompanied by differentiation of prechondrogenic cells into chondrocytes. In turn, starting from the articular region and progressing towards the middle of the cartilage element, chondrocytes within the cartilage elements further undergo a progressive program of maturation

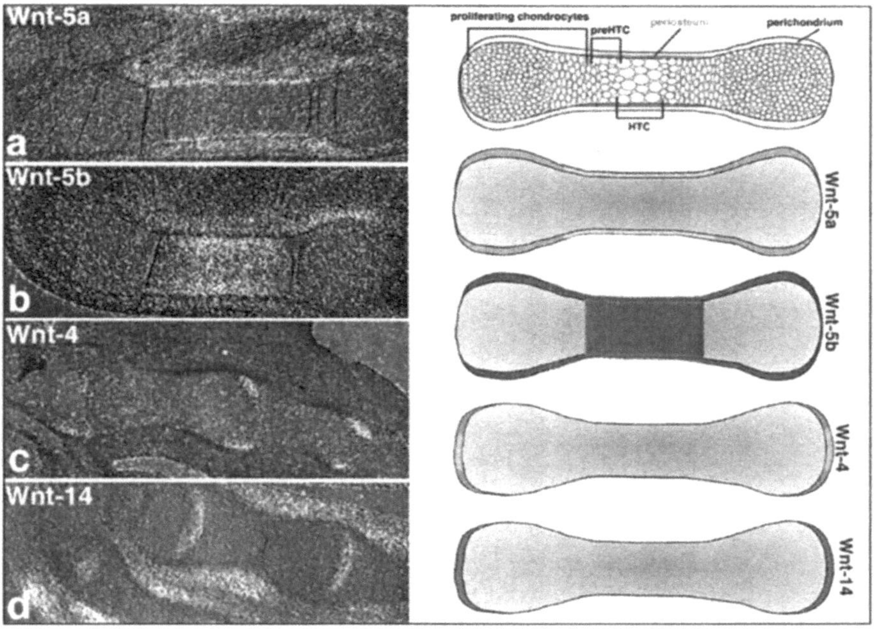


Figure 3. Expression of Wnt-genes in chondrogenic regions of the appendicular skeleton. Left panel: radioactive section in situ hybridizations showing expression of the Wnt-genes, *Wnt-5a*, *Wnt-5b*, *Wnt-4* and *Wnt-14* (expression signal is white). Right panel: schematic overview of the different maturation zones within a cartilage element and of the Wnt expression patterns (*Wnt-5a*: green; *Wnt-5b*: blue; *Wnt-4*: purple; *Wnt-14*: red). From top to bottom: a) Expression of *Wnt-5a* in the central region of the perichondrium, the future periosteum, surrounding the cartilage element (shown: ulna). b) Expression of Wnt-5b in the mesenchyme surrounding the cartilage element and in prehypertrophic chondrocytes within the cartilage element (shown: ulna). c) Expression of *Wnt-4* in the lateral areas of the joints between adjacent skeletal elements (shown: digits in the foot). d) Expression of *Wnt-14* throughout the joint interzone (shown: digits in the foot). preHTC: prehypertrophic chondrocytes. HTC: hypertrophic chondrocytes.

from proliferating to postmitotic prehypertrophic to hypertrophic chondrocytes, the terminally differentiated cell type, which eventually is replaced by bone. This maturation program has to be tightly controlled by multiple factors, including negative regulators, whose role is to prevent premature maturation (which would result in shorter bones), and positive regulators, whose role is to ensure that a sufficient number of cells undergoes maturation. In addition, the maturation of the chondrocytes within the cartilage element is accompanied by differentiation of the perichondrium (the mesenchyme immediately surrounding the cartilage element), whereby cells within in the so-called periosteum (the region of the perichondrium surrounding the domain of prehypertrophic and hypertrophic chondrocytes) differentiate into preosteoblasts. In turn, the innermost preosteoblasts adjacent to the hypertrophic cells subsequently mature into osteoblasts, which secret a bone matrix (osteoid) that becomes progressively calcified, forming a membranous bone collar around the diaphysis of the cartilage element.[47]

Many different signaling molecules have been implicated in regulating the maturation of perichondrial cells and chondrocytes. Among these, recent work by several groups has implicated Wnt molecules in the regulation of chondrocyte maturation (Fig. 4).[48-53] In particular, gain-of-function experiments in the chick have suggested that Wnt-5a and Wnt-4 have opposing

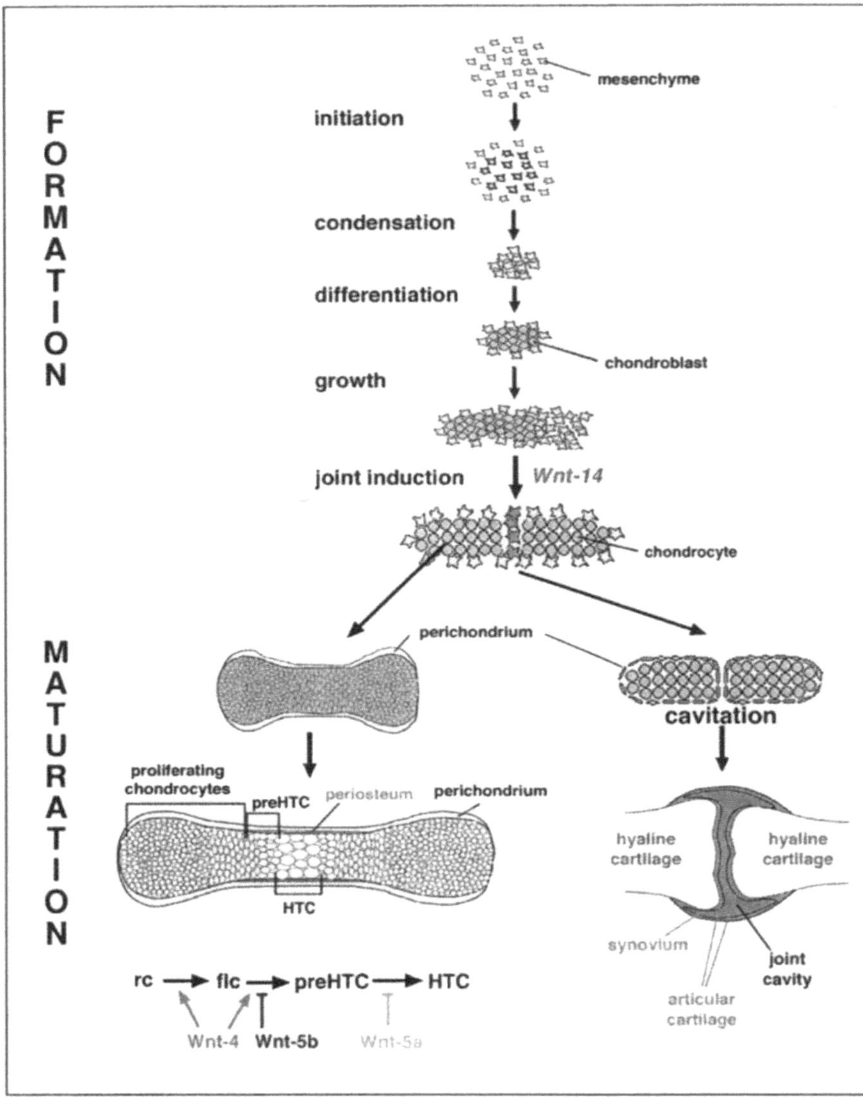


Figure 4. Schematic diagram showing the different steps involved in the formation of cartilage elements and joints as well as their subsequent maturation and at which steps the various Wnt-genes are involved. rc: round proliferating chondrocytes; flc: flattened proliferating chondrocytes; preHTC: prehypertrophic chondrocytes; HTC: hypertrophic chondrocytes

activities on chondrocyte maturation.[50,51] In these studies, while *Wnt-5a* misexpression causes a delayed appearance of hypertrophic chondrocytes (resulting in shortening of the infected long bones), Wnt-4 misexpression results in an acceleration of chondrocyte maturation (also accompanied by a shortening of the long bones). Further analysis on the histological and the molecular level suggested that Wnt-5a negatively regulates the transition from prehypertrophic to hypertrophic chondrocytes. However, *Wnt-5a* expression in the chick is restricted in the perichondrium to the region which will develop into the periosteum (Fig. 3)[51], and thus, its effect on chondrocyte maturation must be indirect. In this regard, it is interesting to note that Wnt-5a has been shown to regulate the expression of a closely related Wnt gene, *Wnt-5b*, which is expressed in a subpopulation of prehypertrophic chondrocytes.[51] Significantly, Wnt-5b misexpression also results in a delay of chondrocyte maturation (Hartmann and Tabin, unpublished observation).

Consistent with the distinct phenotypes observed following misexpression, there is further evidence that Wnt-5a and Wnt-4 regulate chondrocyte maturation by signaling through different pathways. In particular, gain-of-function experiments using a constitutively active form of β-catenin resulted in acceleration of chondrocyte maturation similar to the effects seen in Wnt-4 misexpression experiments. Conversely, misexpression of dominant-negative molecules such as truncated forms of the Frizzled receptors that lack the C-terminal intracellular domain[54] or a dominant-negative form of the transcription factor LEF-1 both delay chondrocyte maturation (C. Hartmann, unpublished observation).[51] On the basis of these observations, it has been proposed that Wnt-4, but not Wnt-5a, signals through the β-catenin pathway.[51] Future studies are necessary to elucidate through which signaling pathway(s) Wnt-5a and Wnt-5b convey their effects on chondrocyte maturation.

To summarize the current model, *Wnt-4*, which is expressed in mesenchymal cells adjacent to the joint region (Fig. 3), is thought to act on chondrocytes in the articular region to positively control their progression from proliferating to the post-mitotic, prehypertrophic state (Fig. 4). This notion is supported by the finding that misexpression of either Wnt-4 or activated β-catenin resulted in an upregulation of LEF-1 expression in the articular regions.[51] Significantly, LEF-1 has previously been shown to be a transcriptional target of the β-catenin pathway,[20] further supporting the notion that Wnt-4 in this context signals through the canonical pathway. At the cell surface, Wnt-4 may bind to the receptor Frizzled 7, which is expressed in chondrocytes of the articular regions and has been shown in other systems to act predominantly through the β-catenin pathway.[29] In contrast, *Wnt-5a* signals from the periosteum surrounding the cartilage element, negatively controlling the transition from prehypertrophic to hypertrophic chondrocytes. This Wnt-5a activity, in turn, may be conveyed, at least in part, through another Wnt-gene, *Wnt-5b*, which is expressed in prehypertrophic chondrocytes (Fig.3) and is upregulated in response to Wnt-5a.

In vitro experiments by Bergwitz and colleagues using rat calvarial cell lines have further demonstrated that the two Wnt genes, *Wnt-5a* and *Wnt-4*, have opposite effects in their assay system.[53] However, the effects in the culture system are different from those observed in the in vivo gain-of-function experiments in chick. In particular, while Wnt-5a overexpression in vivo delayed chondrocyte maturation, it promoted proliferation in the in vitro experiment.[53] No effect of Wnt-5a on proliferation had been observed in the chick *Wnt-5a* misexpression studies assaying for BrdU incorporation (C. Hartmann, unpublished observation). Nevertheless, studies of Wnt-5a loss-of-function mice has revealed a mitogenic activity for Wnt-5a.[26] In contrast, *Wnt-4*, among other Wnt genes (*Wnt-1, Wnt-3a, Wnt-7a*, and *Wnt-7b*), represses both the formation of colonies and the expression of the *Collagen II* chondrogenic reporter-gene construct in the cell culture system.[53] The growth arrest of the colonies could be explained by a model in which this group of Wnt genes, including *Wnt-4*, stimulates exit of proliferating chondrocytes from the cell cycle, leading to their subsequent differentiation. In this regard, the differentiation status of those chondrocytes remains to be determined; however, the suppression of *Collagen II* expression, which is a marker for proliferating and prehypertrophic chondrocytes,[55] suggests that these cells may be terminally differentiated hypertrophic chondrocytes. Bergwitz and colleagues[53] also addressed whether the protein kinase C (PKC) pathway might act downstream of Wnt-5a in the chondrocyte culture system, as Wnt-5a has been shown to activate PKC in other systems.[29] Surprisingly, experiments using specific inhibitors of PKC had similar effects to co-culturing the cells with *Wnt-5a* expressing fibroblasts. This result would indicate that the Wnt-5a signal in this context is not mediated through PKC, but some other pathway.

A number of Frizzled genes have been reported to be expressed in chondrogenic cells.[56] However, their expression patterns and functions in developing skeletal elements remain to be determined. Very recently, mutations in the human LRP5 gene have been identified as the cause of the autosomal recessive disorder osteoporosis-pseudoglioma syndrome (OPPG).[57] Affected individuals have very low bone mass and are therefore prone to develop multiple fractures and deformations. *LRP5* is expressed in immature and mature osteoblasts and its expression profile in cell lines changes upon induction of osteoblastic differentiation.[57] These observations, together with the bone phenotype of OPPG patients suggest that LRP5 regulates osteoblastic proliferation and/or differentiation. *LRP5* is closely related to the vertebrate LRP6 gene and the *Drosophila arrow* gene, which encode transmembrane proteins that are involved as co-receptors during Wnt-signaling.[44,45] Previous studies have implicated a role for *LRP5* in Wnt signaling through the β-catenin pathway,[44,58] and in vitro studies by Gong and colleagues using dominant-negative forms of the receptor have added further support to this hypothesis.[57] These studies also provided evidence that Wnt-signaling may play a role in bone growth in the mouse, since bone growth in calvarial explants can be inhibited by a secreted dominant-negative form of the LRP5 receptor, presumably by binding and interfering with Wnts.[57]

Recent gain-of-function studies in the chick have disclosed yet another function for Wnt signaling during skeletogenesis, specifically, in the segmentation process of the appendicular skeleton. During limb outgrowth, a mesenchymal condensation forms first in the proximal region of the limb, forming the anlagen of the humerus (or femur) and proceeds distally, branching as it extends to form the ulna and radius (or fibula and tibia). Significantly, these early branched, prechondrogenic condensations appear spatially continuous and only subsequently segment into individual elements. Most of the appendicular skeletal elements are thought to arise from this process of branching and segmentation of preexisting prechondrogenic elements.[6,7] Mechanistically, this process is thought to require a signal which in the presumptive joint region prevents early prechondrogenic cells from differentiating into chondrocytes and may instruct those cells to become joint-interzone cells instead. *Wnt-14*, which was originally isolated by Bergstein et al.[59] is expressed, like *Wnt-4*, in the joint regions (Fig. 3).[60] However, *Wnt-14* does not accelerate chondrocyte maturation in a manner similar to Wnt-4; rather its misexpression induces morphological and molecular changes characteristic of the early joint. Furthermore, Wnt-14 has been shown to be capable of arresting and reversing chondrogenic differentiation in vitro. These observations, together with the early expression of *Wnt-14* in the presumptive joint region, make *Wnt-14* an excellent candidate to be involved in the joint induction processes (Fig. 4).[60]

Wnt signaling has also been implicated recently in pathological alterations of the mature joint such as rheumatoid arthritis. Certain Wnt genes, along with certain members of the Frizzled gene family, have been shown to be expressed in the RA synovium, and moreover, some of these genes are specifically upregulated in synovia of rheumatoid arthritis patients.[61] A follow-up study by Sen and colleagues has further shown that Wnt-5a/Frizzled 5 signaling is involved in the activation of rheumatoid arthritis synoviocytes.[62] Additional clues for a role for Wnt signaling in joint disease etiology come from the finding that mutations in *WISP-3*, a member of the CCN family of secreted, cystein-rich growth factors, cause progressive pseudorheumatoid dysplasia (PPD), a disease that resembles RA phenotypically.[63] The highly related *WISP* genes, *WISP1* and *WISP2*, were originally identified as *Wnt-1*-responsive genes.[64,65]

Significantly, Wnt-14 continues to be expressed in cells of the mature chick joint, including the joint capsule and the synovial membrane lining the inside of the joint capsule.[60] Thus, the early developmental role of Wnt-14 in joint induction and its persistent expression in structures of the mature joint allude to a possible role for Wnt-14 in maintaining joint integrity during postnatal development.

Wnts are key signals at various stages of limb development; however, because of the discrepancies in the expression patterns of certain Wnts observed between mouse and chick it remains to be uncovered to what extend the different roles for the Wnt molecules in limb bud development are evolutionary conserved. Future studies will hopefully shed more light on the signaling pathways involved in these different processes as well as on the evolutionary conservation of Wnt signaling in limb development.

### *Acknowledgments*

I would like to thank Dr. Clifford Tubin for helpful comments on this manuscript.

## References

1. Chevallier A, Kieny M, Maugner A. Limb-somite relationship: origin of the limb musculature. J Embryol Exp Morphol 1977; 41:245-258.
2. Christ B, Jacob HJ, Jacob, M. Experimental analysis of the origin of the wing musculature in avian embryos. Anat Embryol 1977; 150:171-186.
3. Saunders JW Jr. The proximal-distal sequence of origin of the parts of the chick wing and the role of the ectoderm. J Exp Zool 1948; 108:363-404.
4. Summerbell D. A quantitative analysis of the excision of the AER from the chick limb-bud. J Embryol Exp Morphol 1974; 32:651-660.
5. Rowe DA, Fallon JF. The proximodistal determination of skeletal parts in the developing chick leg. J Embryol Exp Morphol 1982; 68:1-7.
6. Shubin NH, Alberch P. A morphogenetic approach to the origin and basic organization of the tetrapod limb. Evolutionary Biol 1986; 20:319-387.
7. Oster GF, Shubin N, Murray JD, Alberch P. Evolution and morphogenetic rules: The shape of the vertebrate limb in ontogeny and phylogeny. Evolution 1988; 42:862-884.
8. Hall BK, Miyake T. The membranous skeleton: the role of cell condensations in vertebrate skeletogenesis. Anat Embryol (Berl.) 1992; 186:107-124.
9. Stephens TD, McNulty TR. Evidence of metameric pattern in the development of the chick humerus. J Embryol Exp Morphol 1981; 61:191-205.
10. Geduspan JS, Solursh M. A growth-promoting influence from the mesonephros during limb outgrowth. Dev Biol 1992; 151:242-250.
11. Strecker TR, Stephens TD. Peripheral nerves do not play a trophic role in limb skeletal morphogenesis. Teratology 1983; 27(2):159-167.
12. Fernandez-Teran, M, Piedra, ME, Simandl, BK et al. Limb initiation and development is normal in the absence of the mesonephros. Dev Biol 1997; 189(2):246-255.
13. Crossley PH, Minowada G, MacArthur CA, Martin GR. Roles for FGF-8 in the induction, initiation, and maintenance of chick development. Cell 1996; 84:127-136.
14. Cohn MJ, Izpisua-Belmonte JC, Abud H et al. Fibroblast growth factors induce additional limb development from the flank of chick embryos. Cell 1995; 80:739-746.
15. Vogel A, Rodriguez C, Izpisua-Belmonte, JC. Involvement of FGF-8 in initiation, outgrowth and patterning of the vertebrate limb. Development 1996; 122(6):1737-1750.
16. Ohuchi H, Nakagawa T, Yamamoto A et al. The mesenchymal factor, FGF-10, initiates and maintains the outgrowth of the chick limb bud through interaction with FGF-8, an apical ectodermal factor. Development 1997; 124(2):2235-2244.
17. Min H, Danilenko DM, Scully SA et al. FGF-10 is required for both limb and lung development and exhibits striking functional similarity to Drosophila branchless. Genes Dev 1998; 12(20):3156-3161.

18. Sekine K, Ohuchi H, Fujiwara M et al. FGF 10 is essential for limb and lung formation. Nat Genetics 1999; 21(1):138-141.
19. Kawakami Y, Capdevila J, Buscher D et al. WNT signals control FGF-dependent limb initiation and AER induction in the chick embryo. Cell 2001; 104(6):891-900.
20. Kengaku M, Capdevila J, Rodriguez-Esteban, C et al. Distinct WNT pathways regulating AER formation and dorsoventral polarity in the chick limb bud. Science 1998; 280(5367):1274-1277.
21. Takada S, Stark KL, Shea MJ et al. Wnt-3a regulates somite and tailbud formation in the mouse embryo. Genes Dev 1994; 8(2):174-189.
22. Yoshikawa Y, Fujimori T, McMahon AP, Takada S. Evidence that absence of Wnt-3a signaling promotes neuralization instead of paraxial mesoderm development in the mouse. Dev Biol 1997; 183(2):234-242.
23. Galceran J, Farinas I, Depew MJ et al. Wnt3a-/- -like phenotype and limb deficiency in LEF-1(-/-)TCF-1(-/-) mice. Genes Dev 1999; 13(6):709-717.
24. Dealy CN, Roth A, Ferrari D et al. Wnt-5a and Wnt-7a are expressed in the developing chick limb bud in a manner suggesting roles in pattern formation along the proximodistal and dorsoventral axes. Mech Dev 1993(2-3); 43:175-186.
25. Parr BA, Shea MJ, Vassileva G, McMahon AP. Mouse Wnt genes exhibit discrete domains of expression in the early embryoinc CNS and limb buds. Development 1993; 119(1):247-261.
26. Yamaguchi TP, Bradley A, McMahon AP, Jones S. A Wnt5a pathway underlies outgrowth of multiple structures in the vertebrate embryo. Development 1999; 126(6):1211-1223.
27. He X, Saint-Jeannet JP, Wang Y et al. A member of the frizzled protein family mediating axis induction by Wnt-5A. Science 1997; 275:1652-1654.
28. Slusarski DC, Corces, VG, Moon, RT. Interaction of Wnt and a Frizzled homologue triggers G-protein-linked phosphatidylinositol signalling. Nature 1997; 390(6658):410-413.
29. Sheldahl LC, Park M, Malbon CC, Moon RT. Protein kinase C is differentially stimulated by Wnt and Frizzled homologs in a G-protein-dependent manner. Curr Biol 1999; 9(13):695-698.
30. Ng M. Diaz-Benjumea, FJ, Vincent JP et al. Specification of the wing by localized expression of wingless protein. Nature 1996; 381(6580):316-318.
31. Neumann CJ, Cohen SM. Distinct mitogenic and cell fate specification functions of wingless in different regions of the wing. Development 1996; 122(6):1781-1789.
32. Dickinson ME, Krumlauf R, McMahon, AP. Evidence for a mitogenic effect of Wnt-1 in the developing mammalian central nervous system. Development 1994; 120(6):1453-1471.
33. MacCabe JA, Errick J, Saunders JW Jr. Ectodermal control of the dorsoventral axis in the leg bud of the chick embryo. Dev Biol 1974; 39(1):69-82.
34. Gavin BJ, McMahon JA, McMahon AP. Expression of multiple novel Wnt-1/int-1-related genes during fetal and adult mouse development. Genes Dev 1990; 4(12B):2319-2332.
35. Parr BA, McMahon AP. Dorsalizing signal Wnt-7a required for normal polarity of D-V and A-P axes of mouse limb. Nature 1995; 374(6520):350-353.
36. Riddle RD, Ensini M, Nelson C et al. Induction of the LIM homeobox gene Lmx-1 by Wnt-7a establishes dorsoventral pattern in the vertebrate limb. Cell 1995; 83:631-640.
37. Vogel A, Rodriguez C, Warnken W, Izpisua-Belmonte JC. Dorsal cell fate specified by chick Lmx-1 during vertebrate limb development. Nature 1995; 378(6558):716-720.
38. Cygan JA, Johnson RL, McMahon AP. Novel regulatory interactions revealed by studies of murine limb pattern in Wnt-7a and En-1 mutants. Development 1997; 124(24):5021-5032.
39. Loomis CA, Harris E, Michaud J et al. The mouse Engrailed –1 gene and ventral limb patterning. Nature 1996; 382:360-363.
40. Logan C, Hornbruch A, Campbell I, Lumsden A. The role of Engrailed in establishing the dorsoventral axis of the chick limb. Development 1997; 124(12):2317-2324.
41. Loomis CA, Kimmel RA, Tong CX et al. Analysis of the genetic pathway leading to formation of ectopical ectodermal ridges in mouse Engrailed-1 mutant limbs. Development 1998; 125(6):1137-1148.

42. Yang Y, Niswander L. Interaction between the signaling molecules WNT-7a and SHH during vertebrate limb development: dorsal signals regulate antero-posterior patterning. Cell 1995; 80(6):939-947.
43. Kawakami Y, Wada N, Nishimatsu S, Nohno T. Involvement of frizzled-10 in Wnt-7a signaling during chick limb development. Dev Growth Differ 2000; 42(6):561-569.
44. Wehrli M, Dougan ST, Caldwell K et al. Arrow encodes a LDL-receptor-related protein essential for Wingless signalling. Nature 2000; 407(6803):527-530.
45. Tamai K, Semenov M, Kato Y et al. LDL-receptor-related proteins in Wnt signal transduction. Nature 2000; 407(6803):530-535.
46. Pinson KI, Brennan J, Monkley S et al. An LDL-receptor-related protein mediates Wnt signalling in mice. Nature 2000; 407(6803):535-538.
47. Caplan AI, Pechak DG. The cellular and molecular embryology of bone formation. In: Peck WA, ed. Bone and mineral research. New York: Elsevier, 1987:117-183.
48. Zakany J, Duboule D. Correlation of expression of Wnt-1 in developing limbs with abnormalities in growth and skeletal patterning. Nature 1993; 362(6420):546-549.
49. Rudnicki J A, Brown AM. Inhibition of chondrogenesis by Wnt gene expression in vivo and in vitro. Dev Biol 1997; 185(1):104-118.
50. Kawakami Y, Wada N, Nishimatsu SI et al. Involvement of Wnt-5a in chondrogenic pattern formation in the chick limb bud. Dev Growth Differ 1999; 41(1):29-40.
51. Hartmann C, Tabin CJ. Dual roles of wnt signaling during chondrogenesis in the chicken limb. Development 2000; 127:3141-3159.
52. Ladher RK, Church VL, Allen S et al. Cloning and expression of the Wnt antagonists Sfrp-2 and Frzb during chick development. Dev Biol 2000; 218(2):183-198.
53. Bergwitz C, Wendlandt T, Kispert A, Brabant G. Wnts differentially regulate colony growth and differentiation of chondrogenic rat calvaria cells. Biochim Biophy Acta 2001; 1538(2-3):129-140.
54. Nasevicius A, Hyatt T, Kim H et al. Evidence for a frizzled-mediated wnt pathway required for zebrafish dorsal mesoderm formation. Development 1998; 125(21):4283-4292.
55. Nah HD, Rodgers BJ, Kulyk WM et al. In situ hybridization analysis of the expression of the type II collagen gene in the developing chicken limb bud. Coll Relat Res 1988; 8:277-294.
56. Xu L, Tan L, Goldring MB, Olsen BR, Li Y. Expression of frizzled genes in mouse costochondral chondrocytes. Matrix Biol 2001; 20(2):147-151.
57. Gong Y, Slee RB, Fukai N et al. LDL Receptor-Related Protein 5 (LRP5) affects bone accrual and eye development. Cell 2001; 107(4):513-523.
58. Mao J, Wang J, Liu B et al. Low-density lipoprotein receptor-related proten-5 binds to Axin and regulates the canonical Wnt signaling pathway. Mol Cell 2001; 7(4):801-809.
59. Bergstein I, Eisenberg LM, Bhalerao J et al. Isolation of two novel WNT genes, WNT14 and WNT15, one of which (WNT15) is closely linked to WNT3 on human chromosome 17q21. Genomics 1997; 46:450-458.
60. Hartmann C, Tabin C. Wnt14 plays a pivotal role in inducing synovial joint formation in the developing appendicular skeleton. Cell 2001; 104:341-351.
61. Sen M, Lauterbach K, El-Gabalawy H et al. Expression and function of wingless and frizzled homologs in rheumatoid arthritis. Proc Natl Acad Sci USA 2000; 97(6):2791-2796.
62. Sen M, Chamorro M, Reifert J et al. Blockade of Wnt-5a/frizzled 5 signaling inhibits rheumatoid synoviocyte activation. Arthritis Rheum 2001; 44(4):772-781.
63. Hurvitz JR, Suwairi WM, Van Hul W et al Mutations in the CCN gene family member WISP3 cause progressive pseudorheumatoid dysplasia. Nat Genet 1999; 23:94-98.
64. Pennica D, Swanson TA, Welsh JW et al. WISP genes are members of the connective tissue growth factor family that are up regulated in wnt 1 transformed cells and aberrantly expressed in human colon tumors. Proc Natl Acad Sci USA 1998; 95:14717-14722.
65. Xu L, Corcoran RB, Welsh JW et al. WISP 1 is a Wnt 1 and beta catenin responsive oncogene. Genes Dev 2000; 14(5):585-595.

# Chapter 11

# Wnt/Wg and Heart Development

**Petra Pandur and Michael Kühl**

## Abstract

The heart is the first functional organ in a developing embryo, which ensures the distribution of vital nutrients within the growing organism. For decades, scientists have been intrigued by the well-orchestrated morphological and molecular events that result in the formation of this complex organ, which is composed of differently specialized cells derived from lateral plate mesoderm. Dissecting the genetic pathways involved in cardiogenesis revealed crucial roles for members of various growth factor families in this process. This chapter highlights major events in invertebrate and vertebrate cardiogenesis and focuses on the role of Wnt/Wg signaling in the induction of cardiac fate during early development. In contrast to the essential function of Wg in *Drosophila* cardiogenesis, the role of *Wnt* genes in vertebrate heart development has not yet been fully elucidated. To this date major contributions regarding Wnt function in vertebrate cardiogenesis have come from studies in avian embryos and in Xenopus and therefore will be discussed in more detail.

## Introduction to *Drosophila* Cardiogenesis

Despite its obvious morphological and functional differences, the *Drosophila* heart has proven to be a rich source of knowledge about early molecular events in the determination of cardiac fate. Its tubular morphology, which is why the *Drosophila* heart is also referred to as the dorsal vessel, resembles the primary heart tube in vertebrates. Another feature of cardiogenesis shared by invertebrates and vertebrates is the formation of the heart from two cardiac primordia. The mesodermal progenitors that differentiate into cardiac tissue move towards the dorsal midline where they fuse (because of the reversal of the dorsal-ventral body plan between vertebrates and invertebrates the *Drosophila* heart forms dorsally). There are two major cell types that define the *Drosophila* heart: the cardial and the pericardial cells. The latter form the outer layer of the dorsal vessel and do not express muscle-specific structural genes. The contractile cardial cells form the inner layer and generate the lumen of the heart (Fig. 1A). In contrast to the vertebrate heart, the *Drosophila* heart lacks an endocardium because its hemolymph is pumped through the body in an open circulatory system and blood vessels are absent. Although major progress has been made in the elucidation of cardiogenic molecular events, we are still far from understanding how various signaling pathways act together to initiate cardiac differentiation in a specific subset of mesodermal cells.

One of the crucial molecules in cardiogenesis is the homeobox-containing transcription factor tinman. *Tinman* (*tin*) is initially expressed in the entire presumptive mesoderm and becomes restricted to the dorsal mesoderm after gastrulation.[1,2] Since the dorsal mesoderm gives rise to the heart, gut muscles and dorsal skeletal muscle, flies that are mutant for *tin* not

*Wnt Signaling in Development*, edited by Michael Kühl.

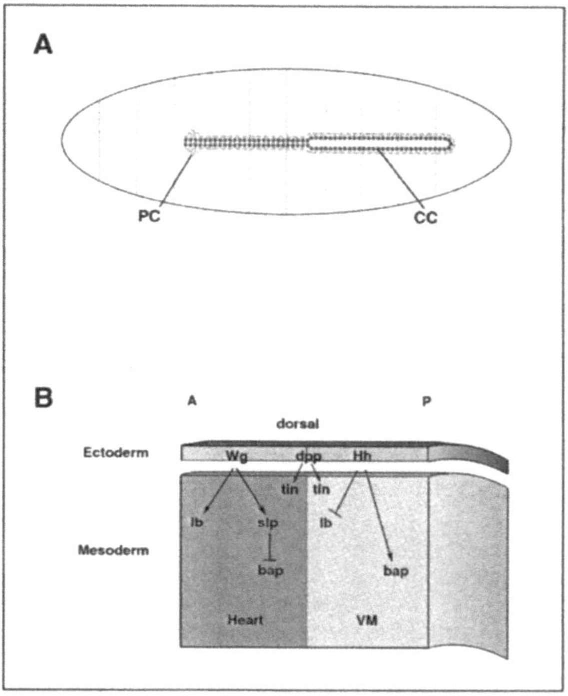


Figure 1. (A) As in vertebrates, the *Drosophila* heart arises from two cardiac primordia, which, in the fly, fuse at the dorsal midline to form a linear tube (dorsal vessel). The *Drosophila* heart consists of two cell types. The cardial cells (CC), which form the contractile inner layer of the heart are surrounded by the pericardial cells (PC), which do not express proteins of the contractile machinery.
(B) The fate of mesodermal cells to enter the cardiac lineage is determined by the coordinated spatiotemporal expression pattern of various growth factors. This simplified cartoon shows that cardiac induction requires the combined expression of *Wingless* (*Wg*), *decapentaplegic* (*dpp*), and *tinman* (*tin*). The Wg-dependent expression of *sloppy paired* (*slp*) promotes the formation of cardiac mesoderm by inhibiting *bagpipe* (*bap*), which is required for the formation of visceral mesoderm (VM). *Ladybird* (*lb*) functions in the process of heart precursor diversification. It needs to be noted that slp and lb are part of feedback loops, which are important for the maintenance of *Wg* expression in the ectoderm. A: anterior, Hh: Hedgehog, P: posterior.

only lack a heart but in addition fail to form visceral mesoderm. Hence, *tin* alone is not sufficient to confer cardiac identity to the cells in which it is expressed and additional factors are required. One of these factors is decapentaplegic (*dpp*), which belongs to the TGFβ family of secreted growth factors. The protein encoded by *dpp* is secreted from the dorsal ectoderm and is required for the maintenance of *tin* expression in the underlying mesoderm (Fig. 1B).[2,3] The restricted expression of *dpp* within the dorsal ectoderm is maintained by DJun and consequently, *tin* expression in flies that are mutant for the effector of DJun kinase signaling is either reduced or absent.[4,5] It is of further importance to note that activation of the JNK pathway in the dorsalmost cells is regulated by Wg.[6] Flies that do not express *dpp* exhibit the same phenotype as *tin* mutants, that is, they lack cardiac as well as visceral mesoderm. Moreover, although *dpp* and *tin* are necessary for heart formation, they are not sufficient as they can not significantly expand the cardiac primordia when ectopically expressed.[2,7,8] These results indicate that multiple signaling pathways are involved in the specification of cardiac fate. Since a detailed description of additional regulatory proteins involved in *Drosophila* heart development can not be considered here, the reader is referred to the corresponding literature.[9-13]

## Wg Signaling in *Drosophila* Cardiogenesis

In addition to Dpp, the dorsal ectoderm is the source of another secreted signaling molecule, the segment polarity gene *wingless* (*wg*), which is the homolog of vertebrate *Wnt-1*. *Wg* is expressed in 15 transversal stripes, perpendicular to the *dpp* expression domain, in the trunk region of the embryo.[14] Wg, like Dpp, is involved in a plethora of patterning events during *Drosophila* development. However, with respect to mesoderm specification, *wg* mutants fail to form cardiac progenitor cells whereas formation of visceral mesoderm is unaffected. Some somatic mesoderm also forms, albeit abnormally patterned. By using a temperature-sensitive *wg* allele Wu et al.[15] showed that the cardiogenic function of zygotic *wg* is distinct from its role in segmentation and neurogenesis and that it is required for heart precursor formation at the stage when *tin* is restricted to dorsal mesoderm. Hence, *wg* provides additional spatial cues leading to a further subdivision of the *dpp* and *tin* expressing dorsal mesoderm (Fig. 1B). Although *wg* is also transiently expressed in the dorsal mesoderm during gastrulation, the crucial positional information as to where the heart progenitors are formed is provided by the wg signal emitted from the dorsal ectoderm.[16] A more detailed investigation of the spatiotemporal relationship between *Wg*, *dpp*, and *tin* expression demonstrates very nicely that their concerted dynamic expression pattern determines which cells become specified to enter the cardiac lineage. Misexpression studies showed that only *tin* expressing cells that have been exposed to simultaneous Wg and Dpp signaling at defined stages during development undergo cardiac specification.[5] As one would predict, examination of components of the Wg signal transduction cascade identified dishevelled (*dsh*), armadillo (*arm*), and dTCF to also positively regulate heart development.[17,18] Embryos that are mutant for both maternal and zygotic *dsh* or *arm* lack any heart precursors.[17] A similar dramatic effect was observed in flies overexpressing a dominant-negative form of dTCF.[18] Additionally, overexpression of Dsh results in an increase of heart precursor cells.[17] The fact that Dsh becomes hyperphosphorylated either in response to Wg signaling or when it is overexpressed has been opportune to show that *dsh* is epistatic to wg and is required autonomously within the mesoderm for heart development.[17,19]

The situation is somewhat more complex for zeste-white3/shaggy (zw3), which is regulated by Dsh. Zw3 acts antagonistically to Dsh and arm and therefore, one would expect that overexpression of *zw3* suppresses heart precursor formation and loss of *zw3* function results in an increase thereof. Neither is the case though. A more detailed investigation of the role of *zw3* in cardiogenesis showed that zw3 has two distinct functions in mesoderm specification and thus the observed phenotypes require careful interpretation.[18] The initial requirement of zw3 function to specify dorsal mesoderm is obvious in *zw3* mutants, which are characterized by severely reduced tin and fasciclin III expression domains, the latter being a marker for visceral mesoderm.[18] Hence, the loss of cardiac progenitor cells is primarily due to a failure of dorsal mesoderm formation. To further elucidate the role of zw3, Park et al.[18] performed overexpression experiments using a zw3 construct, which is under the control of a heat-shock promoter. Excess zw3 function elicited both cardiac hypertrophy and failure of heart formation. The first observation is consistent with the function of *zw3* as specifier of dorsal mesoderm, the latter indicates an inhibitory role for *zw3* in cardiogenic wg signaling. These important findings suggest two temporally distinct functions for *zw3*, which would be consistent with studies on the temporal requirement for wg itself.[15] In vertebrates, a similar scenario regarding the role of GSK3 and β-catenin (*zw3* and *arm* in *Drosophila*) in patterning the dorsal-ventral axis and in heart formation has to be considered and will be discussed in more detail later.

For a better understanding of the role of *wg* in heart formation it is necessary to put *wg* signaling in the context of other segment patterning genes. Initial molecular events during

*Drosophila* development subdivide the ectoderm into segmentally repeated units, each of which is composed of an anterior (A) and a posterior (P) compartment. Within each segment, *wg* expression is confined to the A compartment, whereas the P compartment is characterized by the expression of *hedgehog* (*hh*), another secreted segmentation gene product that positively regulates Wg transcription (Fig. 1B).[20,21] Like wg, hh is required for heart formation during early gastrula.[14] In contrast to wg null mutants, however, flies lacking *hh* expression possess some cardiac precursor cells. Furthermore, overexpression of wg in a *hh* mutant background demonstrated that *wg* signaling determines cardiac fate directly and does not exert its function by regulating other segmentation genes.[17] It is also noteworthy that the mesodermal heart precursors form right beneath the ectodermal *wg* expression domains, which is consistent with the notion that the secreted *wg* gene product acts on neighboring cells. For example, expression of the homeobox transcription factor *ladybird* (*lb*) requires high Wg concentrations and as a consequence is restricted to heart cells that are located underneath dorsal ectodermal cells expressing *wg* (Fig. 1B).[22] Within the P compartment, *hh* signaling suppresses *lb* expression.[22] The gene products of *lb* (lb early and lb late) also play a role in the diversification of cardiac lineages where during the segregation of heart precursors *lb* genes act to specify a subset of cardioblasts and pericardial cells.[22] In addition to wg, the forkhead transcription factor *sloppy paired* (*slp*) is expressed in the ectodermal and mesodermal portions of each A compartment. The identification of putative binding sites for the Wg effector Pangolin (the *Drosophila* homolog of vertebrate LEF-1/TCF) within the 5'-flanking region of the *slp* gene, implicates that *slp* is a direct target of Wg signaling.[20] Sloppy paired plays an important role in the specification of cardiac and visceral mesoderm. It is required for the formation of heart progenitor cells, as flies mutant for *slp* do not form a heart.[17] A likely explanation for this effect is the fact that ectodermal *wg* expression requires slp and hence, the loss of cardiac precursors results from the lack of *wg* signaling.[20,23] It is known that in the ectoderm *wg* and *slp* act in an autoregulatory feed back loop.[20] *slp* function is also required to suppress visceral mesoderm formation by inhibiting *bagpipe* (*bap*) expression, which is consistent with the observation that loss of *wg* signaling is accompanied by an expansion of visceral mesoderm.[24] Furthermore, *wg* and *slp* cooperate in cardiogenesis as neither wg or slp alone can rescue the loss of heart progenitors in embryos mutant for *slp* and *wg*, respectively.[20] Consistent with this finding, only coexpression of *wg* and *slp* in the mesoderm resulted in a significant increase of heart progenitor cells.[20] The experiments also revealed the complexity of the positive regulatory role of Wg in heart development. Although slp is required for mediating the Wg signal it is not sufficient. This suggests that cells expressing *slp* are rendered competent to respond to subsequent *wg* signals.[20]

The *Drosophila* mutant *heartless* (*htl*) demonstrates the requirement of FGF signaling in heart formation. *heartless* codes for an FGF receptor, which is coexpressed with *tin* throughout the mesoderm. Unlike the factors discussed so far, *htl* is not required for the processes that specify cardiac mesoderm but rather for the migration of mesodermal cells towards the dorsal side of the embryo.[25,26] Previous findings indicate that domains of the dorsal mesoderm with overlapping activities of Dpp, tin, and Wg could create an environment that makes those cells competent to respond to FGF receptor signaling.

In summary, these findings demonstrate the crucial role of Wg signaling in *Drosophila* cardiogenesis and they outline the position of Wg within this complex regulatory network of molecules involved in cardiac differentiation.

## Introduction to Vertebrate Cardiogenesis

As in *Drosophila*, the vertebrate heart derives from distinct patches of mesodermal cells that, under the influence of multiple signaling events, have been specified to form the cardiac primordia. These primordia are initially located on either side of the embryonic organizer (Fig. 2A,B). As development progresses they eventually reach their final destination, the ventral

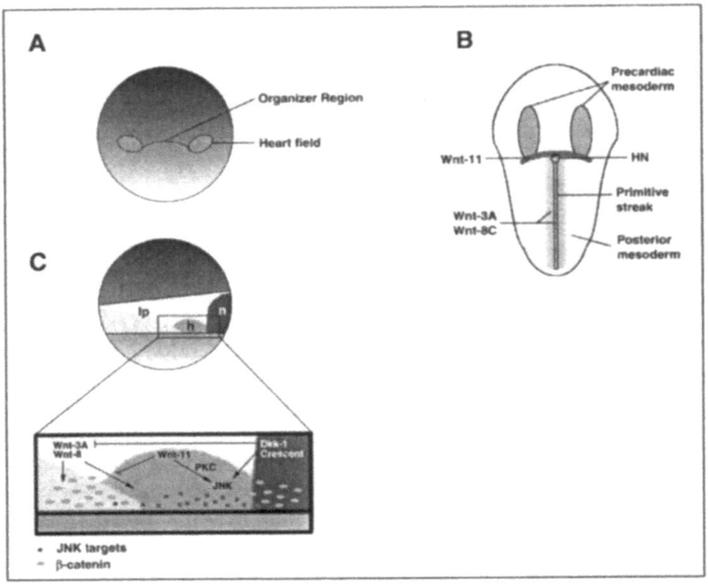


Figure 2. (A) Schematic drawing of a gastrula stage *Xenopus* embryo, depicting the dorsal blastoporus and the bilaterally established cardiac primordia adjacent to the Organizer region.
(B) Schematic drawing of a gastrula stage chicken embryo. The cardiac primordia are located on either side of Hensen's node (HN). Wnt-11 is expressed in the posterior region of the cardiac primordia, whereas expression of Wnt-3A and Wnt-8C is predominantly detected in Hensen's node and the primitive streak. In contrast to Wnt-3A, which becomes restricted to the dorsal midline, Wnt-8C expression is more diffuse within the posterior mesoderm. As discussed in the text, Wnt-3A and Wnt-8C function as signals that suppress cardiac induction in posterior mesoderm and could serve to refine the boundaries of the initial heart fields.
(C) In *Xenopus*, cardiac specification of mesodermal cells requires low activities of signaling through β-catenin and simultaneous activation of Jun-N-terminal kinase (JNK) signaling. This is accomplished by the inhibition of the Wnt/β-catenin pathway at various levels. As in chicken, the cardiac-inhibiting function of XWnt-3A and XWnt-8 could be required to establish the boundaries of the heart field.

midline where they fuse to form the primary heart tube. The heart tube is composed of an outer, contractile layer, the myocardium, which encloses the inner endocardium. During development, the latter becomes continuous with the endothelium of the blood vessels. The heart tube is composed of the outflow tract (bulbus cordis), ventricle, atrium, and sinus venosus, which are arranged in a linear order along the anteroposterior axis of the embryo. Soon after its formation, the primary heart tube undergoes major morphological rearrangements, which place the distinct functional units at their proper positions to give rise to the multichambered heart. Explant studies and transplantation experiments have provided initial clues as to the sources of the molecules involved in cardiogenesis. These areas comprise the organizer region including more lateral domains as well as the underlying endoderm.[27-29]

The combinatorial expression of cardiac transcription factors characterizes cells that are specified to form cardiac tissue. Signaling molecules that act upstream of these transcription factors have been identified by their ability to induce cardiac differentiation in non-cardiogenic mesoderm. For example, the combination of bone morphogenetic proteins 2/4 (BMP-2/4), which belong to the TGFβ-family, and FGF-4 is able to convert avian posterior mesoderm, which normally does not contribute to the heart, into cardiac tissue in chicken.[30,31] In *Xenopus*, BMP signaling is necessary for the maintenance of *XNkx2.5* expression, terminal differentiation, and normal cardiac morphogenesis.[32-34] The most dramatic effect of a secreted molecule on early cardiogenesis has been observed with cripto-1, an EGF-like signaling protein. Mice that are mutant for *cripto-1* lack cardiomyocytes.[35]

Although the cardiac-promoting role of some growth factors has been thoroughly investigated, our current knowledge about regulatory events leading to heart formation is mostly based on previously characterized transcription factors that belong to very different gene families. Exploiting the advantages of different model organisms, studies of these transcription factors have provided major insight into various steps of cardiogenesis. These range from the specification of cardiac fate within a particular population of mesodermal cells to the establishment of diverse structures, which constitute the adult heart. One of the biggest challenges up to this date is to identify molecules that regulate these cardiac transcription factors and to understand their intertwined regulatory relationships in this context.

The crucial function of *tinman* during *Drosophila* heart development initiated the search for its vertebrate homolog and was found in *Nkx2.5*. This homeobox transcription factor of the NK2 family is one of the earliest cardiac marker genes expressed in vertebrate embryos. RNA injections of a dominant-negative Nkx2.5 construct inhibit formation of the linear heart tube and terminal differentiation of cardiomyocytes in *Xenopus* embryos.[36] Nkx2.5 null mutant mice produce a lethal phenotype due to aberrant heart development.[37] The phenotype is characterized by defective looping morphogenesis and reduced or eliminated cardiac marker gene expression.[37-40] It needs to be noted though, that none of the phenotypes elicited by negative interference with *Nkx2.5* expression in any vertebrate embryo is as dramatic as the complete failure of heart formation observed in *Drosophila*. This is likely due to the presence of additional Nkx2 genes in vertebrates whose functions appear to be redundant. It also implicates that the action of various genes is required for cardiac specification. In fact, a regulatory circuit has been demonstrated for Nkx2.5 and GATA factors in the heart.[41-44] The GATA family of zinc finger transcription factors comprises six members, three of which, GATA-4, -5, and -6, have been shown to function in cardiogenesis.[43,45] Like Nkx2.5, expression of GATA-4/5/6 commences during gastrulation when cardiac progenitors are specified. GATA proteins can regulate the transcription of multiple cardiac genes, for example, overexpression of GATA-4 in *Xenopus* embryos results in a premature activation of genes encoding *α-myosin heavy chain* and *cardiac α-actin*.[45] Introduction of antisense GATA-4 transcripts into the mouse embryonic carcinoma stem cell line P19 blocks the DMSO-inducible expression of cardiac muscle markers.[46] Targeted mutations in the GATA-4 gene in mice disturb normal migration of cardiac cells, which results in cardia bifida.[47,48]

A thorough description of all known factors involved in vertebrate heart development would extend beyond the topic of this chapter and therefore we refer the reader to refs. 49-54.

## Wnt Signaling in Vertebrate Cardiogenesis

In contrast to the well-investigated roles of the signaling molecules described above, the function of members of the *Wnt* gene family in vertebrate cardiogenesis has only started to be under thorough examination. But already, the previously published data on the role of Wnt proteins in heart development has revealed the complexity of this subject, which arises from the existence of different Wnt signal transduction cascades.

Among the known Wnt proteins, several are expressed in a spatiotemporal pattern that implicates their involvement in heart development. The conserved expression pattern of *Wnt-11* in precardiac mesoderm in birds, mouse, and *Xenopus* makes this gene the best candidate molecule for playing a role in signaling events leading to cardiac differentiation.[55-58] In mouse and chick, *Wnt-11* expression is also conserved in the primary heart tube and transcripts of human *Wnt-11* are detected in the adult heart.[59-61] In the early mouse gastrula, Wnt-2 expression overlaps with that of Wnt-11.[62] Later in development mWnt-2 expression becomes restricted to the pericard.[63] Mouse *Wnt-8* is only weakly expressed during early embryogenesis, followed by a peak of expression at stage 10.5 when its transcripts are detected in the myocardium at high levels.[63] Additional avian and mouse *Wnt* genes expressed within or lateral to the primitive streak during early gastrula are *Wnt-3a, Wnt-5a, Wnt-5b,* and *Wnt-8c.*[64-66] Four other Wnt genes are expressed in the tubular heart of the chick: *Wnt-2b, Wnt-14, Wnt-5a,* and *Wnt-7a.*[60,67] To this date, only Wnt-11 has been ascribed a promoting role in cardiac differentiation. This does not exclude positive regulatory roles for other Wnts expressed in cardiac tissue unfortunately there is no well-documented experimental evidence for this hypothesis yet available. Targeted gene disruptions of different Wnt genes in mice suggest that the functions of these proteins are interchangeable.

The Wnt signal is transduced after the ligand has bound to its putative Frizzled (Fz) receptor. Transcripts of different Fz have been detected in cardiac tissue including cardiac mesoderm and adult heart as well as in cardiac neural crest cells. These are human *Fz-1, -2, -7, -8, -9,* mouse *Fz-2, -4, -9,* and Xenopus *Fz-7, -8, -10a, -10b.*[68-75]

## Vertebrate Cardiogenesis Involves Different Wnt Signal Transduction Cascades

Secreted signals from the embryonic organizer are essential for the formation of the vertebrate heart. For example, *Xenopus* embryos lacking the organizer region fail to form cardiac tissue despite the presence of presumptive cardiac mesoderm.[27] Recent studies in chick and *Xenopus* demonstrated that cardiac tissue formation can be initiated in non-cardiogenic posterior ventral mesoderm by overexpression of the Wnt inhibitors dickkopf-1 (Dkk-1) and crescent, which are secreted by the embryonic organizer.[76,77] Ectopic expression of Dkk-1 and crescent not only induce cardiac marker gene expression but also stimulate the formation of beating tissue. In contrast to Dkk-1, crescent induces a contractile phenotype to a much lesser extent. Interestingly, ectopic cardiac differentiation is not elicited by other Wnt inhibitors, such as WIF, FrzA and Sizzled or dominant-negative Wnt-8.[77] A possible explanation could be that these inhibitors differ in their abilities to block signaling of different Wnts.[78-81] Dkk-1, for example, is a very potent inhibitor of Wnts that activate the β-catenin signaling pathway.[82] Overexpression of GSK3β RNA, which inhibits maternal β-catenin-mediated Wnt signaling in *Xenopus* posterior ventral mesoderm, phenocopied the effect of Dkk-1, further suggesting that cardiac induction is initiated simply by blocking β-catenin signaling. This assumption is supported by the finding that activation of zygotic β-catenin signaling in the heart-forming regions of *Xenopus* embryos by Wnt-3a and Wnt-8 inhibits endogenous cardiac gene expression.[77] The conclusion, however, is challenged by two observations. First, it needs to be noted that GSK3β is part of a multi-protein complex, whose regulation is influenced by multiple signals in a temporally regulated manner and in dependence on the cellular context.[83,84] If the mere inhibition of β-catenin signaling was indeed sufficient to induce cardiogenesis, the introduction of a dominant-negative mutant of LEF-1/TCF should phenocopy the effect of Dkk-1 and GSK3β. This specific block of β-catenin signaling on the level of transcription, however, did not induce cardiogenesis in posterior ventral mesoderm suggesting that additional signaling events must be involved in this process.[85] These events could be triggered by Wnts that do not act through β-catenin. This idea is supported by data demonstrating that overexpression of Wnt-5a and Wnt-11, which activate non-canonical signaling pathways, in the presumptive heart region had no inhibitory effect on heart formation in vivo.[77]

Before we discuss the cardiac-promoting role of Wnt-11, another feature of β-catenin signaling during *Xenopus* development needs to be stressed. It is important to distinguish between the role of maternal and zygotic β-catenin-mediated signaling. Maternally provided β-catenin is essential for the formation of dorso-anterior tissue. UV irradiation of *Xenopus* eggs or LiCl treatment of early cleavage stage embryos lead to ventralization or hyperdorsalization of the tissue. Both effects are elicited by interference with β-catenin signaling, which concomitantly leads to changes in the expression of genes characteristic for either tissue type.[86] Hence, with regard to cardiogenesis, the expression domains of GATA-4,-5,-6 are expanded in LiCl-treated embryos, whereas UV-irradiated embryos lack expression of these cardiac genes.[45] Likewise, embryos that were hyperdorsalized by dorsal injections of XWnt-8 RNA had expanded expression domains of XNkx2.5 (own observation). Indeed, evidence for a dual role for β-catenin signaling in *Xenopus* development has already been provided with regard to head formation. Here, maternal β-catenin signaling is initially required to determine dorso-anterior cell fate, subsequently however, Wnt/β-catenin signaling has to be inhibited for head formation to occur.[87] In fact, ectopic activation of zygotic β-catenin signaling by XWnt-8 interferes with head development.[88] Last but not least, the dual role of vertebrate β-catenin resembles the dual role of *Drosophila zw3*, which subsequently regulates arm function in the formation of the invertebrate heart.

With respect to cardiogenesis the data described above suggests that zygotic β-catenin signaling activated by Wnt-8 and/or Wnt-3a may be required to restrict the heart forming domains within the dorsolateral mesoderm. It is known that the initial heart field, which comprises cells that have the potential to become cardiac tissue is broader than the actual heart forming area.[89-91] Additionally, it has been demonstrated that signals from the neural tube and notochord repress cardiogenesis.[90,92-94] Results obtained from ablation and explant studies in chicken are consistent with such a role for Wnt-8c and Wnt-3a, which are expressed in neural tissue (Fig. 2B). In detail, removal of the anterior neural tube initiated cardiogenesis in head mesenchyme, whereas the cardiac inducing property of anterior endoderm in the overlying mesoderm was inhibited in cocultures of anterior mesoderm with neural tube and notochord.[95]

Studying the role of Wnt-11 in cardiac induction was pioneered by Eisenberg et al.[57] who used the quail mesoderm cell line QCE-6 as a culture model. The QCE-6 cell line is representative of early nondifferentiated mesoderm cells and has the potential to differentiate into cardiomyocytes, endothelial or red blood cells.[96] Differentiation of QCE-6 cells is initiated by treating the cells with a combination of retinoic acid, bFGF, TGFβ2, and TGFβ3.[96] Most importantly though, the endogenous expression of Wnt-11 in this cell line is a requisite for myocardial differentiation, as Wnt-11-minus cells did not immunoreact with an antibody against muscle myosin heavy chain.[57] This cardiogenic function of Wnt-11 was further elucidated in quail embryos.[97] Explants of anterior precardiac mesendoderm dissected at stage 5 or 6 differentiate into cardiac tissue, i.e. they become contractile in culture, whereas explants of posterior mesendoderm do not differentiate into beating tissue. Interestingly, explants of the posterior mesendoderm do become contractile and express sarcomeric proteins when dissected at stage 4.[97] To test the ability of Wnt-11 to ectopically activate a cardiac developmental program, posterior non-cardiogenic mesendoderm was exposed to Wnt-11. Indeed, addition of Wnt-11 promoted cardiogenesis in posterior tissue dissected at stage 5 or 6.[97] These results and the fact that Wnt-11 is the only Wnt gene known to this date to be expressed in avian precardiac mesoderm demonstrate the importance of Wnt-11 activated signaling pathways in heart development. As in the avian embryo, *Xenopus Wnt-11* (*XWnt-11*) is present at the right time and in the appropriate areas to play a role in heart development. In *Xenopus*, cardiac specification of mesodermal cells is completed by the end of gastrulation. XWnt-11 RNA is present maternally and is post-transcriptionally regulated, which leads to an accumulation of XWnt-11 protein on the dorsal side of the embryo as early as the 64-cell stage.[58] After the onset of zygotic transcription, Xwnt-11 mRNA first becomes detectable in the dorsal marginal zone where high levels persist during gastrulation.[55] Loss-of-function studies of XWnt-11 finally demonstrate a crucial function for this gene in cardiogenesis, since inhibition of XWnt-11 signaling, for example by morpholino antisense oligonucleotides, dramatically decreases cardiac marker gene expression.[85] Overexpression of XWnt-11 in ectodermal tissue explants demonstrated that this gene is sufficient to induce early cardiac genes, such as GATA-4, GATA-6 and XNkx2.5, as well as genes coding for structural proteins of the myocardium.[85] Most importantly, ectopic expression of XWnt-11 in ventral marginal zone (VMZ) explants is sufficient to convert this non-cardiogenic tissue to contractile tissue.

Although the molecular mechanism of Wnt-11 triggered cardiogenesis in avian embryos remained obscure at the time, several studies regarding gastrulation movements in *Xenopus* have demonstrated in the meantime that Wnt-11 activates non-canonical Wnt signaling pathways.[98,99] Known mediators of non-canonical signaling are PKC, $Ca^{2+}$/calmodulin- dependent kinase II (CamKII) in vertebrates and Jun-N-terminal kinase (JNK) in *Drosophila*. In two independent experimental assays it could be demonstrated that the regulatory function of cardiac genes by XWnt-11 involves PKC and JNK but not CamKII.[85] Decreased cardiac gene expression in dorsal marginal zone (DMZ) explants that include the heart forming region as a result of inhibiting XWnt-11 signaling could be rescued by activation of PKC.[85] Injections of a kinase-dead mutant form of JNK phenocopied the effect of dnXWnt-11 in DMZ explants. Consistent with these findings, JNK-1 activation was shown to be regulated by XWnt-11, as XWnt-11 overexpression leads to increased levels of phosphorylated JNK-1 and injections of XWnt-11 morpholino antisense oligonucleotides resulted in a decrease thereof.[85] The activation of JNK-1 was shown to be specific for XWnt-11 and could not be mimicked by XWnt-8.

The picture that emerges from the data summarized above is that cardiac induction is initiated in a region of the embryo that is characterized by low levels of β-catenin signaling activity and high levels of JNK signaling activity (Fig. 2C). This assumption is substantiated by the findings that inhibitors of the Wnt/β-catenin pathway, Dkk-1 and crescent, as well as GSK3, can also activate JNK signaling. [85]

The conservation of the cardiogenic function of Wnt-11 could be demonstrated using the mouse embryonic carcinoma stem cell line P19. The P19 cell line is a widely used model for investigating molecular events involved in cardiac differentiation. Addition of DMSO to the culture medium activates a cardiac developmental program in these cells and likewise, treatment with Wnt-11 conditioned medium mimics the effect of DMSO.[85]

In summary, the available data on Wnt function in cardiogenesis may be interpreted as follows, currently being most appropriate for *Xenopus* but also likely to apply for chick:

During early cleavage stages maternal β-catenin specifies dorsoanterior tissue and maternally provided XWnt-11, together with proteins of other growth factor families initiate events that are required for the establishment of the bilateral heart fields in presumptive dorsolateral mesoderm. The cardiac promoting processes regulated by these signaling molecules continue through gastrula stages when heart precursor cells become specified. The emergence of zygotically expressed Wnts in the ventrolateral mesoderm and neural tissue, Wnt-8 and Wnt-3a, that counteract cardiac development, would then be necessary to restrict the initial heart field, thereby refining the area of cells that are specified to become cardiac tissue. Obviously, β-catenin-mediated signaling by these Wnts needs to be blocked within the cardiac tissue, which is achieved by the appropriate inhibitors secreted from the organizer. At the same time, components of the non-canonical signaling pathways, PKC and JNK, that are activated, for example, by XWnt-11 and Dkk-1, promote cardiogenesis in the dorsolateral mesoderm.

## Conclusion and Future Perspectives

In conclusion, multiple signaling pathways converge to specify cardiac fate in cells that have been exposed to diverse growth factors at defined stages during development. To this date, the role of Wg in the formation of the heart in *Drosophila* has been studied in more detail and its requirement has been demonstrated in multiple experiments. Several lines of evidence indicate an important function for Wnt genes in vertebrate heart development. Studies conducted so far have primarily demonstrated the roles of various members of this gene family in early steps of cardiogenesis. They also make clear that Wnts belonging to different classes (Wnt-1 or Wnt-5a class) may differ in their function of regulating the expression of cardiac genes. Therefore, detailed analyses of *Wnt* gene expression in combination with gain-and loss-of-function studies of Wnts are required to deepen our understanding of the roles of these secreted factors during early cardiogenesis as well as during later events in heart development. In the same way as findings obtained from embryological studies in various model organisms will be helpful with respect to repair processes after heart injury, the determination of molecules involved in various cardiac diseases will help to further elucidate the molecular pathways in the formation of this complex organ.

### *Acknowledgements*

We would like to thank Daniel Maurus, Marc-Phillip Hitz and Matthias Läsche for discussion and comments on the manuscript. Work in our lab has been supported by the Deutsche Forschungsgemeinschaft (DFG, Junior Group SFB 271).

## References

1. Bodmer R. The gene tinman is required for the specification of the heart and visceral muscles in Drosophila. Development 1993; 118:719-729.
2. Yin Z, Frasch M. Regulation and function of tinman during dorsal mesoderm induction and heart specification in Drosophila. Dev Genet 1998; 22:187-200.
3. Frasch M. Induction of visceral and cardiac mesoderm by ectodermal Dpp in the early Drosophila embryo. Nature 1995; 374:464-467.
4. Riesgo-Escovar JR, Hafen E. Drosophila Jun kinase regulates expression of decapentaplegic via the ETS-domain protein Aop and the AP-1 transcription factor DJun during dorsal closure. Genes Dev 1997; 11:1717-1727.
5. Lockwood WK, Bodmer R. The patterns of wingless, decapentaplegic, and tinman position the Drosophila heart. Mech Dev 2002; 114:13-26.
6. McEwen DG, Cox RT, Peifer M. The canonical Wg and JNK signaling cascades collaborate to promote both dorsal closure and ventral patterning. Dev 2000; 127:3607-3617.
7. Rugendorff A, Younossi-Hartenstein A, Hartenstein V. Embryonic origin and differentiation of the Drosophila heart. Roux's Arch Dev Biol 1994; 203:266-280.
8. Klapper R, Holz A, Janning W. Fate map and cell lineage relationships of thoracic and abdominal mesodermal anlagen in Drosophila melanogaster. Mech Dev 1998; 71:77-87.
9. Bodmer R, Frasch M. Genetic determination of Drosophila heart development. In Rosenthal N, Harvey R (eds). Heart Development. New York; Academic Press 1999:65-90.
10. Bodmer R, Venkatesh TV. Heart development in Drosophila and vertebrates: Conservation of molecular mechanisms. Dev Genet 1998; 22:181-186.
11 .Frasch M. Intersecting signalling and transcriptional pathways in Drosophila heart specification. Cell Dev Biol 1999; 10:61-71.
12. Fossett N, Schulz RA. Conserved cardiogenic functions of the multitype zinc-finger proteins u-shaped and fog-2. Trends Cardiovasc Med 2001; 5:185-190.
13. Gajewski K, Zhang Q, Choi CY et al. Pannier is a transcriptional target and partner of Tinman during Drosophila cardiogenesis. Dev Biol 2001; 233:425-436.
14. Baker N. Molecular cloning of sequences from wingless, a segment polarity gene in Drosophila: the spatial distribution of a transcript in embryos. EMBO J 1987; 6:1765-1773.

15. Wu X, Golden K, Bodmer R. Heart development in Drosophila requires the segment polarity gene wingless. Dev Biol 1995; 169:619-628.
16. Lawrence PA, Bodmer R, Vincent JP. Segmental patterning of heart precursors in Drosophila. Development 1995; 121:4303-4308.
17. Park M, Wu X, Golden K et al. The wingless signaling pathway is directly involved in Drosophila heart development. Dev Biol 1996; 177:104-116.
18. Park M, Venkatesh TV, Bodmer R. Dual role for the zeste-white3/shaggy-encoded kinase in mesoderm and heart development. Dev Gen 1998; 22:201-211.
19. Yanagawa S, Leeuwen F, Wodarz A et al. The Dishevelled protein is modified by wingless signaling in Drosophila. Genes Dev 1995; 9:1087-1097.
20. Lee HH, Frasch M. Wingless effects mesoderm patterning and ectoderm segmentation events via induction of its downstream target sloppy paired. Development 2000; 127:5497-5508.
21. Perrimon N. The genetic basis of patterned baldness in Drosophila. Cell 1994; 76:781-784.
22. Jagla K, Frasch M, Jagla T et al. Ladybird, a new component of the cardiogenic pathway in Drosophila required for diversification of heart precursors. Development 1997; 124:3471-3479.
23. Cadigan KM, Grossniklaus U, Gehring WJ. Localized expression of sloppy paired protein maintains the polarity of Drosophila parasegments. Genes Dev 1994; 8:899-913.
24. Azpiazu N, Lawrence P, Vincent JP et al. Segmentation and specification of the Drosophila mesoderm. Genes Dev 1996; 10:3183-3194.
25. Gisselbrecht S, Skeath J, Doe C et al. heartless encodes a fibroblast growth factor receptor (DFR1/DFGF-R2) involved in the directional migration of early mesodermal cells in the Drosophila embryo. Genes Dev 1996; 10:3003-3017.
26. Michelson AM, Gisselbrecht S, Baek KH et al. Dual functions of the heartless fibroblast growth factor receptor in the development of Drosophila embryonic mesoderm. Dev Genet 1998; 22:212-229.
27. Sater AK, Jacobson AG. The role of the dorsal lip in the induction of heart mesoderm in Xenopus laevis. Development 1990a; 108:461-470.
28. Nascone N, Mercola M. An inductive role for the endoderm in Xenopus cardiogenesis. Development 1995; 121:515-523.
29. Schultheiss TM, Xydas, S, Lassar AB. Induction of avian cardiac myogenesis by anterior endoderm. Development 1995; 121:4203-4214.
30. Lough J, Barron M, Brogley M et al. Combined BMP-2 and FGF-4, but neither factor alone, induces cardiogenesis in non-precardiac embryonic mesoderm. Dev Biol 1996; 178:198-202.
31. Barron M, Gao M, Lough J. Requirement for BMP and FGF signaling during cardiogenic induction in non-precardiac mesoderm is specific, transient, and cooperative. Dev Dyn 2000; 218:383-393.
32. Shi Y, Katsev S, Cai C et al. BMP signaling is required for heart formation in vertebrates. Dev Biol 2000; 224:226-237.
33. Breckenridge RA, Mohun TJ, Amaya E. A role for BMP signalling in heart looping morphogenesis in Xenopus. Dev Biol 2001; 232:191-203.
34. Walters MJ, Wayman GA, Christian JL. Bone morphogenetic protein function is required for terminal differentiation of the heart but not for early expression of cardiac marker genes. Mech Dev 2001; 100:263-273.
35. Xu C, Liguori G, Persico MG et al. Abrogation of the Cripto gene in mouse leads to a failure of postgastrulation morphogenesis and lack of differentiation of cardiomyocytes. Development 1999; 126:483-494.
36. Grow MW, Krieg PA. Tinman function is essential for vertebrate heart development: elimination of cardiac differentiation by dominant inhibitory mutants of the tinman-related genes, XNkx2-3 and XNkx2-5. Dev Biol 1998; 204:187-196.
37. Lyons I, Parsons LM, Hartley L, et al. Myogenic and morphogenetic defects in the heart tubes of murine embryos lacking the homeobox gene Nkx-2.5. Genes Dev 1995; 9:1654-1666.
38. Zou Y, Evans S, Chen J et al. CARP, a cardiac ankyrin repeat protein, is downstream in the Nkx2-5 homeobox gene pathway. Development 1997; 124:793-804.
39. Biben C, Harvey, RP. Homeodomain factor Nkx2-5 controls left/right asymmetric expression of bHLH gene eHand during murine heart development. Genes Dev 1997; 11:1357-1369.

40. Durocher D, Chen CY, Ardati A et al. The atrial natriuretic factor promoter is a downstream target for Nkx2-5 in the myocardium. Mol Cell Biol 1996; 16:4648-4655.
41. Searcy RD, Vincent EB, Liberatore CM et al. A GATA-dependent nkx-2.5 regulatory element activates early cardiac gene expression in transgenic mice. Development 1998; 125:4461-4470.
42. Lien CL, Wu C, Mercer B et al. Control of early cardiac-specific transcription of Nkx2-5 by a GATA-dependent enhancer. Development 1999; 126:75-84.
43. Jiang Y, Drysdale TA, Evans T. A role for GATA-4/5/6 in the regulation of Nkx2.5 expression with implications for patterning the precardiac field. Dev Biol 1999; 216:57-71.
44. Molkentin JD, Antos C, Mercer B et al. Direct activation of a GATA6 cardiac enhancer by Nkx2.5: evidence for a reinforcing regulatory network of Nkx2.5 and GATA transcription factors in the developing heart. Dev Biol 2000; 217:301-309.
45. Jiang Y, Evans T. The Xenopus GATA-4/5/6 genes are associated with cardiac specification and can regulate cardiac-specific transcription during embryogenesis. Dev Biol 1996; 174:258-270.
46. Grepin C, Robitaille L, Antakly T et al. Inhibition of transcription factor GATA-4 expression blocks in vitro cardiac muscle differentiation. Mol Cell Biol 1995; 15:4095-4102.
47. Kuo CT, Morrisey EE, Anandappa R et al. GATA-4 transcription factor is required for ventral morphogenesis and heart tube formation. Genes Dev 1997; 11:1048-1060.
48. Molkentin JD, Lin Q, Duncan SA et al. Requirement of the transcription factor GATA4 for heart tube formation and ventral morphogenesis. Genes Dev 1997; 11:1061-1072.
49. Fishman MC and Chien KR. Fashioning the vertebrate heart: earliest embryonic decisions. Development 1997; 124:2099-2117.
50. Sucov HM. Molecular insights into cardiac development. Annu Rev Physiol 1998; 287-308.
51. Chen JN, Fishman MC. Genetics of heart development. Trends Genet 2000; 16:383-388.
52. Stainier DYR. Zebrafish genetics and vertebrate heart formation. Nature 2001; 2:39-48.
53. Brand T, Andree B, Schlange T. Molecular characterization of early cardiac development. In: Brand-Saberi B et al. (eds). Results and Problems in Cell Differentiation. Heidelberg; Springer Verlag, 2002 (in press).
54. Farrell MJ, Kirby ML. Cell biology of cardiac development. Int Rev Cytol 2001; 202:99-158.
55. Ku M, Melton DA. XWnt-11: a maternally expressed Xenopus wnt gene. Development 1993; 119:1161-1173.
56. Kispert A, Vainio S, Shen L et al. Proteoglycans are required for maintenance of Wnt-11 expression in the ureter tips. Development 1996; 122:3627-3637.
57. Eisenberg CA, Gourdie RG, Eisenberg LM. Wnt-11 is expressed in early avian mesoderm and required for the differentiation of the quail mesoderm cell line QCE-6. Development 1997; 124:525-536.
58. Schroeder KE, Condic ML, Eisenberg LM et al. Spatially regulated translation in embryos: asymmetric expression of maternal Wnt-11 along the dorsal-ventral axis in Xenopus. Dev Biol 1999; 214:288-297.
59. Christiansen JH, Dennis CL, Wicking CA et al. Murine wnt-11 and wnt-12 have temporally and spatially restricted expression patterns during embryonic development. Mech Dev 1995; 51:341-350.
60. Eisenberg LM, Eisenberg CA. Onset of a cardiac phenotype in the early embryo. In: Dube DK (ed). Cardiovascular Molecular Morphogenesis: Myofibrillogenesis. Springer Verlag 2001 (in press).
61. Kirikoshi H, Sekihara H, Katoh M. Molecular cloning and characterization of human WNT11. Int J Mol Med 2001; 6:651-656.
62. Monkley SJ, Delaney SJ, Pennisi DJ et al. Targeted disruption of the Wnt2 gene results in placentation defects. Development 1996; 122:3343-3353.
63. Jaspard B, Couffinhal T, Dufourcq P et al. Expression pattern of mouse sFRP-1 and mWnt-8 gene during heart morphogenesis. Mech Dev 2000; 90:263-267.
64. Hume CR, Dodd J. CWnt-8c: A novel Wnt gene with a potential role in primitive streak formation and hindbrain organization. Development 1993; 119:1147-1160.
65. Takada S, Stark KL, Shea MJ et al. Wnt-3a regulates somite and tailbud formation in the mouse embryo. Genes Dev 1994; 8:174-189.
66. Liu P, Wakamiya M, Shea MJ et al. Requirement for Wnt3 in vertebrate axis formation. Nature Genet 1999; 22:361-365.

67. Dealy CN, Roth A, Ferrari D et al. Wnt-5a and Wnt-7a are expressed in the developing chick limb bud in a manner suggesting roles in pattern formation along the proximodistal and dorsoventral axes. Mech Dev 1993; 43:175-186.
68. Sagara N, Toda G, Hirai M et al. Molecular cloning, differential expression, and chromosomal localization of human frizzled-1, frizzled-2, and frizzled-7. Biochem Biophys Res Commun 1998; 252:117-122.
69. Saitoh T, Hirai M, Katoh M. Molecular cloning and characterization of human Frizzled-8 gene on chromosome 10p11.2. Int J Oncol 2001; 18:991-996.
70. van Gijn ME, Blankesteijn WM, Smits JF et al. Frizzled 2 is transiently expressed in neural crest-containing areas during development of the heart and great arteries in the mouse. Anat Embryol 2001; 203:185-192.
71. DeRossi C, Laiosa MD, Silverstone AE et al. Mouse fzd4 maps within a region of chromosome 7 important for thymus and cardiac development. Genesis 2000; 27:64-75.
72. Wang YK, Sporle R, Paperna T et al. Characterization and expression of the frizzled gene Fzd9, the mouse homolog of FZD9 which is deleted in Williams-Beuren syndrome. Genomics 1999; 57:235-248.
73. Wheeler GN, Hoppler S. Two novel Xenopus frizzled genes expressed in the developing heart and brain. Mech Dev 1999; 86:203-207.
74. Deardorff MA, Tan C, Conrad LJ et al. Frizzled-8 is expressed in the Spemann organizer and plays a role in early morphogenesis. Development 1998; 125:2687-2700.
75. Moriwaki J, Kajita E, Kirikoshi H et al. Isolation of Xenopus frizzled-10A and frizzled-10B genomic clones and their expression in adult tissues and embryos. Biochem Biophys Res Commun 2000; 278:377-384.
76. Marvin MJ, Di Rocco G, Gardiner A et al. Inhibition of Wnt activity induces heart formation from posterior mesoderm. Genes Dev 2001; 15:316-327.
77. Schneider V, Mercola M. Wnt antagonism initiates cardiogenesis in Xenopus laevis. Genes Dev 2001; 15:304-315.
78. Wang S, Krinks M, Moos M. Frzb-1, an antagonist of Wnt-1 and Wnt-8, does not block signaling by Wnts-3A, -5A or -11. Biochem Biophys Res Comm 1997; 236:502-504.
79. Xu Q, D'Amore PA, Sokol SY. Functional and biochemical interactions of Wnts with FrzA, a secreted Wnt antagonist. Development 1998; 125:4767-4776.
80. Dennis S, Aikawa M, Szeto W et al. A secreted Frizzled related protein, FrzA, selectively associates with Wnt-1 protein and regulates Wnt-1 signaling. J Cell Sci 1999; 112:3815-3820.
81. Krupnik VE, Sharp JD, Jiang C et al. Functional and structural diversity of the human Dickkopf gene family. Gene 1999; 238:303-313.
82. Glinka A, Wu W, Delius H et al. Dickkopf-1 is a member of a new family of secreted proteins and functions in head induction. Nature 1998; 391:357-362.
83. Ferkey DM, Kimelman D. GSK3: New thoughts on an old enzyme. Dev Biol 2000; 225:471-479.
84. Dominguez I, Green JBA. Missing links in GSK3 regulation. Dev Biol 2001; 235:303-313.
85. Pandur P, Läsche M, Eisenberg LM et al. Wnt-11 activation of a non-canonical Wnt signaling pathway is required for cardiogenesis. Nature 2002; 418:638-641.
86. Moon RT, Kimelman D. From cortical rotation to organizer gene expression: toward a molecular explanation of axis specification in Xenopus. Bioessays 1998; 20:536-545.
87. Glinka A, Wu W, Onichtchouk D et al. Head induction by simultaneous repression of Bmp and Wnt signalling in Xenopus. Nature 1997; 389:517-519.
88. Christian JL, Moon RT. Interactions between XWnt-8 and Spemann organizer signaling pathways generate dorsoventral pattern in the embryonic mesoderm of Xenopus. Genes Dev 1993; 1:13-28.
89. Rawles ME. A study in the localization of organ-forming areas in the chick blastoderm of the head process stage. J Exp Zool 1936; 138:505-555.
90. DeHaan RL. Morphogenesis of the vertebrate heart. In: Organogenesis. DeHaan RL, Ursprung H, eds. New York: Holt, Rinehart & Winston, 1965:377-419.
91. Jacobson AG, Sater AK. Features of embryonic induction. Development 1988; 104:341-359.
92. Sater AK, Jacobson AG. The restriction of the heart morphogenetic field in Xenopus laevis. Dev Biol 1990b; 140:328-336.

93. Goldstein AM, Fishman MC. Notochord regulates cardiac lineage in zebrafish embryos. Dev Biol 1998; 201:247-252.
94. Raffin M, Ming Leong L, Rones MS et al. Subdivision of the cardiac Nkx2.5 expression domain into myogenic and nonmyogenic compartments. Dev Biol 2000; 218:326-340.
95. Tzahor E, Lassar AB. Wnt signals from the neural tube block ectopic cardiogenesis. Genes Dev 2001; 15:255-260.
96. Eisenberg CA, Bader DM. The establishment of the mesodermal cell line QCE-6: A model system for cardiac cell differentiation. Circ Res 1996; 78:205-216.
97. Eisenberg CA, Eisenberg LM. WNT11 promotes cardiac tissue formation of early mesoderm. Dev Dyn 1999; 216:45-58.
98. Tada M Smith JC. XWnt-11 is a target of Xenopus Brachyury: regulation of gastrulation movements via Dishevelled, but not through the canonical Wnt pathway. Development 2000; 127:2227-2238.
99. Djiane A, Riou JF, Umbhauer M et al. Role of frizzled 7 in the regulation of convergent extension movements during gastrulation in Xenopus laevis. Development 2000; 127:3091-3100.

# Chapter 12

# Wnt Signaling in *C. elegans*

**Michael A. Herman**

## Abstract

The Wnt signaling pathway controls many aspects of *Caenorhabditis elegans* development A canonical Wnt pathway functions to control cell migrations and determine certain cell fates. Noncanonical Wnt pathways function primarily to control the polarity of specific cells. *C. elegans* shares the involvement of canonical and noncanonical Wnt pathways in various developmental processes with many other animals. However, other aspects of Wnt signaling appear to be unique to *C. elegans*. This chapter will review the major process in *C. elegans* development that involve Wnt signaling, discuss the interactions of Wnt pathway components involved in each process and compare and contrast Wnt signaling pathways in the worm with those found in other species.

## Introduction

Reading this book, or perusing the literature, should leave one with the impression that Wnt signaling controls many aspects of the development of most animals. The same is true of the development of the nematode *Caenorhabditis elegans*. Like most animals, *C. elegans* has developmental processes that are controlled by both canonical and noncanonical Wnt signaling pathways (see Chapter 1). These processes include the control of cell polarity, cell migrations and cell fate determination involved in the generation of multiple organ systems.

During *C. elegans* embryogenesis blastomeres divide to generate cells of all the major organ systems. Cell migration and other morphogenic events occur so that at hatching the worm body plan is essentially complete. Development then proceeds through four larval stages, each separated by a molt during which the animal sheds its cuticle. Postembryonic larval development is characterized by additional cell divisions, cell migrations and inductions, some of which differentiate the two sexes: self-fertile hermaphrodites and males. Approximately 64 hours after fertilization the life cycle is complete and the worm begins to reproduce. All of the cellular and developmental processes can be easily monitored because the animal is transparent; which allowed the determination of the complete cell lineage.[1,2] Although the cell lineage is essentially invariant, cell-cell interactions are involved in many developmental processes, many of which involve Wnt signaling. This chapter will review the major processes in which Wnt signaling is involved and discuss what we know about the components that are used in each signaling event (Tables 1 and 2).

## Catalog of *C. elegans* Wnt Pathway Components

PCR screens for Wnt-1 homologs identified the first *C. elegans* Wnt genes *cwn-1* and *cwn-2*.[3,4] Genetic screens for mutations that affected particular developmental processes identified

*Wnt Signaling in Development*, edited by Michael Kühl.

***Table 1.* C. elegans *Wnt pathway parts list***

| Component | Number of Orthologs in: *Drosophila* | Human | *C. elegans* Orthologs | Function Determined by: Mutation | RNAi |
|---|---|---|---|---|---|
| Porcupine | 1 | 0 | *mom-1* | X | X |
| Wnt | 7 | 19 | *mom-2* | X | |
| | | | *lin-44* | X | |
| | | | *egl-20* | X | |
| | | | *cwn-1* | X | |
| | | | *cwn-2* | | |
| FrzB | 0 | 5 | X73B6BL.21 | | |
| Frizzled | 4 | 12 | *mom-5* | X | X |
| | | | *lin-17* | X | |
| | | | *mig-1* | X | |
| | | | *CFz2* | | |
| Dishevelled | 1 | 3 | *mig-5* | X | X |
| | | | *dsh-1* (C34F11.9) | | |
| | | | *dsh-2* (C27A2.6) | X | |
| Casein Kinase I | 1 | 1 | *kin-19* | | X |
| GSK-3 | 1 | 1 | *sgg-1* | | X |
| | | | C44H4.6 | | |
| Axin | 1 | 2 | *pry-1* | X | |
| APC | 2 | 2 | *apr-1* | X | X |
| β-catenin | 1 | 2 | *wrm-1* | | X |
| | | | *bar-1* | X | X |
| | | | *hmp-2* | X | X |
| TCF/LEF | 1 | 3 | *pop-1* | X | X |
| Groucho | 1 | 4 | *unc-37* | X | X |
| NLK | 1 | 1 | *lit-1* | X | X |
| TAK1 | 1 | 1 | *mom-4* | X | X |
| ? | ? | ? | *mig-14/mom-3* | X | |

Orthologs of Wnt pathway components not found in *C. elegans*: Cerberus, WIF, Dickkopf, GBP/Frat, Naked. Numbers of *Drosophila*, Human and *C. elegans* orthologs from Ruvkun[10], the Wnt gene Home Page (http://www.stanford.edu/~rnusse/wntwindow.html) or determined by querying WormPD and HumanPSD databases (Proteome, Inc). Orthologs of Wnt pathway components found in *C. elegans*, but for which a Wnt-related function has not been reported: LRP/Arrow, proteoglycans, Phosphatase PP2C, PAR-1, Slimb/βTrCP (*lin-23*). X, indicates whether a component's function was determined by mutation, RNAi, or both.

many other genes encoding *C. elegans* Wnt signaling components. The first gene found in this way was *lin-44*, which was identified by mutations that affected the orientation of certain asymmetrically dividing cells in the tail to the body axis.[5,6] The identification of other components involved in embryonic endoderm induction by both genetic and sequence-based methods soon followed.[7,8]

The completion of the *C. elegans* genome sequence has allowed the complete cataloging of the *C. elegans* Wnt pathway component orthologs.[9,10] Those that have been identified are listed in Table 1. Briefly, *C. elegans* has five Wnts, compared to seven in *Drosophila* and 19 in Humans; four Frizzleds (Fz), as does *Drosophila*, but compared to 12 in humans; and three Dishevelleds (Dsh) compared to one in *Drosophila* and three also in Humans. Comparisons for

***Table 2. Summary of C. elegans Wnt controlled processes and the components involved***

| Canonical Component | QL.d Migration | Endoderm Induction | Spindle Orientation | P12 Cell Fate | VPC Fate | T Cell Polarity | Gonad Polarity | V5 Polarity | Ray Formation |
|---|---|---|---|---|---|---|---|---|---|
| **Porcupine** | | MOM-1[7,8] | MOM-1[40] | | MOM-1† | | MOM-1[77] | | |
| **Wnt** | EGL-20[25] | MOM-2[7,8] | MOM-2[40] | LIN-44[55] | | LIN-44[5,6] | none?[77] | EGL-20[78] | EGL-20[90] |
| **Frizzled** | LIN-17[25] | MOM-5[7,8] | MOM-5[40] | LIN-17[55] | | LIN-17[75,76] | LIN-17[75,77] | LIN-17[78] | LIN-17[90] |
| | MIG-1* | | | | | | | | |
| **Disheveled** | MIG-5[24] | MIG-5[42] | | MIG-5[24] | MIG-5[24] | | | | |
| | | DSH-2 | | | | | | | |
| **GSK-3** | SGG-1[11] | SGG-1[40] | SGG-1[40] | | | | | | |
| **Casein Kinase I** | | KIN-19[41] | | | | | | | |
| **Axin** | PRY-1[40,11] | | | PRY-1† | PRY-1[72] | no** | | PRY-1[78] | |
| **APC** | | APR-1[7] | | | APR-1[72,73] | | | | |
| **β-catenin** | BAR-1[26] | WRM-1[7] | | BAR-1[28] | BAR-1[71] | none[27] | WRM-1[77] | | BAR-1[90] |
| **TCF/LEF** | POP-1[17,27] | POP-1[35] | | | POP-1[35,72] | POP-1[27] | POP-1[77] | | |
| **target gene** | *mab-5*[22] | *end-1*[44] | | *egl-5*[55] | *lin-39*[71,72] | *tlp-1*[81] | | | *mab-5*[90] |
| **Noncanonical Component** | | | | | | | | | |
| **NLK** | | LIT-1[19,36,37] | | no† | LIT-1[19] | LIT-1[77] | | | |
| **TAK1** | | MOM-4[8,37] | | | | | no[77] | | |
| **MIG-14/MOM-3** | yes[25] | yes[8] | yes[40] | yes[28] | yes[28] | no** | | yes[78] | |

Genes and the proteins involved in the major processes that are controlled by, or involve, Wnt signaling are listed. References that provide evidence for the involvement of each component are indicated. *S. Clark, personal communication, **M.H. unpublished, †D. Eisenmann, personal communication. If a protein is named, it has been shown to be involved in the process listed in that column. "no" indicates that a function for a particular component was assayed, but found not to be involved. In the case of MIG-14/MOM-3, "yes" indicates that it is involved in the process listed in that column. A blank indicates that the involvement of the component in the process listed has not been assayed. Modified from http://www.stanford.edu/~rnusse/pathways/wormsignal.htm, original by D. Eisenmann.

other components are listed in Table 1. Interestingly, *C. elegans* has three β-catenins compared to only one in Drosophila and two in humans. *C. elegans* orthologs were not found for several Wnt pathway components identified in other animals. Like *Drosophila*, *C. elegans* does not appear to employ Wnt inhibitors such as Cerberus, WIF, Dickkopf or sFRPs (R. Nusse, World Wide Wnt Window: http://www.stanford.edu/~rnusse/pathways/pathwleg.html). However, an apparent FrzB homolog, Y73B6BL.21, is present in the *C. elegans* genome; although its function has not yet been determined. The *C. elegans* genome also does not appear to contain an LRP/Arrow ortholog that functions in Wnt signaling, suggesting Wnts in *C. elegans* bind to Frizzled receptors without the aid of a co-receptor. In addition, orthologs of GBP/Frat and Naked have not been identified in C. elegans. However it should be noted that standard BLAST analyses did not identify an Axin homolog, although one exists.[11] Further, although *apr-1/APC, pry-1/Axin, bar-1/β-catenin* and *pop-1/TCF* are functional orthologs of Wnt pathway components, they are only distantly related to their *Drosophila* and vertebrate counterparts. This suggests that more sophisticated database searches may identify additional *C. elegans* orthologs of Wnt pathway components. A Wnt related function for some *C. elegans* orthologs of Wnt pathway components has not been reported. For example, a role for proteoglycans in *C. elegans* Wnt signaling has not been reported. In addition, *C. elegans* homologs of Phosphatase PP2C which binds Dishevelled and can dephosphorylate Axin in other animals[12] exist, but have not been reported to play a role in Wnt signaling. PAR-1 acts as a positive regulator of Wnt signaling in several animals[13] and functions in establishing the anteroposterior polarity of the *C. elegans* zygote,[14] but its function in *C. elegans* Wnt signaling events has not been established. *lin-23*, the *C. elegans* ortholog of Slimb/βTrCP which functions in the degradation of β-catenin in other animals,[15] is a negative regulator of cell division that functions throughout *C. elegans* development [16] but appears to have no role in Wnt signaling. Thus the *C. elegans* genome contains many, but not all, of the Wnt signaling components found in other animals. Finally, several novel Wnt signaling components have been identified by genetic analyses in *C. elegans.*

## The Three *C. elegans* β-Catenins Are Involved in Distinct Processes

Curiously, *C. elegans* has three β-catenin homologs whereas a single protein in *Drosophila* performs both cell adhesion and Wnt signaling functions of β-catenin, and vertebrates have two proteins, β-catenin and plakoglobin, which have partially overlapping roles. Study of the interactions of the three C. elegans β-catenin homologs with POP-1/TCF revealed that the *C. elegans* β-catenins have divergent functions.[17] First, it was established that POP-1 binds to a consensus TCF target sequence, indicating that POP-1 can behave like a canonical TCF. In addition, in mammalian tissue culture cells, POP-1 can complex with the *Drosophila* β-catenin, Armadillo, and activate transcription of a reporter gene containing several upstream TCF binding sites. BAR-1 is the only β-catenin homolog that interacts strongly with POP-1. When expressed together, BAR-1 and POP-1 form a complex that can activate transcription of a TCF-reporter gene. An amino-terminally deleted version of POP-1 that removed the putative β-catenin interaction domain, ΔN-POP-1, did not interact with BAR-1,[17,18] did not activate a TCF-reporter gene and behaved as a dominant negative protein in phenotypic assays. Thus like other TCF proteins, POP-1 interacts with a β-catenin via an amino-terminal domain.

Weak interactions between WRM-1/β-catenin and POP-1/TCF in yeast hybrid assays have been reported,[18,19] but stable interactions were not observed in co-immunoprecipitation assays.[17] Furthermore, WRM-1 also interacted weakly with ΔN-POP-1, suggesting that WRM-1 interacts with POP-1 in a manner different from BAR-1 and canonical β-catenin-TCF interactions. The third *C. elegans* β-catenin homolog, HMP-2, does not interact with POP-1 in any assay, but does interact with the cadherin homolog HMR-1 and the α-catenin HMP-1.[17,18]

One interesting conclusion from this work is that the functions of the *C. elegans* β-catenins in adhesion and in signaling are performed by separate proteins. This is a major difference between *C. elegans* and other systems. Of the three *C. elegans* β-catenins, BAR-1 appears to function in signaling pathways as a canonical β-catenin. In addition, of the two ß-catenins involved in signaling, BAR-1 is involved in canonical β-catenin signaling whereas WRM-1 is involved in unusual Wnt signaling pathways (see below).

Why might the functions of the β-catenin orthologs be separated among different processes in *C. elegans*? It should be noted that these proteins are quite divergent; the Hydra β-catenin ortholog[20] is more similar to vertebrate β-catenins that are the *C. elegans* orthologs. Perhaps this arrangement is unique to nematodes, although why this might be is unclear. One possibility is that this arrangement separates the adhesion, canonical signaling and non-canonical signaling, β-catenin pools to prevent pathway cross-talk.

## A Canonical Wnt Pathway Controls the Migration of the Descendants of the QL Neuroblast

QL and QR are left/right homologous neuroblasts that migrate and each generate three neurons and two cells that undergo programmed cell death during the first larval stage (L1). The QR cell and its descendants (known as the QR.d) migrate anteriorly, whereas the QL neuroblast and its descendants (known as the QL.d) migrate posteriorly and divide. The posterior QL daughter, QL.p, stops migrating and generates two neurons, and a cell that undergoes programmed cell death. The anterior QL daughter, QL.a, continues a posterior migration and divides to generate a cell that undergoes programmed cell death and a neuron, which continues to migrate posteriorly until it reaches its characteristic position in the tail (Fig. 1).[1] The direction Q cell descendants migrate is controlled by the Hox gene *mab-5*. Cells that express *mab-5* migrate posteriorly, whereas cells that do not, migrate anteriorly. Ectopic expression of *mab-5* in both the QL.d and QR.d descendants caused them to migrate posteriorly, just as the QL.d do in wild-type animals.[21,22] Loss of mab-5 expression (as in *mab-5* mutants) cause the QL.d and QR.d to migrate anteriorly.[22,23]

A canonical Wnt signaling pathway controls the migration of the QL.d by regulating the expression of *mab-5*.[17,24-27] Mutations in *egl-20/Wnt, lin-17/Fz, mig-5/Dsh* and *bar-1/β-catenin*, all cause the loss of *mab-5* expression in the QL lineage and anterior migration of the QL.d. The inhibition of *pop-1/TCF* function by RNAi also causes anterior migration of the QL.d, as well as loss of *mab-5* expression in the QL lineage [27] as does the expression of the dominant negative ΔN-POP-1 construct (see below).[17] Furthermore, overexpression of *pop-1* also causes anterior migration of the QL.d, antagonizing Wnt signaling in a manner similar to observations of TCFs in other Wnt pathways.[27] In addition, mutations in *pry-1* cause posterior migration of the QR.d and the ectopic expression of *mab-5* in the QR lineage, suggesting that it functions as a negative regulator of the Wnt pathway.[26] Genetic analysis demonstrated that pry-1 functions downstream of egl-20/Wnt and upstream of bar-1 β-catenin, pop-1 and mab-5; and it was recently shown that pry-1 encodes the *C. elegans* Axin homolog.[11,26] Overexpression of *pry-1* and *sgg-1*, the *C. elegans* GSK-3 ortholog, caused anterior migration of the QL.d, indicating that they function negatively in the pathway. Finally, PRY-1 interacts with BAR-1, SGG-1, APR-1/APC and MIG-5/Dsh in a canonical manner in a yeast two-hybrid assay.[11] Although a role for APR-1 in QL.d migration has not been established, it appears to function negatively in vulval development (see below). Taken together, current evidence indicates that the Wnt pathway that controls QL.d migration and *mab-5* expression is a canonical Wnt pathway (Table 2, Fig. 9). In addition to these components, *mig-14* mutants also cause anterior migrations of the QL.d as well as a loss of *mab-5* expression, suggesting that MIG-14 functions in the canonical Wnt pathway.[25,28] The molecular identity of MIG-14 has not yet been deter-

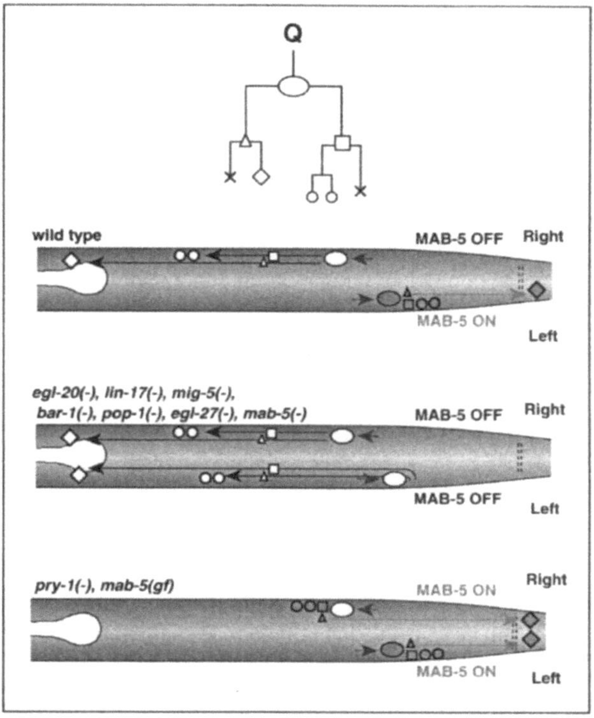


Figure 1. A canonical Wnt pathway controls QL.d migration. The Q neuroblast lineage is shown on top. Both QL and QR (ovals) divide to generate three neurons (diamond and circles) and two cells that undergo programmed cell death (X). Migrations of the QL.d and QR.d in wild-type and mutant animals are shown below. A dorsal view is depicted so that both left and right sides are visible. Soon after hatching QL begins to migrate posteriorly, expresses *mab-5* (shaded) and divides. The QL.d continue to express *mab-5* and migrate posteriorly. Soon after hatching, QR begins to migrate anteriorly and divides. The QR.d do not express *mab-5* and continue to migrate anteriorly. Final positions of the QL.d and QR.d are indicated. In *egl-20, lin-17, mig-5, bar-1, pop-1, egl-27* and *mab-5* mutants, QL migrates posteriorly, but does not express *mab-5*. Subsequently, the QL.d do not express *mab-5* and migrate anteriorly. In *pry-1* and *mab-5* mutants QL, QR and their descendants all express mab-5 and migrate posteriorly. After Whangbo, J and Kenyon, C. (1999) A Wnt signaling system that specifies two patterns of cell migration in C. elegans. Mol Cell. 4: 851-8, with permission from Elsevier Science

mined. This should turn out to be an important discovery as, *mom-3* mutations, which also have defects in endoderm specification and spindle orientation (see below), were shown to be allelic to *mig-14* mutations; indicating that this unknown protein may function more generally in Wnt signaling. I will subsequently refer to this protein as MIG-14/MOM-3.

Another gene that functions in the control of QL.d migration is *egl-27*. Mutations in *egl-27* also cause the anterior migration of the QL.d as well as the loss of *mab-5* expression in the QL lineage.[29] EGL-27 contains a region of similarity to MTA, a factor overexpressed in metastatic cells.[30] Further analysis of the EGL-27 sequence has revealed several regions of similarity to proteins involved in chromatin remodeling.[31] MTA1 has been shown to be a subunit of the NURD complex that displays ATP-dependant chromatin-remodeling and histone deacetylase activities.[32] Although *egl-27* appears to function upstream of *mab-5*, it is not clear whether it functions upstream, downstream or in parallel with *pop-1*.

## Embryonic Endoderm Introduction

At the four-cell stage of *C. elegans* embryogenesis, the posterior-most blastomere, $P_2$, polarizes it's anterior neighbor, EMS, and induces it to form endoderm (Fig. 2). After reorienting its spindle (see below), the EMS blastomere subsequently divides to generate the MS cell, which generates mesoderm, and the E cell, that generates all the endoderm, or intestine, of the animal. Blastomere isolation and reconstitution experiments demonstrated that the $P_2$ blastomere induces the EMS blastomere to generate the E cell fate[33] and that the position of contact between $P_2$ and EMS establishes which side of the EMS cell is partitioned in the E cell.[34]

The nature of the $P_2$-to-EMS signal was revealed by genetic screens that identified several maternally expressed genes required for differentiated E and MS cell fates.[7,8,35,36] These screens identified five *mom* genes (for *more* mesoderm), *lit-1* (for *loss* of *intestine*) and *pop-1* (for *posterior pharynx* defective). Mutations in the *mom* genes and in *lit-1* cause the EMS blastomere to generate two MS-like daughter blastomeres, thus more mesoderm is produced and endodermal cell fates are lost. Mutations in *pop-1* cause the opposite effect, the EMS blastomere generates two E-like blastomeres, thus mesodermal fates (including the posterior pharynx) are lost and excess endodermal cell fates are produced. Three of the mom genes and *pop-1* encode components of a Wnt pathway: *mom-1* encodes the *C. elegans* Porcupine homolog, *mom-2* encodes a Wnt, *mom-5* encodes a Frizzled homolog and *pop-1* encodes the TCF/LEF homolog[7,8,35] (Tables 1 and 2). Blastomere isolation and recombination experiments demonstrated that *mom-1*, *mom-2* and *mig-14/mom-3* function in the signaling cell, $P_2$, whereas *mom-4* functions in the receiving cell EMS.[8] This makes sense for *mom-1/Porc* and *mom-2/Wnt* and suggests that *mig-14/mom-3* functions in expression or secretion of the Wnt signal. The genes *mom-4* and *lit-1* encode components of a mitogen-activated protein kinase (MAPK) pathway: *mom-4* encodes a MAPK kinase kinase related to vertebrate TAK1 and *lit-1* encodes a MAPK-like protein that is 70% identical to the *Drosophila* tissue-polarity gene *nemo* and its vertebrate homolog nemo-like kinase (NLK).[19,37,38] This MAPK pathway functions with the Wnt pathway to downregulate POP-1 protein levels (see below). As mentioned above, the molecular identity of *mig-14/mom-3* has yet to be determined.

Other Wnt pathway genes required for embryonic endoderm induction in *C. elegans* were identified by sequence similarity and their roles in endoderm induction were revealed by RNA-mediated interference (RNAi) experiments.[39] *wrm-1/β-catenin* (for wormadillo), *apr-1/APC* (for *APC*-related), *sgg-1/GSK-3* (for *shaggy*-like) and *kin-19* (a homolog of casein kinase I) were all shown to be required for endoderm induction by RNAi.[7,40,41] However, one cautionary note is required here: with the exception of *wrm-1*, *lit-1* and *pop-1*, mutation or RNAi of any of these genes only produces partially penetrant defects in endoderm specification that ranges from only 2—5% for *mom-5* to 88% for *mom-2*.[7,8,40] Genetic redundancies exist for several components *(mom-2, mom-5, sgg-1* and possibly *mom-4*), which might explain why mutations in these genes do not cause fully penetrant defects. Along these lines, it is not yet known whether any of the three *C. elegans* Dishevelled proteins is involved in endoderm induction. However, *mig-5* and *dsh-2* (C27A2.6) messages are enriched in oocytes, as are the other Wnt components involved in endoderm induction,[42] suggesting that they might function together in endoderm induction. Genetic redundancies do not appear to exist for other components (*apr-1* and *mom-1*) that do not have fully penetrant Mom defects, however. The low penetrance Mom defects observed in several *mom* mutants might be caused by the existence of a redundant pathway that specifies endoderm development.

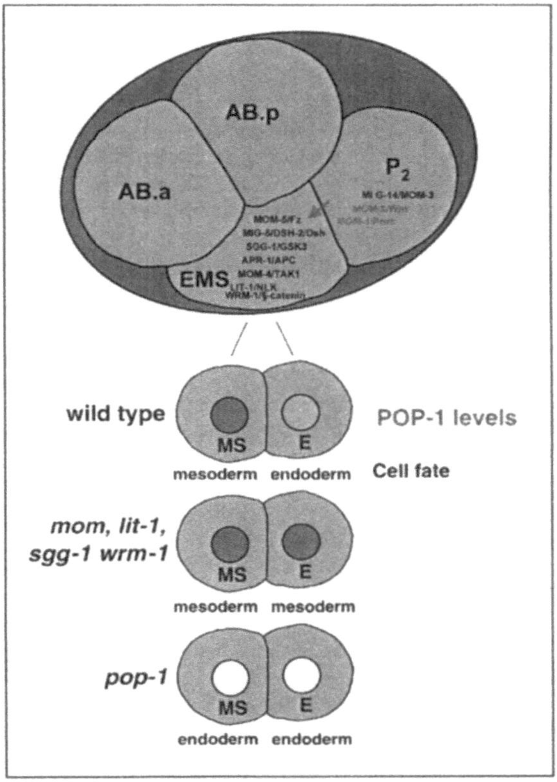


Figure 2. Endoderm induction. In the four-cell *C. elegans* embryo a Wnt signal from the $P_2$ blastomere polarizes the EMS blastomere and induces endoderm fate. Genes are listed inside the blastomere in which they have been shown to function (top). In wild type embryos, this results in a high level of POP-1 protein in the anterior MS blastomere nucleus and a low level in the E blastomere nucleus. Effects of mutations or RNAi on POP-1 localization and EMS daughter fates are indicated (bottom). See text for details.

## Unusual Aspects of Wnt Signaling during Endoderm Induction

In the absence of MOM-2/Wnt signal, POP-1/TCF appears to repress the E cell fate. The level of POP-1 has been shown to be higher in the anterior MS cell than in the posterior E cell, which receives a Wnt signal and in *pop-1* mutants both daughters adopt the E cell fate (Fig. 2).[35,43] Thus endoderm induction appears to be achieved by inhibition of a repressor. In the MS cell POP-1 is required to represses endoderm fate. At least for the endoderm-specific GATA-like transcription factor gene, *end-1*,[44] this appears to be achieved by POP-1 functioning to recruit a complex of the histone deacetylase (HDAC) HDA-1 and UNC-37/Groucho.[45] In this respect, POP-1 functions as a canonical TCF/LEF factor by recruiting HDAC and Groucho to function as corepressors in the absence of a Wnt signal.[46,47] Interestingly, the *C. elegans* p300 homolog CBP-1 is required positively in the E cell for *end-1* transcription and endoderm fate.[45,48] Although in mammalian cells CBP/p300 can interact with β-catenin/TCF to activate transcription,[49] neither POP-1 nor WRM-1/β-catenin has been reported to interact with CBP-1 in *C. elegans*. The involvement of the MAPK pathway components MOM-4/TAK and LIT-1/NLK make this pathway unusual. In the E cell, the Wnt and MAPK components function together to repress POP-1; although exactly how this is achieved is not yet clear. In vertebrate cells NLK has been shown to phosphorylate TCF-4, which interfered with DNA binding by the TCF-4/β-catenin complex.[38] The *C. elegans* proteins, WRM-1 and LIT-1 have been shown to interact in tissue culture cells and together can phosphorylate POP-1. Thus Wnt and MAPK pathways might collaborate to phosphorylate POP-1, which could lead to its degradation, accounting for the repression of POP-1 function. However, it is unlikely that POP-1 and WRM-1 form a complex that binds DNA as their vertebrate counterparts do, as WRM-1 and POP-1 have only been shown to interact weakly in yeast two-hybrid assays [18,19] and failed to interact in a co-immunoprecipitation assay.[17] Interestingly, when POP-1 is expressed alone in vertebrate COS cells it is localized to the nucleus; but when POP-1 is coexpressed with WRM-1 and LIT-1, it becomes cytoplasmic.[19] Thus WRM-1 and LIT-1 appear to regulate POP-1 nuclear localization which might cause a perceived lowering of POP-1 levels by dilution in the cytoplasm or alternatively, selective degradation in the cytoplasm. Thus POP-1 must be regulated differently during endoderm induction in *C. elegans* than is TCF-4 in vertebrate cells. WRM-1 and LIT-1 appear to interact, somehow leading to a lowering of POP-1 levels in the E blastomere by an unknown mechanism.

Another very unusual aspect of the Wnt pathway involved in endoderm induction is that APR-1/APC and SGG-1/GSK3 appear to act positively in the transduction of the Wnt signal leading to the negative regulation of POP-1/TCF (Fig. 9). In the absence of Wnt signaling, the *Drosophila* and mammalian homologs of APC and GSK3 function negatively by promoting the degradation of β-catenin. A WRM-1/β-catenin/POP-1 complex might be targeted for degradation by the functions of APR-1 and SGG-1. How Wnt signaling might activate rather than inhibit SGG-1/GSK3 function is not clear;[50] especially in light of the possibility that two Dsh proteins might function redundantly in endoderm induction. The positive role of APR-1 is also curious. However, APR-1 is only 1186 amino acids, much smaller than the 2843 amino acid human APC protein,[7] suggesting APR-1 might have a divergent function in endoderm induction. Other data suggest a more canonical, negative role, for APR-1 in the Wnt signaling pathway involved in vulval development (see below). Determination of WRM-1 localization patterns as well as the investigation of the role of Dsh in this pathway is needed to learn how this unusual Wnt pathway establishes endoderm fates.

## Wnt Signaling Might Directly Target the Cytoskeleton to Control Mitotic Spindle Orientation

During EMS division, the nucleus and centrioles are initially aligned along the left-right axis. Just before mitosis, the EMS nucleus and centrioles rotate about 90° and align along the anterior-posterior axis (Fig. 3). Signaling from $P_2$ also induces this spindle rotation. If EMS is isolated, no rotation occurs. However, when placed in contact with $P_2$, the EMS nuclear complex rotates and aligns along an axis perpendicular to the plane of contact.[51] Spindle rotation and endoderm specification by the signal from $P_2$ can be uncoupled, however. EMS loses competence to respond to $P_2$ to orient its spindle before it loses competence to polarize endoderm fate. For example, the spindle orients randomly when $P_2$ is placed in contact with EMS late in the EMS cell cycle.

In the 8-cell *C. elegans* embryo, there are four descendants of the AB blastomere. The mitotic spindle of three of these descendants are oriented in parallel, but the spindle of the fourth descendant, AB.ar, is oriented perpendicular to the others. The orientation of the AB.ar spindle is defective in some, but not all, of the *mom* mutants. Specifically, mutations in *mom-1, mom-2, mom-3, mom-5* and *sgg-1(RNAi)* cause defects in AB.ar spindle orientation, whereas mutations in *pop-1, apr-1* and *wrm-1* do not.[7,8,40] By contrast with the effects on endoderm specification, the defects in AB.ar spindle orientation are highly penetrant. Although it was not

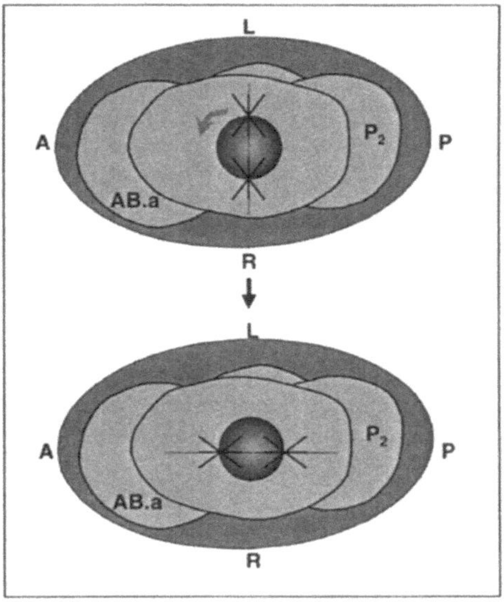


Figure 3. In the four-cell embryo, a Wnt signal from the $P_2$ blastomere orients the EMS mitotic spindle. A ventral view is depicted. The EMS spindle is initially aligned along the left (L)-to-right (R) axis (top). Approximately 30 seconds prior to nuclear envelope breakdown the EMS spindle rotates about 90° and becomes aligned along the Anterior (A)-to-posterior (P) axis (bottom).

clear whether the defects in AB.ar spindle orientation were direct effects of Wnt signaling in $P_2$ or any other blastomere, they suggested that Wnt signals might also be involved in the orientation of the EMS spindle.[40]

The same genes that affect AB.ar spindle orientation also affect EMS spindle orientation. The results of blastomere isolation and reconstitution experiments using wild-type and mutant blastomeres indicated that upstream components *mom-1/Porc, mom-2/Wnt, mom-5/Fz* and *sgg-1/GSK3* affect both EMS spindle orientation and endoderm specification; whereas downstream components *apr-1/APC, wrm-1/β-catenin* and *pop-1/TCF* only affect endoderm induction.[40] Thus *sgg-1* appears to be a branch point in these pathways (Fig. 4).[50] In addition, because the EMS spindle rotates quickly after $P_2$ signals and still rotates if transcription is blocked, it appears that Wnt signaling in *C. elegans* may directly target the cytoskeleton.[50] This may also occur in other systems as there is evidence of Wnt signaling affecting axonal cytoskeletal elements (reviewed in ref. 52) as well as gap junction patterns.[53]

## Wnt and Ras Pathways Control P12 Cell Fate

The P11 and P12 cells are located in the laterally in the animal at hatching and are the most posterior pair of postembryonic ventral nerve cord precursors. The P12 cell usually is on the right side and P11 on the left. Midway through the first larval stage they migrate into the ventral nerve cord and divide.[1] Before they enter the ventral cord, either cell can adopt the P12 cell fate, as killing either with a laser microbeam prior to its migration results in the remaining cell adopting the P12 cell fate.[54] Multiple pathways and genes can affect this cell fate determination. The genes of the *lin-3/let-23/let-60* pathway, the *C. elegans* Ras signaling pathway that functions in vulval development (see below), the Wnt pathway and the Hox genes *mab-5* and *egl-5* are all involved. Mutations in several genes including *let-23, lin-44/Wnt* and

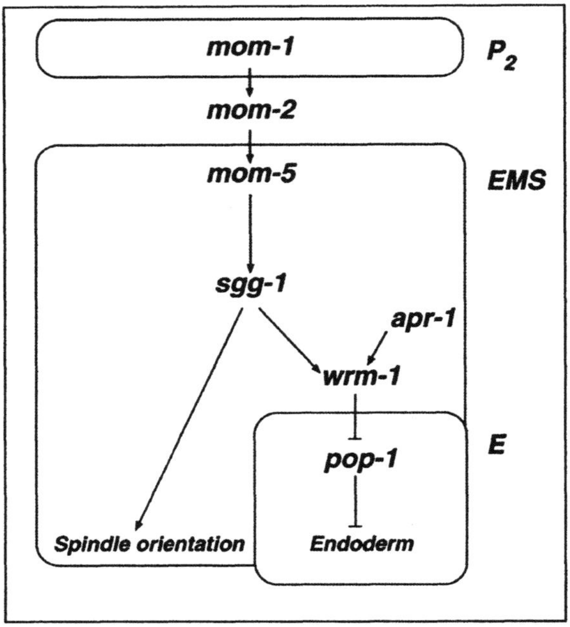


Figure 4. A Wnt signal from the $P_2$ blastomere induces both endoderm fate and spindle orientation in the EMS blastomere. See text for details. Reprinted with permission from Elsevier Science from Thorpe, CJ, et al. (2000) Wnt signalling in *Caenorhabditis elegans*: regulating repressors and polarizing the cytoskeleton. Trends Cell Biol. 10: 10-17.

*egl-5* cause a P12 to P11 cell fate transformation. Mutations in the Wnt pathway genes *mig-14/mom-3*, *mig-5/Dsh*, *bar-1/β-catenin* and *lin-17/Fz* also cause this cell fate transformation,[28,55] while mutations in *pry-1* cause the opposite transformation (D. Eisenmann, personal communication). Interestingly, a moderate reduction in *lin-3/EGF* function enhances the P12 to P11 cell fate transformation of *lin-44* mutants, but a *lin-17* mutation does not. This indicates that *lin-3* and *lin-44* function in separate pathways and lin-44 and lin-17 function in the same pathway for P12 cell fate specification.[55] Additional genetic evidence suggests that LIN-44 and LIN-3 functions are required at different times during development for P12 cell fate specification. LIN-44 appears to be required during late embryogenesis and LIN-3 is required during early L1 stage, thus LIN-44 may function before LIN-3. One possibility is that the LIN-44/Wnt pathway functions to make both cells competent to receive the LIN-3/EGF signal.[55] It is also possible that these two pathways both function to regulate the expression of the target gene, *egl-5*, which is required for P12 cell fate (Fig. 9). Finally, although it appears that P12 cell fate competence/specification is controlled by a canonical Wnt pathway, it is not clear what other Wnt signaling components are involved. The examination of recently isolated *pop-1* alleles that affect postembryonic development will more definitively establish whether this is a canonical Wnt pathway.

## Wnt Signaling in Vulval Development

*C. elegans* vulval development has been extensively studied as an excellent example of induction of cell fates in one tissue by a cell in an overlying tissue (reviewed by Greenwald[56]).

For proper vulval cell fate specification to occur, cells must integrate information from both a Ras/MAPK pathway and a Wnt pathway. Vulval development can be broken down into four processes: the generation of the vulval precursor cells (VPCs) that will divide to give rise to the vulva; determination of particular vulval cell fates; execution of a particular vulval cell fate; and morphogenesis of the vulva, where the 22 cells that are produced organize, migrate and rearrange to form the functional vulva that connects the uterus to the outside (Fig. 5).[57]

The generation stage begins when six cells from the each side of the animal (the 12 P cells) migrate to the ventral midline, interdigitate and divide. The posterior daughters become the 12 Pn.p cells, P1.p-P12.p. The six central cells, P3.p-P8.p express the Hox gene *lin-39* and become VPCs, forming the Vulval Equivalence Group.[58-60] The other cells, P1.p, P2.p, P9.p-P11.p do not express *lin-39*, adopt the F (fused) fate and fuse with the surrounding epidermal syncytium in the L1 stage. Later, in the L3 stage, P3.p adopts the F fate about 50 percent of the time.

The six VPCs, P3.p-P8.p, are multipotent, able to adopt any of the three vulval fates: 1°, 2° or 3°.[54,61,62] During the late L2 and early L3 individual VPC fates are determined based on extracellular signals. The pattern ofVPC fates adopted by P3.p-P8.p is usually 3°-3°-2°-1°-2°-3°. The anchor cell (AC) in the somatic gonad secretes a LIN-3/EGF signal, that induces P6.p, which is closest to the AC, to adopt the 1° fate. [62-64] Activation of a Ras pathway that includes LET-23/RTK, LET-60/Ras and MPK-1/MAPK specifies the 1° cell fate. [65-67] P6.p then signals laterally to its immediate neighbors, P5.p and P7.p, causing them to adopt the 2° fate [68] via activation of the LIN-12/Notch receptor. [69] VPCs that do not receive a signal from the anchor cell or a neighbor adopt the uninduced 3° fate.

The role of Wnt signaling in vulval development appears to be to maintain LIN-39 expression during VPC determination. LIN-39 is required twice during vulval development: first during the generation step to specify VPC versus F cell fate and second during the determination step to specify 1° versus 2° versus 3° cell fates, where it might make P3.p-P8.p competent to receive signals from the anchor cell. For example, mutations that eliminate *lin-39* function cause P3.p-P8.p to adopt the F fate rather than VPC fate;[58,59] while mutations that partially reduce *lin-39* function allow some VPC fates to be specified, but these VPCs often adopt 3°, F or 1°/2° hybrid cell fates.[60,70] LIN-39 levels in the VPCs are regulated by the Ras pathway: reduction-of-function mutations in *lin-3/EGF, let-23/RTK* and *let-60/Ras* cause a decrease in LIN-39 expression, while gain-of-function mutations in *let-60* cause an increase in LIN-39 expression. [60] Like in *lin-39* mutants, P3.p-P8.p in *bar-1/β-catenin* mutants often adopt F cell fates and VPC fate specification is defective, indicating a role for Wnt signaling in vulval development. [71] In addition, there is a loss of LIN-39 expression during the late L2 stage in bar-1 mutants. Furthermore, *lin-39* expressed from the *bar-1* promoter rescues *bar-1* mutants, suggesting that *bar-1* function is required for *lin-39* expression. Thus Wnt signaling may also specify the competence of these cells to respond Ras pathway signaling as with P12 cell fate. It is likely that the Wnt and Ras pathways both impinge upon LIN-39 expression during VPC fate determination (Fig. 5C). Double mutant analyses suggest *bar-1* functions after *let-60, mpk-1/MAPK* and *mek-2/MAPKK* but prior to, or in parallel with, the transcription factors *lin-1/ETS* or *lin-31/WH* (Winged helix protein).[71] Interestingly, when expressed from the *bar-1* promoter, both *wrm-1* and *hmp-2* can rescue a *bar-1* mutant, suggesting the other *C. elegans* β-catenins can also function in the vulval development Wnt pathway when overexpressed in the VPCs. This is quite surprising given that HMP-2 does not normally function in any signaling pathways and does not interact with POP-1 in yeast two-hybrid or co-immunoprecipitation assays. Perhaps HMP-2 can interact with POP-1 when overexpressed in worm cells or interacts with POP-1 via another factor not present in yeast or cultured cells.[18]

Other factors implicated in the Wnt pathway that functions during VPC determination include *mig-14/mom-3, pry-1/Axin* and *apr-1/APC. mig-14/mom-3* mutants display defects in VPC generation and determination similar to those observed in bar-1 mutants, indicating that

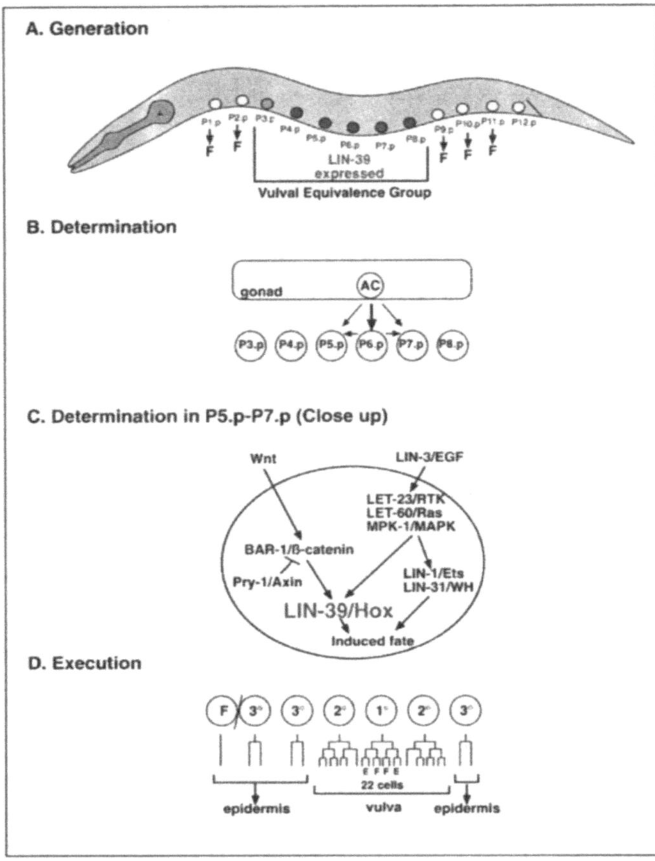


Figure 5. Vulval development in *C. elegans*. Three of the four stages of vulval development are diagramed. A) Generation of P1.p –P12.p during the L1 stage. Expression of LIN-39/Hox in P3.p-P8.p causes them to take on the VPC fate instead of fusing with the epidermal syncytium, or F fate, as do P1.p, P2.p and P9-11.p. B) Determination of particular vulval cell fates occurs in response to several extracellular signals: an apparently graded signal from the anchor cell (AC) in the overlying gonad, lateral signals between the Pn.p cells and an inhibitory signal from the surrounding epidermal syncytium (not shown). As a result, P6.p adopts the 1° vulval cell fate, P5.p and P7.p adopt the 2° vulval cell fates and P3.p, P4.p and P8.p adopt non-vulval 3° cell fates. P3.p adopts the non-VPC F fate and fuses with the epidermal syncytium 50% of the time in wild-type animals, C) An interaction between and Wnt and Ras pathways impinge on LIN-39 expression, which is required for P5.p-P7.p to take on induced vulval cell fates. Neither a Wnt ligand nor a Fz receptor have yet to be identified. The RTK/MAPK pathway also induces the expression of LIN-1/Ets and LIN-31/WH, transcription factors required for the induced vulval cell fate. D) Execution of vulval cell fates by the vulval precursor cells occurs in the L3 stage. The patterns of cell division of P3.p-P8.p are shown. The vulE (E) and vulF (F) pattern of cell fates among 1° descendants is also indicated. This pattern appears to be generated by Ras signaling from the AC to the inner 1° descendants and Wnt signals via LIN-17/Fz between the inner and outer 1° descendants. After refs. 28 and 71.

it is also involved in regulating LIN-39 expression. However, it is not known whether *mig-14/mom-3* mutations affect LIN-39 levels as do *bar-1* mutations.[18] *pry-1* mutants display an overinduction of vulval cell fates, where too many cells adopt 1° or 2° VPC fates. This VPC determination defect is opposite to that observed in *bar-1* mutants,[72] suggesting that *pry-1/Axin* functions as a negative regulator of Wnt signaling involved in vulval development, as it does in other animals. In addition, expression of a stabilized BAR-1 protein, produced by removing the amino-terminal consensus GSK-3 phosphorylation sites (ΔN-BAR-1) also caused an overinduction of vulval cell fates. Furthermore, double mutant analyses indicated that the Overinduced defect of *pry-1* mutants was dependant upon *bar-1, pop-1* and *lin-39*, but not upon *mig-14/mom-3* (Fig. 9). Interestingly, mutations that reduced Ras signaling did not affect the overinduction of vulval cell fates caused by loss of *pry-1* function and BAR-1 stabilization, suggesting that Wnt signaling can directly affect LIN-39 expression.[72] Thus the Wnt pathway that functions during VPC determination appears to be canonical, and this pathway interacts with a RTK/Ras pathway to regulate *lin-39* Hox gene expression (Fig. 12.5C).

The role of *apr-1/APC* in this pathway is less clear, however. Loss of *apr-1* function specifically in P3.p-P8.p by expression of an antisense *apr-1* transgene driven by the VPC-specific *lin-31* promoter causes defects in the regulation of LIN-39 expression and function similar to those observed in *bar-1* mutants,[73] suggesting that both *apr-1* and *bar-1/β-catenin* functions are required positively for LIN-39 expression. This would be similar to the situation in the EMS blastomere, where the loss of *apr-1* and *wrm-1* function have similar effects. However, the overinduction of vulval cell fates caused by loss of *pry-1/Axin* function or BAR-1 stabilization was enhanced by *apr-1(RNAi)*, indicating that the overinduction does not require *apr-1* function, and that APR-1 may negatively regulate Wnt signaling, as it does in the canonical pathway.[72] Further experiments are required to reconcile these results on the role of *apr-1* in vulval development.

In summary, during vulval development, the Wnt pathway collaborates with the Ras pathway, to specify which vulva cell fate, 1°, 2° or 3° each VPC adopts. Thus, as in QL.d migration and P12 cell fate determination processes in the worm, the Wnt pathway that functions in vulval development regulates the expression of a Hox gene. At present, it appears that this Wnt pathway is canonical, however the functions of many components have yet to be determined. For example, no Wnt gene has been found to affect this process nor has a Fz receptor been identified. Exactly how this Wnt pathway interacts with the Ras pathway is also not clear. Both Wnt and Ras signaling may function together to instruct VPC fates (Fig. 5C). Alternatively, Wnt signaling may function permissively for the VPC to be competent to receive Ras signals, for example to achieve a certain level of LIN-39 expression that is increased by Ras signaling, leading to induced vulval cell fates.[72]

Finally, Wnt signaling has also been implicated in the execution and morphogenesis of the 1° cell fate. The 1° cell generates a specific pattern of vulval cells: four outer vulE cells and four inner vulF cells (Fig. 5D). This pattern appears to be generated by Ras signaling from the AC to the inner 1° descendants and Wnt signals via LIN-17/Fz between the inner and outer 1° descendants ensure the proper differentiation of vulE and vulF cell fates.[74]

## An Unusual Wnt Pathway Controls T Cell Polarity

Wnt signals control the polarities of several cells during *C. elegans* development. Mutations in *lin-44/Wnt* cause the polarities of the B and T cells in the tail to be reversed (Fig. 6A).[5] Mutations in *lin-17/Fz* cause the loss of polarity in these same cells;[75,76] suggesting that LIN-17 is the LIN-44 receptor. LIN-44 is expressed in the epidermal cells at the tip of the developing tail, which is posterior to the cells affected by *lin-44* mutations .[6] Although both *lin-44* and *lin-17* mutations affect the polarities of the same cells, the phenotypes are slightly different, which is unexpected for a receptor-ligand pair. One possibility is that a second signal, probably another Wnt, emanates from an anterior source, causing the reversal of polarity observed in *lin-44* mutants. Unfortunately, as of yet, these is not evidence for this second signal. *egl-20/Wnt* is expressed in cells just anterior to the T cells that are in a perfect position to the source of such an anterior signal. [26] However, *egl-20* mutants have no effect on T cell polarity even when in combination with *lin-44* mutations, indicating that EGL-20 is not the anterior signal (M.H., unpublished).

POP-1 is also involved in the control of T cell polarity. Recall that in the embryo, Wnt signaling functions with components of a MAPK pathway to lower the levels of POP-1 in the posterior daughter of the EMS blastomere. The level of POP-1 is lower in the posterior daughters of many anteroposterior asymmetric cell divisions during development [43] and is also true of the T cell divisions. In wild-type animals, the level of POP-1 is lower in the posterior T cell daughter. Furthermore, in *lin-44* mutants, the level of POP-1 was lower in the anterior T cell daughter, reflecting the reversal of T cell polarity. Interfering with *pop-1* zygotic function by RNAi or expression of the dominant negative ΔN-POP-1 construct causes a loss of T cell polarity similar to that observed in *lin-17* mutants (Fig. 6A).[27] Further, mutations in *pop-1* that cause defects in gonad development (see below) also cause T cell polarity defects.[77] Therefore, a Wnt, Frizzled and TCF are required for proper T cell polarity. However, the pathway that controls T cell polarity appears to be unusual. *lit-1*, which functions in the control of EMS polarity, also appears to act in the control of T cell polarity. Mutations in *lit-1* cause the loss of T cell polarity and *lit-1* is expressed in the T cells.[19] Thus there are some similarities in the control of T cell and EMS polarities. Surprisingly, none of the three *C. elegans* β-catenin homologs appear to function with POP-1 to control T cell polarity. [27] This is a major difference and it is not clear what, if any, other proteins might interact directly with POP-1 in the control of T cell polarity. The identification of which of the other known Wnt signaling components function in T cell polarity may help sort this out.

*lin-44* expressed from a heat-shock promoter, which presumably causes near uniform expression, can rescue the T cell polarity defects of *lin-44* mutants, suggesting that LIN-44 is a permissive rather than an instructive signal for T cell polarity.[6] This is similar to how *egl-20/Wnt* is thought to function in the control of V5 polarity [78] (see below). However, whereas expression of *egl-20* in the head of the animal can rescue V5 polarity defects of *egl-20* mutants,[78] similar expression of *lin-44* does not rescue the T cell polarity defects of *lin-44* mutants (M.H. unpublished). This suggests that LIN-44 acts as an instructive signal. *lin-44* expressed from the heat-shock promoter might not be processed or secreted uniformly, perhaps due to asymmetric expression of *mom-1/Porc* or *mig-14/mom-3*.

The Wnt signaling pathway that controls T cell polarity may exert its effects on transcription via chromatin remodeling complexes. Specifically, mutations in *egl-27*, which encodes a protein with similarity to a factor isolated as a component of the NURD complex (see above), cause the loss of T cell polarity. However it is not clear whether EGL-27 functions in the NURD complex during the T cell division or at any other time. In addition, mutations in genes encoding *C. elegans* homologs of components of the SWI/SNF complex also cause a loss of T cell polarity.[79] The SWI/SNF complex was originally identified in yeast as being required for activation of genes involved in mating–type switching and sucrose fermentation. *Drosophila* and vertebrate SWI/SNF complexes can be involved in either activation or repression of transcription (reviewed by Vignali [80]). Interestingly, components of the *C. elegans* SWI/SNF complex interact genetically with *egl-27*, suggesting that both play roles; it is not clear whether they have the same or distinct targets, however.

Mutations in *tlp-1* also cause defects in T cell asymmetry. *tlp-1* is asymmetrically expressed in the posterior T cell daughter and encodes a C2H2 Zinc finger protein that is likely to function as a transcription factor required for neural and other cell fates. This suggests that

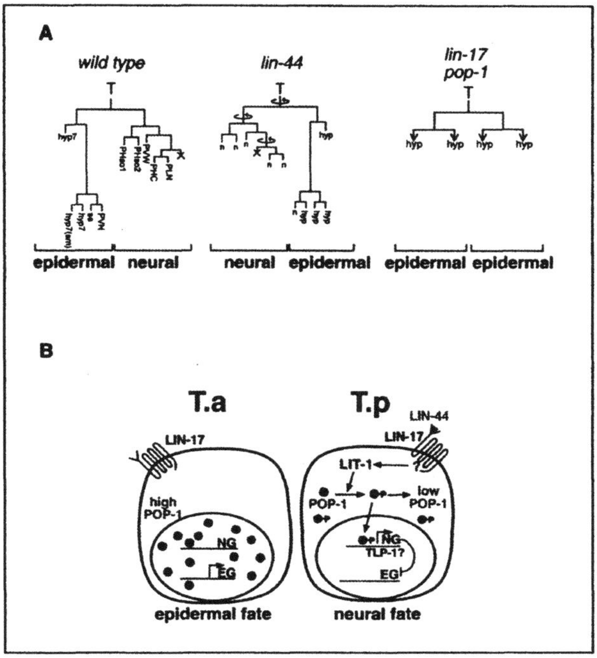


Figure 6. Control of T cell polarity by an unusual Wnt pathway. A) Representative hermaphrodite T cell lineages in wild type, *lin-44*, *lin-17* and *pop-1* mutants. In wild-type animals T.a generates epidermal cell fates and T.p generates neural cell fates. This is reversed in *lin-44* mutants. Cell fates are distinguished by nuclear morphologies.[1,5] hyp7 cells join the epidermal syncytium, including T.apaa which has a smaller nucleus (sm). Phso1, Phso2, PVW, PHC and PLN have neural nuclear morphologies. X indicates programmed cell death. Seam (se) is a specialized epidermal cell. Downward arrows indicate variable numbers of epidermal cells can be generated. B) A model for T cell polarity. LIN-44/Wnt (triangle) binds to LIN-17/FZ on the posterior surface of the T cell before division, although for simplicity it is shown binding to T.p. The signal is transduced through unknown factors (?) leading to the activation of LIT-1, which functions to phosphorylate POP-1 (circles), perhaps in combination with an unknown factor. Some of the modified POP-1 is degraded leading to lower POP-1 levels, and some may activate expression of neural-specific genes (NG). A neural-specific gene then represses epidermal-specific (EG) genes in T.p. TLP-1 might be a target gene that gets activated in T.p. The T.a cell takes on a default epidermal fate. This might occur through the constitutive expression of epidermal-specific genes that promote epidermal fate. Without modification, the high levels of POP-1 in T.a may be nonfunctional. Modified from Herman, M.A. (2001) Development, 128: 581-590, *C. elegans* POP-1/TCF function in a canonical Wnt pathway that controls cell migration and in a noncanonical Wnt pathway that control cell polarity, with permission from Company of Biologists, Ltd.

TLP-1 functions in the execution of cell polarity decisions. *lin-44* and *lin-17* mutants cause *tlp-1::gfp* expression to be reversed or lost, respectively. Thus *lin-44* and *lin-17* must act upstream of *tlp-1*, suggesting that *tlp-1* may be a target of the Wnt pathway that functions to control T cell polarity.[81]

## *A Model for T Cell Polarity*

In wild-type animals T.a generates epidermal cell fates and T.p generates neural cell fates (Fig. 6B). Since the polarity of the T cell is determined before it divides,[5] there is a segregation of cell fates at the T cell division: epidermal fate to T.a and neural fate to T.p. Each cell fate is correlated with a particular level of POP-1/TCF protein: high levels and epidermal fate in T.a and lower levels and neural fate in T.p. Mutations in both *lin-44* and *lin-17* affect POP-1 distribution, but interference with *pop-1* function causes both cells to adopt epidermal fates. One model to reconcile these data is shown in Fig. 6B.[27] LIN-44/Wnt might act through LIT-1/NLK to modify POP-1, which decreases POP-1 levels and at the same time converts POP-1 to an activator that activates neural-specific genes in T.p. The high levels of POP-1 in T.a could be nonfunctional. Specifically, LIN-44 binds to LIN-17/Fz on the posterior end of the T cell before its division and on T.p and its descendants. LIN-44 does not bind to T.a and it accumulates high levels of POP-1 and expresses epidermal-specific genes. This could be the default state, since loss of *pop-1* function also causes T.a to adopt an epidermal fate, and might occur by the constitutive expression of epidermal-specific genes in T.a. Upon binding of LIN-44 to LIN-17, LIT-1 is activated by transduction through unknown components. Activated LIT-1 could then phosphorylate POP-1, leading to the degradation of POP-1, as occurs in the E cell.[50] Some POP-1 function must be required for the neural cell fate since loss of *pop-1* function causes T.p to take on an epidermal fate. A low level of modified, perhaps phosphorylated, POP-1 might function positively without a β-catenin ortholog to activate neural-specific genes, one or more of which could then repress epidermal-specific genes in T.p. The activation of neural-specific genes might require the N-terminal domain of POP-1 as overexpression of ΔN-POP-1 causes a loss of neural cell fates. Further work is required to identify other factors that are involved and how they interact in this unusual Wnt pathway.

## Gonad Polarity

Wnt signaling also controls asymmetric cell divisions that occur during the development of the *C. elegans* gonad. In this case the plane of asymmetry is proximal-distal (P-D), rather than anterior-posterior as with the EMS blastomere and the T cell. *C. elegans* gonadogenesis starts with the specification of the somatic gonad precursor cells, then the gonad primordium forms and gonad polarity is established (reviewed by Hubbard and Greenstein [82]). The gonad primordium contains four cells: the somatic gonadal precursor cells, Z1 and Z4 on each end and the germline precursor cells, Z2 and Z3 in the middle. The hermaphrodite gonad develops two arms, each containing an ovo-testes with its own (P-D) axis. The male gonad develops into a single testis with a single P-D axis. [83] In each sex, the P-D axes are set up by the first divisions of Z1 and Z4, descendants with distal fates lying at the ends and descendants with proximal fates lying in the middle. This arrangement is maintained in hermaphrodites, but in males cell migrations result in the generation of a single P-D axis (Fig. 7).

Mutations in *lin-17/Fz* cause a loss of asymmetry such that both Z1 and Z4 daughters generate proximal cell fates [75], implicating Wnt signaling in this asymmetric cell division. This defect has been termed Sys (for symmetric sisters) and additional mutants with this defect have been isolated, some of which turned out to be alleles of *pop-1/TCF.* [77] The Wnt pathway that controls Z1 and Z4 polarity includes *mom-1/Porc*, *lin-17* (and no other *C. elegans* Frizzleds), *lit-1/NLK*, *wrm-1/β-catenin* (and not bar-1) and *pop-1* since interference with these gives the same Sys phenotype. None of the *C. elegans* Wnt genes appeared to be involved, although the possibility of redundancy has not been ruled out.

The fact that *pop-1* appears to play a positive role in Wnt signaling in the gonad is similar to the situation in the T cell (although *wrm-1* may not be involved in the T cell) but contrasts the situation in the embryo where *pop-1* has a phenotype opposite to that of *lit-1* and *wrm-1*. This suggests that the regulation of polarity in Z1, Z4 and T may be similar, but different, from EMS. Siegfried and Kimble[77] propose two models for how WRM-1 and LIT-1 might function to positively regulate POP-1 function in the gonad. In the first, WRM-1 binds to POP-1 in a

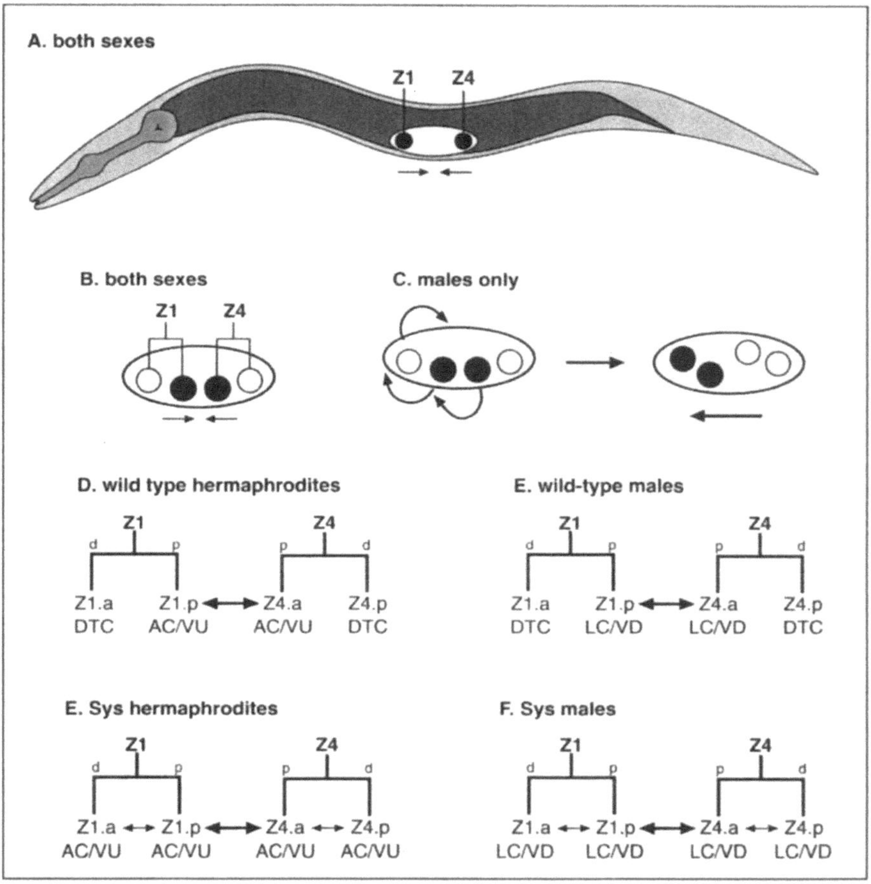


Figure 7. Gonadogenesis in wild-type and sys (*pop-1*) mutants. A) The gonad primordium in males and hermaphrodites appears identical. Arrows indicate primordium symmetry. The somatic gonad precursors, Z1 and Z4, are present at each end of the primordium and germline precursors, Z2 and Z3 (not shown) are in the middle. B) Z1 and Z4 divide asymmetrically along the proximal (p) to distal (d) axis in both sexes. This symmetry is retained throughout gonadogenesis in hermaphrodites. C) In males, an asymmetric structure is generated by Z1/Z4 daughter rearrangements. D) In wild-type hermaphrodites Z1.a and Z4.p generate distal tip cells (DTCs), whereas Z1.p and Z4.a generate AC/VU cells, with one anchor cell (AC) and the other as a ventral uterine precursor (VU) as a result of lateral signaling. E) In wild-type males Z1.a and Z4.p generate DTCs, whereas Z1.p and Z4.a generate bipotential cells that can adopt a linker cell (LC) or a vas deferens (VD) fate. One becomes an LC and the other becomes a VD as a result of lateral signaling. F) In sys mutant hermaphrodites all four Z1/Z4 daughters can generate AC/VU cells. G) In *sys* mutant males Z1 and Z4 generate bipotential LC/VD cells and no DTCs are made. Modified from Siegfried, KR and Kimble, J. (2002) POP-1 controls axis formations during early gonadogenesis in *C. elegans*. Development. 129: 443-453, with permission from Company of Biologists, Ltd.

canonical β-catenin-TCF relationship to activate gene expression leading to the establishment of different proximal-distal cell fates. Although WRM-1 does not bind to the amino terminal domain of POP-1,[17] it appears to bind weakly to full length POP-1. This interaction may be stabilized by LIT-1, which binds to POP-1.[18,19] The second is similar to that proposed for T cell polarity (Fig. 6B).[27] Specifically, WRM-1 and LIT-1 function to modify POP-1, which results in the activation of gene expression. It is not clear whether the action of WRM-1 or LIT-1 also results in decreased POP-1 levels as is proposed in the posterior T cell daughter, or whether a particular level of POP-1 is correlated with a Z1 or Z4 daughter cell fate, as POP-1 antibodies do not stain the gonads of L1 animals.

## Wnt Pathways Might Interact to Control V5 Cell Polarity

*egl-20/Wnt* also controls the polarities of asymmetric cell divisions in the lateral epidermis of the animals. One each side of the animal six epidermal V cells divide asymmetrically along the anteroposterior axis during postembryonic development to generate cuticular and sensory structures. The anterior V cell daughter, Vn.a, becomes a syncytial cell which fuses with the epidermal syncytium (hyp7) that covers most of the body of the animal. The posterior V cell daughters, Vn.p, becomes a seam cell and continues to divide (Fig. 8). Sensory structures are generated from V5.p in both sexes and from V6.p in males. Like most anterior-posterior asymmetric cell divisions, POP-1 is expressed at high levels in the Vn.a cells and at low levels in Vn.p cells.[27,43] The polarity of the V5 division is reversed in 50 percent of *egl-20* mutants [25] and seven percent of *mig-14/mom-3* mutants. Like *lin-44/Wnt*, *egl-20* is expressed in a group of cells that are posterior to the cells whose polarity it controls.[84] *egl-20* expressed from a heat-shock promoter, which presumably causes near uniform expression, can rescue the V5 polarity defects of *egl-20* mutants. Surprisingly, expression of *egl-20* from a pharynx-specific promoter (at the anterior end of the animal) can also rescue the V5 polarity defects of *egl-20* mutants, even thought the source of EGL-20 signal is on the other side of the target cells and over half the body length away![78] This suggests that EGL-20/Wnt must be secreted and can diffuse a distance from its source to function and that EGL-20 is a permissive rather that an instructive signal for V5 polarity.[78] What controls the polarity of the other V cells? Mutations in *cam-1*, that encodes a Ror orphan receptor kinase cause polarity reversals in V1.[85] Thus, some signaling system involving Ror kinases may be involved in V cell polarity. Long periods of heat-shock induced higher levels of *hs::egl-20* expression and also caused the polarity defects in the V1-V4 and V6 cells. This indicates that, although these cells do not normally require *egl-20* to establish their polarities, the system that determines their polarities can be disrupted by high levels of EGL-20.[78]

What could be responsible for the V5 polarity reversals observed in *egl-20* mutants? Killing the V6 and T cells with a laser microbeam suppressed the V5 polarity reversals in *egl-20/Wnt* mutants; whereas killing anterior neighboring cells did not. This suggests that a lateral signal from the cells posterior and adjacent to V5 appears to be responsible for the V5 polarity reversals in *egl-20* mutants. LIN-44/Wnt is not involved in this lateral signaling system as mutations in *lin-44* could not suppress the V5 reversals observed in *egl-20* mutants and did not influence the effects of killing posterior cells. Normally, the suppressive effect of killing the posterior cells in *egl-20* mutants is restricted to the side of the animal on which the cells were killed. Curiously, in *lin-44*; *egl-20* double mutants, this side restriction became uncoupled, suggesting that LIN-44 functions somehow to restrict polarity signals from one side of the animal from affecting the other. LIN-17 does appear to be involved, however, as mutations in *lin-17* suppress the polarity reversals seen in *egl-20* mutants. PRY-1/Axin is also involved as *pry-1* mutations enhance the V5 polarity reversals of *egl-20* mutants. Thus the lateral signal from the cells posterior and adjacent to V5 that appears to be responsible for the V5 polarity reversals in *egl-20* mutants requires *lin-17* and *pry-1*, but not *lin-44*.[78] It appears that lateral

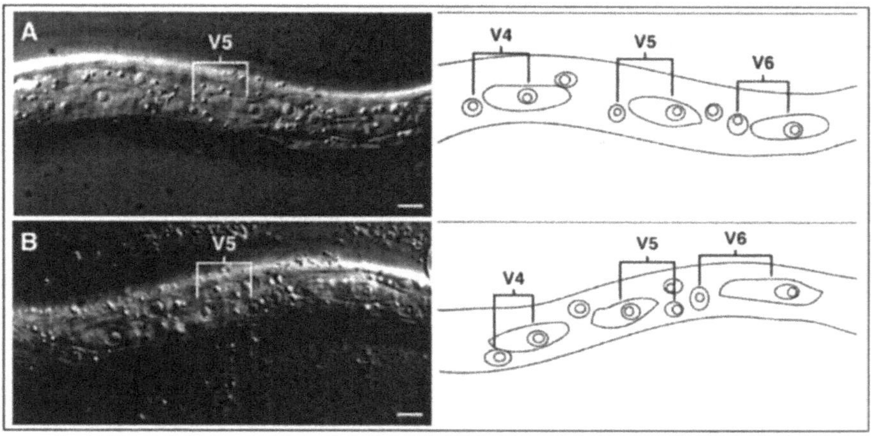


Figure 8. EGL-20 orients the V5 cell division. Nomarski micrographs of L1 larvae are shown on the left, with tracings of the on the right. A) Wild type. The outline of the posterior V5 daughter (V5.p) is visible, which is the seam cell the morphology. The anterior daughter (V5.a) has a syncytial morphology: its nucleus is slightly ventral to the V5.p nucleus, and has a grainy nucleoplasm. B) *egl-20* mutant. The polarity of the V5 cell division is reversed: V5.a has the morphology and position of a seam cell and V5.p has the morphology and position of a syncytial nucleus. Bars, 10 µm. Reprinted from Whangbo, J, et al. (2000) Multiple levels of regulation specify the polarity of an asymmetric cell division in *C. elegans*. Development. 127: 4587-98, with permission from Company of Biologists, Ltd.

signaling occurs through a canonical Wnt pathway, but with only three components identified this is not certain. Whangbo et al[78] hypothesized that EGL-20 function is required to override the lateral signal. This pathway may involve MIG-14/MOM-3, but additional components have not been identified. They further hypothesize an underlying global system may exist that functions to orient anteroposterior cell polarities in the worm. This system may be responsible for the asymmetric distribution of POP-1 at many anterior-posterior asymmetric cell divisions as well as for the polarities of the EGL-20 and LIN-44 dependant cell divisions. A prediction of this model is that LIN-44 would also function as a permissive signal, this does not appear to be the case, however (M.H., unpublished observations). Finally, the targets of the V5 polarity Wnt pathways are not known.

## Sensory Ray Formation

Wnt signaling also plays a role in interactions that occur among the lateral epidermal cells that generate sensilla. In hermaphrodites and males, a descendant of the V5 cells, V5.pa, generates the postderid sensillum. In males, a different type of sensilla, the sensory rays, are produced such that each side of the animal generates nine rays: the V5.pp, V6 and T cells generate one, five and three rays, respectively. Interactions among the lateral epidermal cells regulate the numbers of rays that are generated. If the V5 and V6 cells in males were killed with a laser microbeam, the next anterior V cell, V4, generates a number of rays.[54] If either V4 or V6 is killed in a male, V5 makes a number of rays, but no postderid.[86-88] Thus cell interactions regulate cell fates. *mab-5* is required for generation of the V5- and V6-derived rays, but represses postderid formation. If V6 is killed or if *mab-5* is expressed from a heat-shock promoter, the V5 cell takes on the V6 cell fate, producing extra rays but no postderid.[89] However, following the killing of V6, Wnt signaling involving *egl-20Wnt, lin-17/Fz* and *bar-1/β-catenin* is required for V5 to take on the V6 fate.[90] Furthermore, *pry-1* mutations cause ectopic *mab-5*

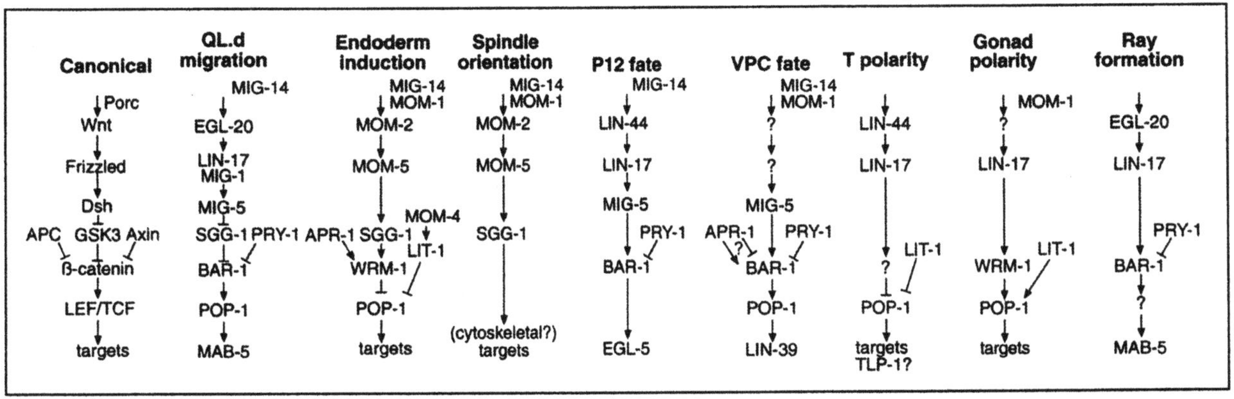


Figure 9. Comparison of the canonical and *C. elegans* Wnt pathways. Known genes involved in QL.d migration, Endoderm induction, EMS spindle orientation, P12 cell fate, VPC cell fate, T cell polarity, gonad polarity and male ray formation and the inferred regulatory relationships are shown. The Wnt pathways that interact the control V5 cell polarity are not shown, as the nature of the interaction is not well understood. MIG-14/MOM-3 is simplified to MIG-14. Only the canonical pathway components with *C. elegans* homologs are shown. APR-1 is shown as influencing WRM-1 independently of SGG-1 in the Endoderm induction pathway, since SGG-1 influences EMS spindle orientations whereas APR-1 does not. "?" indicates that multiple or unknown homologs of a component function in a particular pathway. In the T cell polarity pathway "?" indicates that either a ß-catenin homolog is not involved or that an unknown factor may be involved and "TLP-1?" indicates that tlp-1 might not be a direct target. In the gonad polarity pathway "?" indicates the Wnt involved has not been identified. Similarly, in the VPC fate pathway "?" indicates that a Wnt or a Fz involved has not been identified.

expression (which requires *bar-1*) leading to loss of the postderid and ectopic ray formation.[26] Thus a canonical Wnt signaling pathway activates *mab-5* expression in the lateral epidermal cells, but is normally inhibited by cell contacts by a mechanism that is not clear. In the V6 cell, the cell contact inhibition is overcome by the caudal homolog, *pal-1*, which functions to activate *mab-5* expression to specify V6 cell fate.[90]

## Themes and Remaining Questions in Worm Wnt Signaling

Wnt signaling pathways are involved in three major types of process during *C. elegans* development: control of asymmetric cell divisions and cell polarity, cell fate decisions and cell migration. Both canonical and noncanonical Wnt pathways are used in the various processes. The Wnt pathway that controls the migrations of the QL descendants is clearly canonical. The pathways that control P12 cell fate, VPC fate, V5 cell polarity and signaling in the lateral sensilla also appear to be canonical, although several "gaps" remain.

The Wnt pathways involved in the control of cell polarity of EMS (endoderm induction), T and Z1 and Z4 are each unusual, but in different respects. Whereas the pathway that controls V5 polarity may be canonical, the involvement of many of the canonical pathway components have yet to be assessed. POP-1/TCF and LIT-1/NLK function in the EMS, T and Z1/Z4 cell polarity Wnt pathways, but POP-1 is negatively regulated in the EMS blastomere. Just how POP-1 levels and function are regulated in the EMS daughters, as well as the determination the role of Dsh in this regulation, remain as important questions for the control of EMS polarity. In addition, WRM-1/β-catenin functions in EMS and Z1/Z4 cell polarity but may not function in T cell polarity; although the latter is a negative RNAi result that will have to be reexamined once a *wrm-1* mutation is isolated. The isolation of additional components of these pathways and further examination of the regulatory relationships of the components is required to determine the exact extent to which these pathways are similar or different.

For several processes it is difficult to discern whether or not the pathway is canonical. Some of this uncertainty is caused by the lack of information about the involvement of particular Wnt pathway component in each of the processes. This is particularly true of the Wnt pathway involved in vulval cell fate determination, which appears to canonical in many respects in that PRY-1/Axin, BAR-1/β-catenin and POP-1/TCF appear to be regulated in a canonical manner, but the role of APR-1/APC is unclear as is the identity of the Wnt ligand and Fz receptor that are involved. In order to get a complete picture of Wnt signaling in *C. elegans* the "gaps" that exist in Table 2 need to be filled.

Aside from the types of processes that are regulated by either canonical or noncanonical Wnt signaling, a couple other themes emerge. As is the case in other animals, a target gene of many of the C. elegans Wnt pathways is a Hox gene: *mab-5* in the QL.d migration and lateral sensilla pathways, *egl-5* in the P12 cell fate pathway and *lin-39* in the VPC fate pathway. Another possible theme is that for both P12 cell fate and VPC fate, Wnt signaling might control the competence of cells to respond to other signals involving the Ras pathway that control cell fates. However, in the case of VPC fate it is also possible that Wnt and Ras pathways collaborate to specify cell fates.

Finally, MIG-14/MOM-3 is involved in multiple C. elegans Wnt signaling pathways (Table 2, Fig. 9), yet the molecular identity of MIG-14/MOM-3 has not been determined. *mig-14/mom-3* maps to a genetic region that appears to be devoid of known Wnt signaling components or regulators, suggesting that it may encode a novel Wnt signaling component. Since *mom-3* functions in P2, the signaling cell, for endoderm induction, it is likely to function in the signaling cell in other signaling events. Thus, MIG-14/MOM-3 appears to function generally in the synthesis, secretion or processing of Wnt proteins that function in many, if not all, *C. elegans* Wnt pathways and may also be conserved in Wnt signaling pathways in other organisms.

## *Acknowledgements*

I thank David Eisenmann, Rik Korswagen, Judith Kimble and Joel Rothman for communication of results prior to publication and Dave Eisenmann for Table 12.2. I also thank David Eisenmann and Rik Korswagen for critical reading of this manuscript. Work in the Herman Lab is supported by NIH grant GM56339.


## References


1. Sulston JE, Horvitz HR. Post-embryonic cell lineages of the nematode, Caenorhabditis elegans. Dev Biol 1977; 56(1):110-156.
2. Sulston JE, Schierenberg E, White JG et al. The embryonic cell lineage of the nematode Caenorhabditis elegans. Dev Biol 1983; 100(1):64-119.
3. Kamb A, Weir M, Rudy B et al. Identification of genes from pattern formation, tyrosine kinase, and potassium channel families by DNA amplification. Proc Natl Acad Sci USA 1989; 86:4372-4376.
4. Shackleford GM, Shivakumar S, Shiue L et al. Two wnt genes in Caenorhabditis elegans. Oncogene 1993; 8:1857-1864.
5. Herman MA, Horvitz HR. The Caenorhabditis elegans gene lin-44 controls the polarity of asymmetric cell divisions. Development 1994; 120:1035-1047.
6. Herman MA, Vassilieva LL, Horvitz HR et al. The C. elegans gene lin-44, which controls the polarity of certain asymmetric cell divisions, encodes a Wnt protein and acts cell nonautonomously. Cell 1995; 83(1):101-110.
7. Rocheleau CE, Downs WD, Lin R et al. Wnt signaling and an APC-related gene specify endoderm in early C. elegans embryos. Cell 1997; 90(4):707-716.
8. Thorpe CJ, Schlesinger A, Carter JC et al. Wnt signaling polarizes an early C. elegans blastomere to distinguish endoderm from mesoderm. Cell 1997; 90(4):695-705.
9. The C. elegans Sequencing Consortium. Genome sequence of the nematode C. elegans: A platform for investigating biology. Science 1998; 282(5396):2012-2018.
10. Ruvkun G, Hobert O. The taxonomy of developmental control in Caenorhabditis elegans. Science 1998; 282(5396):2033-2041.
11. Korswagen HC, Coudreeuse DYM, Betist M et al. The Axin-like protein PRY-1 is a negative regulator of a canonical Wnt pathway in C. elegans. Bioessays 2002; 29(9):801-810.
12. Strovel ET, Sussman DJ. Transient overexpression of murine dishevelled genes results in apoptotic cell death. Exp Cell Res 1999; 253(2):637-648.
13. Sun TQ, Lu B, Feng JJ et al. PAR-1 is a Dishevelled-associated kinase and a positive regulator of Wnt signalling. Nat Cell Biol 2001; 3(7):628-636.
14. Guo S, Kemphues KJ. par-1, a gene required for establishing polarity in C. elegans embryos, encodes a putative Ser/Thr kinase that is asymmetrically distributed. Cell 1995; 81(4):611-620.
15. Marikawa Y, Elinson RP. beta-TrCP is a negative regulator of Wnt/beta-catenin signaling pathway and dorsal axis formation in Xenopus embryos. Mech Dev 1998; 77(1):75-80.
16. Kipreos ET, Gohel SP, Hedgecock EM. The C. elegans F-box/WD-repeat protein LIN-23 functions to limit cell division during development. Development 2000; 127(23):5071-5082.
17. Korswagen HC, Herman MA, Clevers HC. Distinct beta-catenins mediate adhesion and signalling functions in C. elegans. Nature 2000; 406(6795):527-532.
18. Natarajan L, Witwer NE, Eisenmann DM. The Divergent Caenorhabditis elegans beta-Catenin Proteins BAR-1, WRM-1 and HMP-2 Make Distinct Protein Interactions but Retain Functional Redundancy in vivo. Genetics 2001; 159(1):159-172.
19. Rocheleau CE, Yasuda J, Shin TH et al. WRM-1 activates the LIT-1 protein kinase to transduce anterior/posterior polarity signals in C. elegans. Cell 1999; 97(6):717-726.
20. Hobmayer B, Rentzsch F, Kuhn K et al. WNT signalling molecules act in axis formation in the diploblastic metazoan Hydra. Nature 2000; 407(6801):186-189.
21. Hedgecock EM, Culotti JG, Hall DH et al. Genetics of cell and axon migrations in Caenorhabditis elegans. Development 1987; 100(3):365-382.
22. Salser SJ, Kenyon C. Activation of a C. elegans Antennapedia homologue in migrating cells controls their direction of migration. Nature 1992; 355(6357):255-258.

23. Kenyon C. A gene involved in the development of the posterior body region of C. elegans. Cell 1986; 46(3):477-487.
24. Guo C. mig-5, a gene that controls cell fate determination and cell migration in C. elegans, is a member of the Dsh family [Ph.D. Thesis]. Baltimore, MD: Johns Hopkins University; 1995.
25. Harris J, Honigberg L, Robinson N et al. Neuronal cell migration in C. elegans: regulation of Hox gene expression and cell position. Development 1996; 122(10):3117-3131.
26. Maloof JN, Whangbo J, Harris JM et al. A Wnt signaling pathway controls Hox gene expression and neuroblast migration in C. elegans. Development 1999; 126(1):37-49.
27. Herman M. C. elegans POP-1/TCF functions in a canonical Wnt pathway that controls cell migration and in a noncanonical Wnt pathway that controls cell polarity. Development 2001; 128(4):581-590.
28. Eisenmann DM, Kim SK. Protruding vulva mutants identify novel loci and Wnt signaling factors that function during Caenorhabditis elegans vulva development. Genetics 2000; 156(3):1097-1116.
29. Herman MA, Ch'ng Q, Hettenbach SM et al. EGL-27 is similar to a metastasis-associated factor and controls cell polarity and cell migration in C. elegans. Development 1999; 126(5):1055-1064.
30. Toh Y, Pencil SD, Nicolson GL. Analysis of the complete sequence of the novel metastasis-associated candidate gene, mta1, differentially expressed in mammary adenocarcinoma and breast cancer cell lines. Gene 1995; 159(1):97-104.
31. Solari F, Bateman A, Ahringer J. The Caenorhabditis elegans genes egl-27 and egr-1 are similar to MTA1, a member of a chromatin regulatory complex, and are redundantly required for embryonic patterning. Development 1999; 126(11):2483-2494.
32. Xue Y, Wong J, Moreno GT et al. NURD, a novel complex with both ATP-dependent chromatin-remodeling and histone deacetylase activities. Mol Cell 1998; 2(6):851-861.
33. Goldstein B. Induction of gut in Caenorhabditis elegans embryos. Nature 1992; 357(6375):255-257.
34. Goldstein B. Establishment of gut fate in the E lineage of C. elegans: the roles of lineage-dependent mechanisms and cell interactions. Development 1993; 118(4):1267-1277.
35. Lin R, Thompson S, Priess JR. pop-1 encodes an HMG box protein required for the specification of a mesoderm precursor in early C. elegans embryos. Cell 1995; 83(4):599-609.
36. Kaletta T, Schnabel H, Schnabel R. Binary specification of the embryonic lineage in Caenorhabditis elegans. Nature 1997; 390(6657):294-298.
37. Meneghini MD, Ishitani T, Carter JC et al. MAP kinase and Wnt pathways converge to downregulate an HMG-domain repressor in Caenorhabditis elegans. Nature 1999; 399(6738):793-797.
38. Ishitani T, Ninomiya-Tsuji J, Nagai S et al. The TAK1-NLK-MAPK-related pathway antagonizes signalling between beta- catenin and transcription factor TCF. Nature 1999; 399(6738):798-802.
39. Fire A, Xu S, Montgomery MK et al. Potent and specific genetic interference by double-stranded RNA in Caenorhabditis elegans. Nature 1998; 391(6669):806-811.
40. Schlesinger A, Shelton CA, Maloof JN et al. Wnt pathway components orient a mitotic spindle in the early Caenorhabditis elegans embryo without requiring gene transcription in the responding cell. Genes Dev 1999; 13(15):2028-2038.
41. Peters JM, McKay RM, McKay JP et al. Casein kinase I transduces Wnt signals. Nature 1999; 401(6751):345-350.
42. Jiang M, Ryu J, Kiraly M et al. Genome-wide analysis of developmental and sex-regulated gene expression profiles in Caenorhabditis elegans. Proc Natl Acad Sci USA 2001; 98(1):218-223.
43. Lin R, Hill RJ, Priess JR. POP-1 and anterior-posterior fate decisions in C. elegans embryos. Cell 1998; 92(2):229-239.
44. Zhu J, Hill RJ, Heid PJ et al. end-1 encodes an apparent GATA factor that specifies the endoderm precursor in Caenorhabditis elegans embryos. Genes Dev 1997; 11(21):2883-2896.
45. Calvo D, Victor M, Gay F et al. A POP-1 repressor complex restricts inappropriate cell type-specific gene transcription during Caenorhabditis elegans embryogenesis. Embo J 2001; 20(24):7197-7208.
46. Roose J, Molenaar M, Peterson J et al. The Xenopus Wnt effector XTCF-3 interacts with Groucho-related transcriptional repressors. Nature 1998; 395(6702):608-612.
47. Chen G, Fernandez J, Mische S et al. A functional interaction between the histone deacetylase Rpd3 and the corepressor groucho in Drosophila development. Genes Dev 1999; 13(17):2218-2230.

48. Shi Y, Mello C. A CBP/p300 homolog specifies multiple differentiation pathways in Caenorhabditis elegans. Genes Dev 1998; 12(7):943-955.
49. Hecht A, Vleminckx K, Stemmler MP et al. The p300/CBP acetyltransferases function as transcriptional coactivators of beta-catenin in vertebrates. Embo J 2000; 19(8):1839-1850.
50. Thorpe CJ, Schlesinger A, Bowerman B. Wnt signalling in Caenorhabditis elegans: regulating repressors and polarizing the cytoskeleton. Trends Cell Biol 2000; 10(1):10-17.
51. Goldstein B. Cell contacts orient some cell division axes in the Caenorhabditis elegans embryo. J Cell Biol 1995; 129(4):1071-1080.
52. Salinas PC. Wnt factors in axonal remodelling and synaptogenesis. Biochem Soc Symp 1999; 65:101-109.
53. Krufka A, Johnson RG, Wylie CC et al. Evidence that dorsal-ventral differences in gap junctional communication in the early Xenopus embryo are generated by beta-catenin independent of cell adhesion effects. Dev Biol 1998; 200(1):92-102.
54. Sulston JE, White JG. Regulation and cell autonomy during postembryonic development of Caenorhabditis elegans. Dev Biol 1980; 78(2):577-597.
55. Jiang LI, Sternberg PW. Interactions of EGF, Wnt and HOM-C genes specify the P12 neuroectoblast fate in C. elegans. Development 1998; 125(12):2337-2347.
56. Greenwald I. Development of the Vulva. In: Riddle DL, Blumenthal T, Meyer BJ, Priess JR, eds. C. elegans II. Cold Spring Harbor: Cold Spring Harbor Laboratory Press, 1997:519-541.
57. Ferguson EL, Sternberg PW, Horvitz HR. A genetic pathway for the specification of the vulval cell lineages of Caenorhabditis elegans. Nature 1987; 326(6110):259-267.
58. Clark SG, Chisholm AD, Horvitz HR. Control of cell fates in the central body region of C. elegans by the homeobox gene lin-39. Cell 1993; 74(1):43-55.
59. Wang BB, Muller-Immergluck MM, Austin J et al. A homeotic gene cluster patterns the anteroposterior body axis of C. elegans. Cell 1993; 74(1):29-42.
60. Maloof JN, Kenyon C. The Hox gene lin-39 is required during C. elegans vulval induction to select the outcome of Ras signaling. Development 1998; 125(2):181-190.
61. Kimble J. Alterations in cell lineage following laser ablation of cells in the somatic gonad of Caenorhabditis elegans. Dev Biol 1981; 87(2):286-300.
62. Sternberg PW, Horvitz HR. Pattern formation during vulval development in C. elegans. Cell 1986; 44(5):761-772.
63. Hill R, Sternberg PW. The gene lin-3 encodes an inductive signal for vulval development in C. elegans. Nature 1992; 358:470-476.
64. Katz WS, Hill RJ, Clandinin TR et al. Different levels of the C. elegans growth factor LIN-3 promote distinct vulval precursor fates. Cell 1995; 82(2):297-307.
65. Aroian RV, Koga M, Mendel JM et al. The let-23 gene necessary for Caenorhabditis elegans vulval induction encodes a tyrosine kinase of the EGF receptor subfamily. Nature 1990; 348:693-699.
66. Beitel GJ, Clark SG, Horvitz HR. Caenorhabditis elegans ras gene let-60 acts as a switch in the pathway of vulval induction. Nature 1990; 348:503-509.
67. Han M, Sternberg PW. let-60, a gene that specifies cell fates during C. elegans vulval induction, encodes a ras protein. Cell 1990; 63(5):921-931.
68. Simske JS, Kim SK. Sequential signalling during Caenorhabditis elegans vulval induction. Nature 1995; 375(6527):142-146.
69. Greenwald IS, Sternberg PW, Horvitz HR. The lin-12 locus specifies cell fates in Caenorhabditis elegans. Cell 1983; 34(2):435-444.
70. Clandinin TR, Katz WS, Sternberg PW. Caenorhabditis elegans HOM-C genes regulate the response of vulval precursor cells to inductive signal. Dev Biol 1997; 182(1):150-161.
71. Eisenmann DM, Maloof JN, Simske JS et al. The beta-catenin homolog BAR-1 and LET-60 Ras coordinately regulate the Hox gene lin-39 during Caenorhabditis elegans vulval development. Development 1998; 125(18):3667-3680.
72. Gleason J, Korswagen HC, Eisenmann DM. Activation of Wnt signaling bypasses the requirement for RTK/Ras signaling during C. elegans vulval induction. 2002:submitted.
73. Hoier EF, Mohler WA, Kim SK et al. The Caenorhabditis elegans APC-related gene apr-1 is required for epithelial cell migration and Hox gene expression. Genes Dev 2000; 14(7):874-886.

74. Wang M, Sternberg PW. Patterning of the C. elegans 1° vulval lineage by RAS and Wnt pathways. Development 2000; 127(23):5047-5058.
75. Sternberg PW, Horvitz HR. lin-17 mutations of Caenorhabditis elegans disrupt certain asymmetric cell divisions. Dev Biol 1988; 130(1):67-73.
76. Sawa H, Lobel L, Horvitz HR. The Caenorhabditis elegans gene lin-17, which is required for certain asymmetric cell divisions, encodes a putative seven-transmembrane protein similar to the Drosophila frizzled protein. Genes Dev 1996; 10(17):2189-2197.
77. Siegfried KR, Kimble J. POP-1 controls axis formations during early gonadogenesis in C. elegans. Development 2002; 129(2):443-453.
78. Whangbo J, Harris J, Kenyon C. Multiple levels of regulation specify the polarity of an asymmetric cell division in C. elegans. Development 2000; 127(21):4587-4598.
79. Sawa H, Kouike H, Okano H. Components of the SWI/SNF complex are required for asymmetric cell division in C. elegans. Mol Cell 2000; 6(3):617-624.
80. Vignali M, Hassan AH, Neely KE et al. ATP-dependent chromatin-remodeling complexes. Mol Cell Biol 2000; 20(6):1899-1910.
81. Zhao X, Yang Y, Fitch DHA et al. TLP-1 is an asymmetric cell fate determinant that responds to Wnt signals and controls male tail tip morphogenesis in C. elegans. Development 2002; 129(6):1497-1508.
82. Hubbard EJ, Greenstein D. The Caenorhabditis elegans gonad: a test tube for cell and developmental biology. Dev Dyn 2000; 218(1):2-22.
83. Kimble J, Hirsh D. The postembryonic cell lineages of the hermaphrodite and male gonads in Caenorhabditis elegans. Dev Biol 1979; 70(2):396-417.
84. Whangbo J, Kenyon C. A Wnt signaling system that specifies two patterns of cell migration in C. elegans. Mol Cell 1999; 4(5):851-858.
85. Forrester WC, Dell M, Perens E et al. A C. elegans Ror receptor tyrosine kinase regulates cell motility and asymmetric cell division. Nature 1999; 400(6747):881-885.
86. Waring DA, Kenyon C. Selective silencing of cell communication influences anteroposterior pattern formation in C. elegans. Cell 1990; 60(1):123-131.
87. Waring DA, Kenyon C. Regulation of cellular responsiveness to inductive signals in the developing C. elegans nervous system. Nature 1991; 350(6320):712-715.
88. Austin J, Kenyon C. Cell contact regulates neuroblast formation in the Caenorhabditis elegans lateral epidermis. Development 1994; 120(2):313-323.
89. Salser SJ, Kenyon C. A C. elegans Hox gene switches on, off, on and off again to regulate proliferation, differentiation and morphogenesis. Development 1996; 122(5):1651-1661.
90. Hunter CP, Harris JM, Maloof JN et al. Hox gene expression in a single Caenorhabditis elegans cell is regulated by a caudal homolog and intercellular signals that inhibit wnt signaling. Development 1999; 126(4):805-814.

# CHAPTER 13

# The *Wnt* Gene Family and the Evolutionary Conservation of *Wnt* Expression

**Michael Schubert and Linda Z. Holland**

## Abstract

The *Wnt* gene family has undergone extensive gene duplications during metazoan evolution. To date, there is one *Wnt* gene known from the cnidarian *Hydra*, seven from *Drosophila* and nineteen from vertebrates. Phylogenetic and genome analyses group the vertebrate *Wnt* genes into thirteen subfamilies and suggest that the last common ancestor of protostomes and deuterostomes had at least six *Wnt* genes. The present review focuses on the roles of *Wnt* genes in early metazoan development, during the establishment of the major regions and chief axes of the body. *Wnt* genes signal through three pathways: Wnt/b-catenin (the canonical Wnt pathway), Wnt/JNK (involved in convergent extension), and Wnt/$Ca^{2+}$ (involved in tissue separation). The earliest expression of Wnt and other genes of the canonical pathway is around the single gut opening of the cnidarian *Hydra*, at the posterior end of insect embryos, and around the forming blastopore in deuterostomes (sea urchins, amphioxus, and vertebrates). Experimental evidence shows that, in deuterostomes, this posterior Wnt signaling center establishes a gradient of Wnt activity that is high posterior and low anterior. Thus, the major conserved role of the canonical Wnt pathway in Metazoa appears to be patterning the longitudinal embryonic axis. It is likely that this role arose very early in metazoan evolution and that additional roles for Wnt signaling evolved later. Such later roles include specification of neural fate, boundary formation (between segments in insects and at the midbrain-hindbrain junction in vertebrates), somite formation, induction of neural crest, elongation of the body axis in amphioxus and vertebrates, and establishing the dorsoventral axis in amphibians.

## Introduction

*Wnt* genes encode secreted glycoproteins, which mediate intercellular signaling either over short or long distances depending on the tissues where they are being expressed.[1] In animals generally, Wnt signaling is involved in a wide variety of cell interactions from early development through the adult stage. The present review focuses on the early developmental roles of *Wnt* genes. The *Wnt* gene family, which has undergone extensive gene duplications during metazoan evolution, can be subdivided into thirteen subfamilies, which fall into three classes differing in the signal transduction cascades they trigger. Two of the signaling pathways are mediated by Dishevelled. In one, Dishevelled inhibits GSK3, which results in translocation of β-catenin to the nucleus. This pathway is involved in axial patterning and specification of cell fate. In the second, Dishevelled signals through Rho to JNK/SAPK (c-Jun N-terminal kinase/ stress-activated protein kinase). This pathway mediates planar cell polarity and convergent

*Wnt Signaling in Development*, edited by Michael Kühl.

extension movements. The third Wnt signaling pathway involves G-proteins and results in changes in intracellular $Ca^{2+}$ levels, which play roles in cell cycling and tissue separation. Wnt pathways will not be discussed here in detail. For comprehensive reviews see references 2-7.

*Wnt* genes have been cloned from several animal phyla (Fig. 1). The most (nineteen) are known from vertebrates [see Roel Nusse's *Wnt* gene homepage at http://www.stanford.edu/~rnusse/wntwindow.html]. At least five of the vertebrate *Wnt* genes appear to signal predominantly through the canonical Wnt/β-catenin pathway (*Wnt-1, Wnt-2b, Wnt-3a, Wnt-8a*, and *Wnt-8b*), at least two signal through the Wnt/$Ca^{2+}$ pathway (*Wnt-4* and *Wnt-5a*),[4,8] and one (*Wnt-11*) apparently signals preferentially through the Wnt/JNK pathway,[7,9] although it can also signal through the Wnt/$Ca^{2+}$ pathway.[10] The signaling pathways of the remaining *Wnt* genes have not been well studied. Among invertebrate deuterostomes, *Wnt* genes have been studied in sea urchins (four *Wnt* genes cloned),[11-14] ascidians (two cloned),[15-17] and amphioxus (nine cloned).[18-23] Among the protostomes, *Wnt* genes have been most studied in *Drosophila* (seven *Wnt* genes)[24] and in *Caenorhabditis elegans* (five *Wnt* genes).[25-37] Most research on *Drosophila Wnt* genes has concerned *wingless*, the *Wnt-1* subfamily ortholog, although developmental expression is known for five of the others.[38-50] Among other protostomes, one *Wnt* has been cloned from a brachiopod[51] and one from leeches.[52,53] In animals basal to the bilaterians, one *Wnt* gene has been isolated from the cnidarian *Hydra*.[54]

The availability of *Wnt* genes from several species provides a basis for molecular phylogenetic and comparative gene expression analyses. Molecular phylogenetics is useful for revealing the diversification of a gene family in the course of evolution,[55-58] while comparisons of how genetic pathways pattern the embryo can elucidate the evolution of developmental mechanisms. To understand the evolution of the roles of *Wnt* genes in embryonic patterning, it would be most useful if they had been studied in numerous phyla, including their most basal members. Regrettably, both *Drosophila melanogaster* and *Caenorhabditis elegans* are derived within their groups[59-61] and their genes tend to evolve more rapidly compared to those in more basal members of their respective phyla.[62] In addition, it is not known if the one *Hydra Wnt* gene cloned, which belongs to the *Wnt-3* subfamily,[22] is the only *Wnt* gene in cnidarians. Consequently, phylogenetic analyses can only give a minimum value for the number of *Wnt* genes in the ancestral bilaterian. The major conserved role of *Wnt* gene function in early development of metazoan animals is apparently patterning along the longitudinal body axis. Later in development, *Wnt* genes play many roles: in elongation of the body axis, in induction of cell fates, in specification of cell identity, and in establishment of regional boundaries within a tissue.

## Recent Phylogenetic Analyses Fail to Resolve the Evolutionary History of the *Wnt* Gene Family

Signaling of different *Wnt* genes through the three known pathways raises the question of which evolved first. Unfortunately, phylogenetic trees based on available sequences do not provide an answer. Most of the invertebrate *Wnt* genes (mainly from *Drosophila* and *Caenorhabditis*) group within the thirteen subfamilies of vertebrate *Wnt* genes[30] (Table 1), and phylogenetic trees using either complete[63] or partial sequences[64] do not adequately resolve the evolutionary relations among these subfamilies (Fig. 2). Thus, the number of *Wnt* subfamilies in the last common ancestor of both protostomes and deuterostomes remains uncertain and the grouping in the tree of *Wnt* subfamilies signaling via different pathways, e.g., *Wnt-5* with *Wnt-8*, may not reflect the true relationship. What phylogenetic analyses do show is that all the amphioxus sequences mark the base of probable vertebrate gene diversifications, suggesting that gene duplications occurred within the vertebrate lineage.[22] Nevertheless, the overall number of *Wnt* genes in vertebrates is only about 2.7x higher than in *Drosophila* and approximately 2.1x higher than in amphioxus. These are not the expected ratios if two rounds of whole-genome

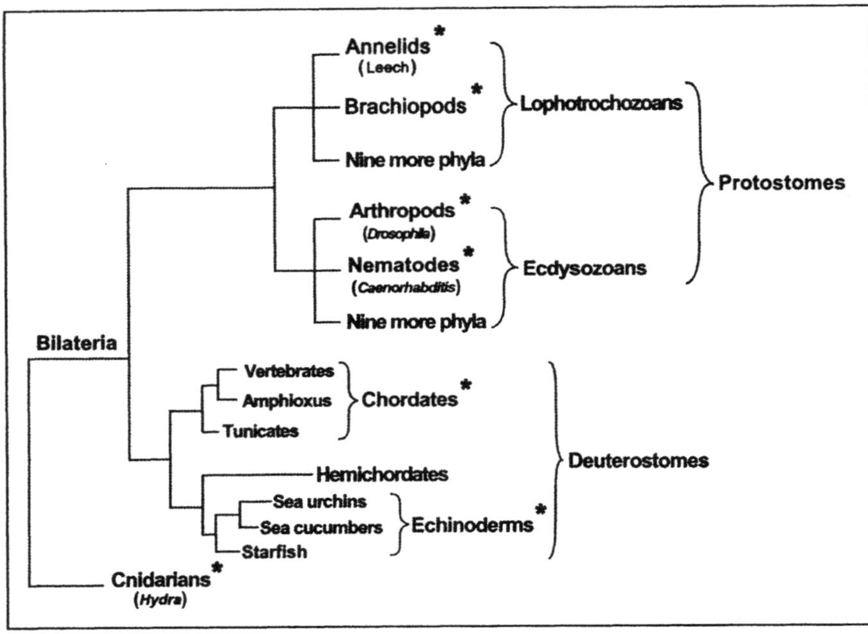


Figure 1. Metazoan phylogeny based on 18S rDNA. Branch lengths are not indicative of phylogenetic distance. Asterisks indicate animal groups from which *Wnt* genes have been isolated. Modified from reference 260.

duplications occurred during early vertebrate evolution,[65,66] unless a massive secondary *Wnt* gene loss has taken place within the vertebrate lineage.

*Wnt* sequences from the protostomes *Drosophila* and *Caenorhabditis* tend to cluster together and form weakly supported clades with very long branches.[63] This clustering results from the small sampling of protostome taxa in the tree and from long-branch attraction due to the high rate of evolution within these two species.[62] Even so, since genome analyses show that the seven *Drosophila Wnt* genes are probably homologous to at least five of the vertebrate *Wnt* genes (*Wnt-1, Wnt-5, Wnt-6, Wnt-7*, and *Wnt-10*),[67,68] it can be concluded that the genome of the last common ancestor of arthropods and deuterostomes probably contained at least these five *Wnt* genes. In addition, the presence of a *Wnt-3* homolog in *Hydra*[54] suggests that a *Wnt-3* gene may have been lost in the *Drosophila* lineage (Table 1) and, therefore, that the last common ancestor of protostomes and deuterostomes probably had at least six *Wnt* genes. The *Hydra Wnt-3* is the only *Hydra Wnt* gene cloned to date and appears to be closely related to the *Wnt-1* subfamily.[22] Thus, unless there are other *Wnt* genes in the *Hydra* genome, the root of the *Wnt* gene family probably falls close to or within the *Wnt-1* or *Wnt-3* subfamilies.

## The Role of *Wnt* Genes in Establishing the Early Anteroposterior Axis in Deuterostomes

Deuterostomes include the echinoderm/hemichordate clade and the chordates, which comprise the vertebrates plus two invertebrate groups (ascidians and amphioxus). In deuterostomes except for ascidians, signaling by the Wnt/β-catenin pathway appears to be involved in early anteroposterior patterning during the late blastula and/or early gastrula stages.

***Table 1. Classification of* Wnt *genes from selected organisms***

| Subfamily | Vertebrates | Amphioxus | *Drosophila* | *Caenorhabditis* | *Hydra* |
|---|---|---|---|---|---|
| *Wnt-1* | • | • | •[a] | | |
| *Wnt-2* | • | | | | |
| *Wnt-3* | • | • | | | •[i] |
| *Wnt-4* | • | • | | •/•[e] | |
| *Wnt-5* | • | • | •[b] | •[f] | |
| *Wnt-6* | • | • | • | | |
| *Wnt-7* | • | • | •[c] | | |
| *Wnt-8* | • | • | • | | |
| *Wnt-10* | • | • | • | | |
| *Wnt-11* | • | • | | | |
| *Wnt-14* | • | | •[d] | | |
| *Wnt-15* | • | | | | |
| *Wnt-16* | • | | | •[g] | |
| | | | | •[h] | |

•Shows that likely orthologs of genes belonging to this *Wnt* subfamily have been isolated from organism indicated.
Classification of amphioxus and *Hydra* sequences according to reference 22; *Drosophila* and *Caenorhabditis* genes were classified by sequence identity scores from BLAST searches. List of genes, which do not adhere to the vertebrate-based *Wnt* subfamily classification:
*Drosophila*: [a]*wingless*; [b]*DWnt-3/5*; [c]*DWnt-2*; [d]*DWnt-4*.
*Caenorhabditis*: [e]Two Caenorhabditis *Wnt* sequences seem to be related to chordate *Wnt-4* genes: *Ce-Wnt-1* and *mom-2*; [f]*Ce-Wnt-2*; [g]*egl-20*; [h]There is no clear relationship between *Caenorhabditis lin-44* and any chordate *Wnt* subfamily.
*Hydra*: [i]*HyWnt*.

## Echinoderms and Hemichordates

Among echinoderms, sea urchins have been most studied in regard to the role of *Wnt* genes in anteroposterior (animal/vegetal) patterning during the late blastula and early gastrula. The first known indication of animal/vegetal polarity in sea urchins is the localization of maternal transcripts of *SpSoxB1*, which encodes a HMG box transcription factor, to animal blastomeres starting at the 4$^{th}$ cleavage.[69] At the 16-cell stage maternal β-catenin becomes localized to vegetal nuclei. Vegetal cells begin to express *Wnt-1* and *Wnt-8* during late cleavage, reinforcing the nuclear localization of maternal β-catenin.[12,13] *Wnt-5*, which preferentially signals through the Wnt/$Ca^{2+}$ pathway, then turns on in the future blastoporal lip at the ectoderm-endoderm boundary.[13] *Wnt-4* has also been cloned from sea urchins, but its expression pattern has not been described.[11] The role of the Wnt signaling pathway in specifying posterior identity in sea urchin embryos is well established.[12,70-77] Nuclear localization of β-catenin in vegetal cells at the late blastula is required for the specification of vegetal cell fates and endoderm formation.[72,73] Moreover, manipulation of the pathway by up-regulation of Wnt signaling (e.g., by lithium treatment or by injection of dominant negative forms of *GSK3β*) vegetalizes (posteriorizes) the embryo.[71] Conversely, blocking Wnt signaling by injection of dominant negative forms of *TCF/LEF* inhibits formation of endoderm.[74]

In sea urchins, Wnt signaling is mediated by the Notch signaling pathway. Nuclear localization of β-catenin in macromeres is required for them to receive inductive signals, mediated by Notch, from the micromeres.[75,78] Moreover, during the late blastula stage Notch becomes localized to the future blastoporal lip in a ring around the vegetal plate at the endoderm/secondary

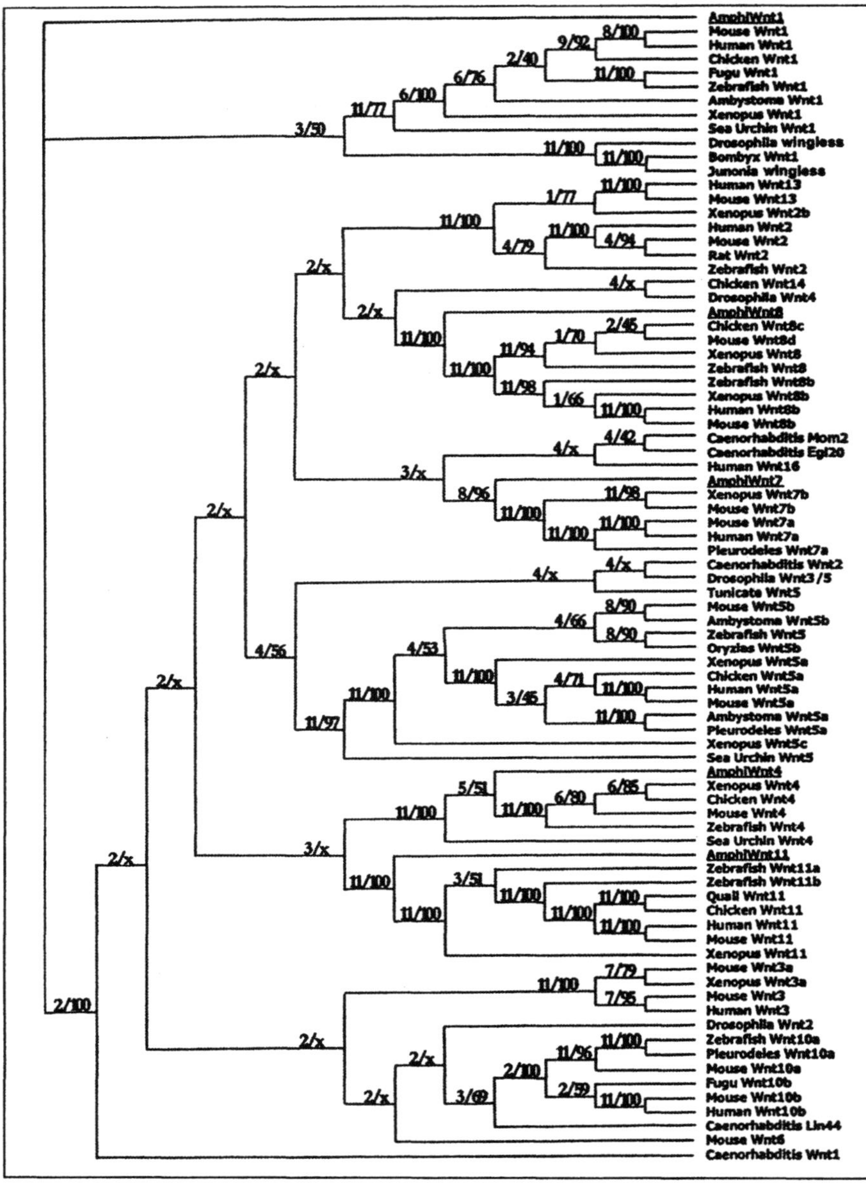


Figure 2. Phylogenetic tree of the *Wnt* gene family based on all full-length sequences available through the end of 1999. The tree includes representatives of twelve of the thirteen *Wnt* subfamilies (*Wnt-15* sequences were not included in the analysis). Bremer support values (ranging from 1 to 11) and jackknife percentages are shown, respectively, before and after the slash (a maximum branch support score would thus be 11/100); "x" indicates jackknife percentages of less than 40. The *Wnt* sequences from the cephalochordate amphioxus are underlined. The phylogenetic structures within most subfamilies are well supported, whereas the branching patterns deeper in the tree are not well resolved. Modified from reference 63.

mesoderm boundary,[79,80] and both the position of this boundary and the localization of nuclear β-catenin in this region are changed by manipulating Notch signaling.[81]

In vertebrates, *brachyury* has been shown to be a target of signaling by Wnt-3a and an upstream activator of *Wnt-11*.[9,82,83] Sea urchin *brachyury* is first expressed in the vegetal plate,[12,84,85] and resolves into a ring around the blastopore as development proceeds.[86] Since vegetal expression of *brachyury* depends on nuclear β-catenin in the macromeres, *brachyury* is probably downstream of Wnt signaling in sea urchins as well as in vertebrates.[86] In starfish and hemichordates, *brachyury* is also expressed in a ring around the blastopore.[87,88] Thus, although *Wnt* genes have not been cloned from starfish and hemichordates, a role in early anteroposterior patterning similar to that in sea urchins seems likely.

## Ascidians

Early development of ascidian tunicates differs considerably from that in other deuterostomes. Ascidian embryos have relatively few cells and determinant cleavage. Depending on the cell lineage, cell fates are restricted between the 64-cell stage and the onset of gastrulation at the 110 cell-stage.[89] Thus, in ascidians, maternal mRNAs play a dominant role in establishing the embryonic axes. The future posterior pole of the embryo is established during ooplasmic segregation, when, shortly after fertilization, the myoplasm migrates to one side of the vegetal pole. This posterior vegetal cytoplasm contains determinants for gastrulation and for posterior fate.[90] Anterior fate evidently results from the absence of posterior cytoplasm.[89] There is no blastocoel, and presumably because there are few cells, gastrulation appears to be highly modified. Cells at the dorsal (often called anterior) lip of the blastopore give rise to endoderm and notochord, while those at the ventral (often called posterior) lip of the blastopore give rise to endoderm and muscle. The anteroposterior axis in ascidian tunicates is not so closely allied to the animal/vegetal axis as it is in echinoderms. Instead, in early embryos it is roughly perpendicular to the animal/vegetal axis,[91] but shifts during gastrulation because of morphogenetic movements.[89] Consequently, at the beginning of gastrulation, the blastoporal lip of ascidian embryos that will involute to give rise to the notochord is defined as "anterior" and the opposite lip as "posterior."

Ascidian *Wnt, Notch*, and *brachyury* genes are not expressed around the blastopore as they are in other deuterostomes. Consequently, the expression of these genes differs from that of their homologs in other deuterostomes. For example, from gastrulation through neurulation, ascidian *Wnt-7* is expressed in a subset of neural precursor cells and subsequently, at the tailbud stage, transcription is detectable exclusively in the dorsal and ventral ependymal cells of the nervous system.[16] For ascidian *Wnt-5*, the early maternal expression is only slightly reminiscent of expression around the blastopore in sea urchins, amphioxus and vertebrates[13,22,92] in that maternal transcripts are initially posterior, being localized first in the myoplasm and then, at the 64-cell stage, in the vegetal blastomeres.[15] However, zygotic *Wnt-5* transcripts, although initially present in cells giving rise to endoderm, notochord, and muscle, subsequently appear in all the animal blastomeres at the 110-cell stage (when gastrulation begins). Expression is very weak during the gastrula and, during the neurula stage, becomes restricted to the notochord and endodermal strand.[15] Moreover, in ascidians, nuclear β-catenin appears to function in endoderm specification and not in axial patterning.[93] Thus, if *β-catenin* is experimentally down-regulated, endoderm does not differentiate and the notochord does not form. Conversely, during early cleavage, treatments up-regulating the Wnt/β-catenin pathway (such as exposure to lithium or injection of *β-catenin* mRNA) transform notochord cells into endoderm,[94] and result in shortened tails. In addition, *brachyury* expression in mesoderm around the blastopore has evidently been lost in ascidians. Expression is limited to the notochord[95] as in other chordates, although a divergent T-box gene (*AsT2*) is expressed at the tip of the tail.[96]

Similarly, posterior expression of *Notch* during the gastrula stage appears to have been lost in ascidians, as has all mesodermal expression. During cleavage stages, maternal *Notch* is concentrated in animal ectodermal cells. Only the relatively late *Notch* expression, which begins at the neural plate stage in the neural folds,[97] is probably evolutionarily related to the neural plate expression of amphioxus and vertebrate *Notch* genes.[98,99]

## Amphioxus

In amphioxus embryos, there is no evidence for maternal expression of *Wnt* genes. However, zygotic expression of amphioxus *Wnt* genes suggests that, as in sea urchins, there is an early posterior Wnt signaling center. The earliest zygotic expression of *Wnt* genes is that of *AmphiWnt-8*—the only *Wnt-8* gene in amphioxus.[21,23] Expression of *AmphiWnt-8* is first detectable at the onset of gastrulation beginning at the same time and in the same pattern as *brachyury*[100]—in a ring around the equator of the late blastula corresponding to the future blastoporal lip (Fig. 3A). However, *AmphiWnt-8* is also expressed somewhat more weakly throughout the mesendoderm.[23] By the early gastrula, expression is down-regulated in the dorsal lip of the blastopore and anterior mesendoderm, and, by the mid-gastrula, the only remaining expression is in the presomitic mesoderm and the posterior ventral endoderm (Fig. 3B, C).[21,23] This pattern persists until the mid-late neurula. The expression pattern of *AmphiWnt-8* around the blastopore and in the presumptive mesendoderm before the completion of invagination tentatively suggests that Wnt signals may be involved in restricting expression of genes such as and *AmphiDll*[101] and *AmphiDRAL*[102] (DRAL has been shown to modulate β-catenin-dependent transcription)[103] to the ectoderm.

Transplantation of the dorsal lip of an amphioxus embryo into the blastocoel of a second embryo produced a secondary axis, suggesting that the dorsal blastoporal lip functions as an organizer in amphioxus as it does in vertebrates.[104] Early expression of *AmphiWnt-8* around the blastopore supports this idea, since in *Xenopus* both transplantation of the blastoporal lip and injection of *Xwnt-8* or other *Wnt* genes signaling through the Wnt/β-catenin pathway induce a secondary axis (see below). However, transplantation experiments in amphioxus must be viewed with caution. Because of the small size of the embryos (about 150μm in diameter), the transplants may have contained more than just the blastoporal lip. Moreover, because the embryos were not marked, it is not certain whether the secondary axes obtained were derived from the transplant or the recipient. Injection of *Wnt* mRNA into amphioxus embryos has not yet been performed, but should help to resolve this question.

In the mid-late gastrula, all the other amphioxus *Wnt* genes described to date are also expressed around the blastopore (Fig. 3A-C). Thus, it seems likely that, as in sea urchins and vertebrates, Wnt signals from the blastoporal region are involved in specifying posterior identity in the neural plate and other tissues. Following *AmphiWnt-8, AmphiWnt-11* expression starts around the blastoporal lip in the very early gastrula, when invagination is half-completed (expression is more intense dorsally) (Fig. 3A).[20] Next, just as invagination is finished, *AmphiWnt-1* turns on uniformly around the blastoporal lip (Fig. 3B).[18] By the mid-gastrula, four other amphioxus *Wnt* genes (*AmphiWnt-3, AmphiWnt-4, AmphiWnt-5,* and *AmphiWnt-6*), together with *Notch*, start to be expressed uniformly around the blastopore.[19,22,99] At the same time, *AmphiWnt-7* transcription commences throughout the mesendoderm (Fig. 3C).[19] *AmphiWnt-4*, although most intensely expressed around the blastopore, is also expressed weakly throughout the mesendoderm.[19] In addition, *AmphiWnt-4, AmphiWnt-7,* and *AmphiWnt-11* are expressed in the ectoderm of the gastrula – *AmphiWnt-11* predominantly in presumptive neuroectoderm,[20] *AmphiWnt-4* and *AmphiWnt-7* most conspicuously in the dorsoposterior ectoderm. It has been proposed that *Wnt* genes signaling via the Wnt/$Ca^{2+}$ pathway may antagonize signals from *Wnt* genes signaling via the Wnt/β-catenin pathway.[4,105] Thus, since

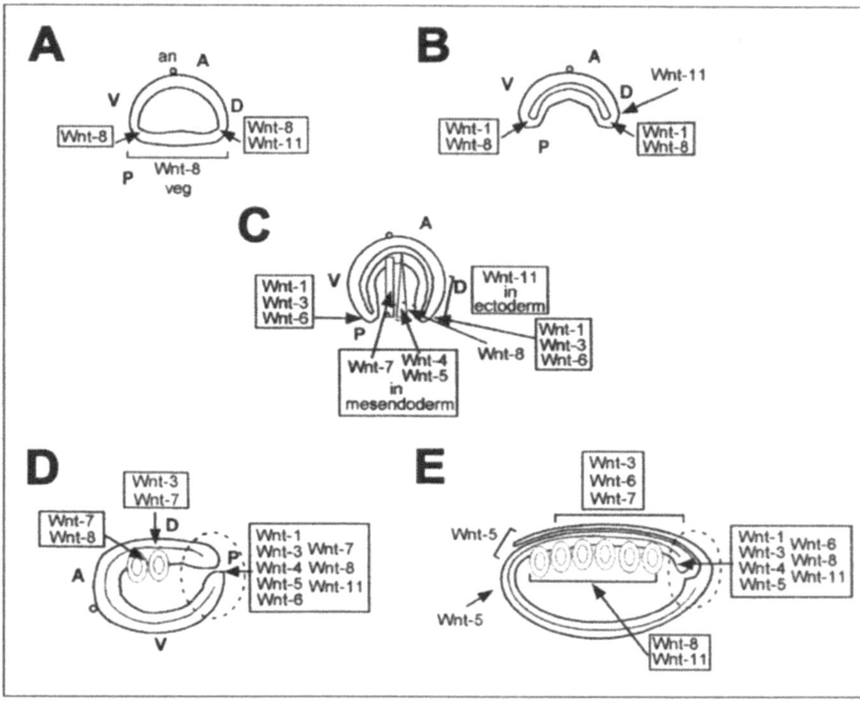


Figure 3. Developmental expression of *Wnt* genes in the cephalochordate amphioxus. The diagrams are not to scale. (A) Zygotic expression of *AmphiWnt-8* and *AmphiWnt-11* begins around the blastopore during the early gastrula stage. (B) *AmphiWnt-1* expression begins around the blastopore as gastrulation proceeds. (C) By the late gastrula, all eight amphioxus *Wnt* genes are expressed posteriorly associated with the blastopore. However, *AmphiWnt-7* is expressed throughout the mesendoderm, while *AmphiWnt-4* and *AmphiWnt-5* are expressed in the mesendoderm in a gradient with strongest expression associated with the blastopore. (D) Transcription of the eight *Wnt* genes remains strong in posterior tissues. Additional expression domains for a few of these genes appear, particularly in the central nervous system and the mesodermal somites. (E) *AmphiWnt-7* ceases to be expressed in posterior tissues by the mid-neurula, while the other amphioxus *Wnt* genes continue to be transcribed posteriorly in and around the tail bud. Some *Wnt* genes are expressed in other tissues, like the central nervous system (*AmphiWnt-3, AmphiWnt-5, AmphiWnt-6,* and *AmphiWnt-7*), the somites (*AmphiWnt-8,* and *AmphiWnt-11*), and the pharynx (*AmphiWnt-5*). List of abbreviations: A, anterior; an, animal pole; D, dorsal; P, posterior; V, ventral; veg, vegetal pole. Modified from reference 261.

*Wnt-11* has been proposed to signal through the Wnt/$Ca^{2+}$ as well as the Wnt/JNK pathway, *AmphiWnt-11* might be antagonizing *AmphiWnt-1, AmphiWnt-3,* and *AmphiWnt-8* to allow expression of neural markers.

In amphioxus, *AmphiWnt-8* is the only *Wnt* gene expressed before the specification of the neuroectoderm as shown by down-regulation of *AmphiDll* and concomitant up-regulation of the pan-neural marker *AmphiSox1/2/3* in the dorsal ectoderm of the early gastrula.[21,23,101,106] Subsequent down-regulation of *β-catenin* in the dorsal ectoderm of the late gastrula and in the early neural plate (our unpublished data) suggests that, as in vertebrates (see below), down-regulation of signaling via the Wnt/β-catenin pathway is necessary for the maintenance of the neuroectoderm and the specification of anterior neural identity. This idea is supported by the finding that up-regulation of the Wnt/β-catenin pathway by a pulse of lithium delivered at the late blastula stage eliminates expression of *AmphiSox1/2/3* in the neural plate (our unpublished data).

## Vertebrates

In developing vertebrates, such as the amphibian *Xenopus*, the canonical Wnt pathway is involved first in establishing the dorsoventral axis and later in patterning the anteroposterior axis. The first of these roles probably has no counterpart in other organisms, except perhaps for amphioxus (see above). Establishment of the dorsoventral axis involves preferential localization of maternal β-catenin to nuclei of cells on the future dorsal side of the embryo.[107] This nuclear localization of β-catenin, which sends a transient dorsal signal to neighboring cells to establish dorsoventral identity,[108] disappears briefly about the time the blastopore begins to form.[109-111] Thus, treatments which stimulate signaling via β-catenin (such as a pulse of lithium, injection of dominant negative forms of *GSK3* or injection of *Xwnt-8*) when applied before the mid-blastula transition, result in embryos that are dorsoanteriorized.[112] Conversely, treatments inhibiting β-catenin signaling block the formation of the dorsoventral axis.[108,113,114] Although several *Xenopus Wnt* genes (*Xwnt-5a, Xwnt-7b, Xwnt-8b,* and *Xwnt-11*) are maternally expressed (Fig. 4A),[115,116] *Wnt* genes are not required for the dorsal β-catenin signal. However, induction of secondary axes by injection of *Wnt* genes signaling through the Wnt/β-catenin pathway into ventral blastomeres at the 4-cell stage[116-118] suggests that transcripts of *Wnt* genes such as *Xwnt-8b*, which are enriched in animal cells of the blastula (Fig. 4B, C),[115] may be involved in maintenance of the dorsal β-catenin signal. This signal is essential for the expression of the early pan-neural markers *chordin, Zicr-1* and *Sox-2*. Expression of these genes shows that specification of the presumptive neuroectoderm occurs before the onset of gastrulation in *Xenopus*[119,120] and does not require the activities of Spemann's organizer.[121] Similarly, in the chick, signals beginning before gastrulation induce transient expression of the early anterior neural markers *Sox-3* and *Otx2*, although signals from the organizer and its derivatives such as the prechordal mesendoderm are subsequently required for maintenance of expression of these genes.[122] It is unlikely that either *Notch* or *brachyury* functions together with β-catenin signaling in this early role of the Wnt/β-catenin pathway in establishing dorsoventral polarity. The *brachyury* gene is not expressed maternally, and although *Xenopus Notch1* (*Xotch*) is expressed maternally, its expression is uniform until the early gastrula, when it becomes slightly enriched in the dorsal mesoderm.[123]

The early role of the Wnt/β-catenin pathway in establishing the dorsoventral axis in vertebrates appears to be quite distinct from its subsequent role in anteroposterior patterning. It is this later role, which begins in *Xenopus* at the mid-blastula transition,[114] that appears to be broadly conserved in evolution. There are several models for how *Xenopus* embryos are patterned along the anteroposterior axis. In recent years, Nieuwkoop's two-signal model involving an anterior activator released from presumptive prechordal plate mesoderm and a transformer released from the notochord, which confers posterior identity, has been modified into a two-inhibitor model.[124] In this model, there is a posterior Wnt signaling center producing an anteroposterior gradient of Wnt activity in the neuroectoderm. This Wnt signal, mediated by members of the Wnt/β-catenin class, posteriorizes neuroectoderm.[125-127]

There is considerable evidence to support such a model for anteroposterior patterning. In *Xenopus*, a posterior Wnt signaling center is established during the late blastula/early gastrula stage. The marginal zone and the blastoporal region express *Wnt* genes of both the Wnt/β-catenin (*Xwnt-3a, Xwnt-8,* and *Xwnt-8b*) and the Wnt/JNK and Wnt/$Ca^{2+}$ classes (*Xwnt-11*) (Fig. 4D, E).[115,128-138] In accord with a change in the role of Wnt/β-catenin signaling at the mid-blastula transition, a new domain of intense nuclear β-catenin expression appears during the late blastula in a ring around the marginal zone, i.e., the future mesoderm.[111] As gastrulation begins, this ring becomes complete around the marginal zone, except for a small region in

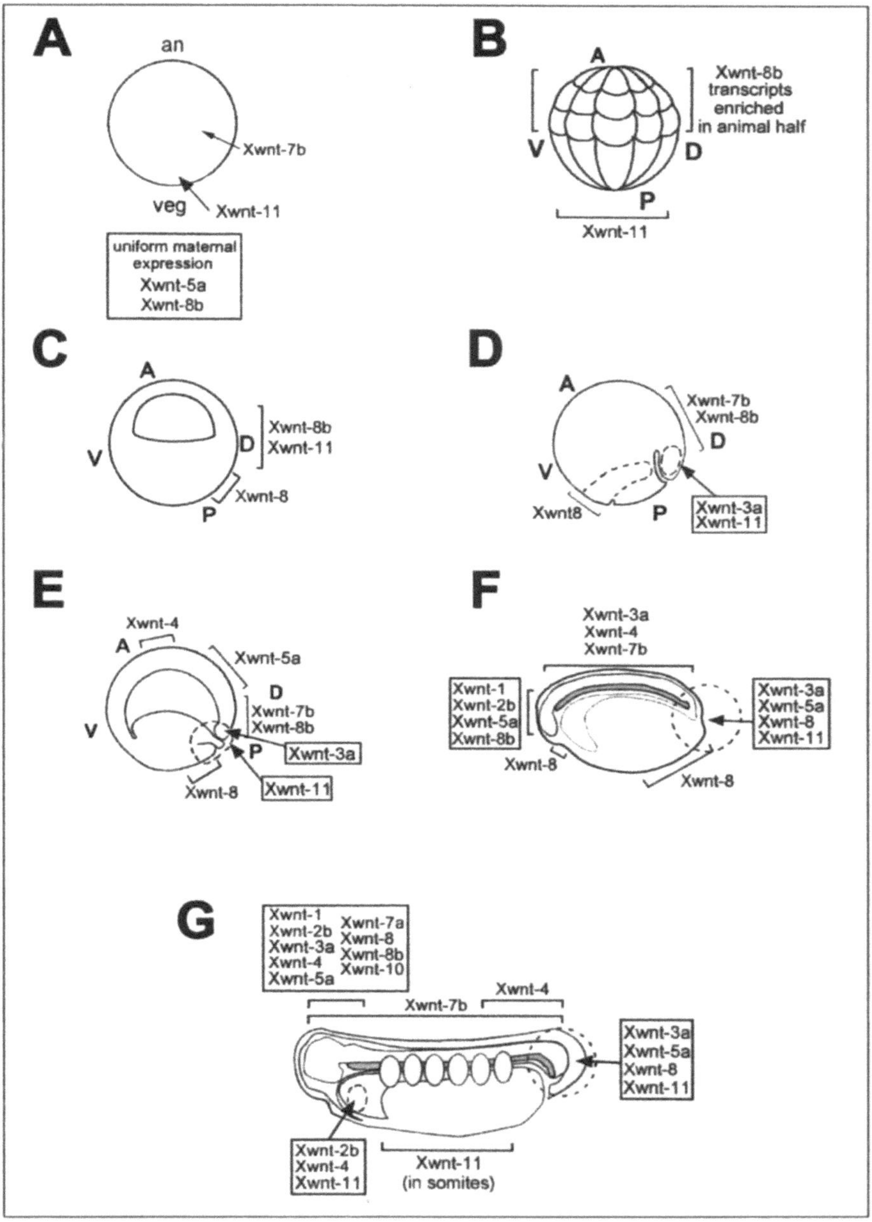


Figure 4. Expression of *Wnt* genes in the vertebrate *Xenopus laevis*. The diagrams are not to scale. (A) Maternal *Xwnt-5a* and *Xwnt-8b* are uniformly distributed throughout the oocyte shortly after fertilization, whereas *Xwnt-11* mRNA is vegetally localized. In addition, *Xwnt-7b* is also maternally expressed. (B) During early cleavage, Xwnt-8b transcripts are enriched in the animal hemisphere, and *Xwnt-11* is still

*Figure legend continued on next page*

*Figure legend continued from previous page*

vegetally expressed. (C) In the late blastula, *Xwnt-8b* and *Xwnt-11* are transcribed in the presumptive dorsal marginal zone region, and *Xwnt-8* is enriched in the future ventral and lateral marginal zone. (D) By the early gastrula stage, *Xwnt-3a, Xwnt-7b, Xwnt-8b* and *Xwnt-11* are expressed in association with the dorsal marginal zone. Xwnt-8 can still be detected in the ventral and lateral marginal zone. (E) In the late gastrula, *Xwnt-3a, Xwnt-7b, Xwnt-8, Xwnt-8b* and *Xwnt-11* are still expressed in close association with the blastopore. In addition, *Xwnt-4* is detectable in the presumptive neuroectoderm, and *Xwnt-5a* expression has commenced in the ectoderm. (F) By the neurula stage, *Xwnt-3a, Xwnt-5a,* Xwnt-8 and *Xwnt-11* are expressed posteriorly. In addition, Xwnt-8 is transcribed in the ventral and lateral mesoderm and lateral to the anterior neural plate. Transcripts of all the other Wnt genes, which are expressed at this stage (*Xwnt-1, Xwnt-2b, Xwnt-3a, Xwnt-4, Xwnt-5a, Xwnt-7b* and *Xwnt-8b*), can be detected in various regions of the central nervous system. (G) During the tailbud stage, *Xwnt-3a, Xwnt-5a, Xwnt-8* and *Xwnt-11* remain to be expressed posteriorly in and around the tail bud. *Xwnt-2b* is transcribed in a region close to the heart primordium, *Xwnt-4* is expressed in the pronephros, and *Xwnt-11* transcripts can be detected in the first branchial arch and the somites. In addition, multiple *Wnt* genes (*Xwnt-1, Xwnt-2b, Xwnt-3a, Xwnt-4, Xwnt-5a, Xwnt-7a, Xwnt-7b, Xwnt-8, Xwnt-8b* and *Xwnt-10*) are expressed in various regions of the central nervous system. List of abbreviations: A, anterior; an, animal pole; D, dorsal; P, posterior; V, ventral; veg, vegetal pole. Modified from reference 261.

the dorsal midline; this zone of nuclear β-catenin approximately coincides with expression of *Xwnt-8.* Later, nuclear β-catenin becomes concentrated at the dorsal and ventral lips of the blastopore. This pattern is similar to that of expression of *Xbra* and *Xwnt-11.*[111]

Posterior Wnt signaling by the Wnt/β-catenin pathway probably is responsible for establishing and/or maintaining posterior identity. First, injection of *Xwnt-8*[113,115,135] after the mid-blastula transition antagonizes signals from the organizer, inhibiting dorsoanterior development and resulting in embryos without head or notochord, but with enlarged somitic muscle. This contrasts with injection of *Xwnt-8* before the mid-blastula transition, which promotes dorsoanterior development. Similarly, lithium applied after the mid-blastula transition, posteriorizes embryos, which develop with malformations of anterior structures.[139,140] Conversely, injection of a dominant negative *Xwnt-8* anteriorizes embryos. Moreover, when either *Xwnt-8* or *Xwnt-8b* is injected into 2- to 4-cell zebrafish embryos, there are dorsoanterior defects including suppression of eye and midbrain-hindbrain boundary formation. This is quite different from the effect of injection of *Xwnt-8* genes into embryos of *Xenopus* itself, and suggests that the primary role of Wnt signaling in zebrafish is in anteroposterior patterning, and that a role for Wnt signaling in establishing the dorsoventral axis as in *Xenopus* may not be general among vertebrates.[141]

*Wnt-3* genes, which signal through the same pathway as *Wnt-8,* have been shown to play a major role in early anteroposterior patterning. Expression of *Xenopus Xwnt-3a* begins in the dorsal marginal zone, chiefly in the ectoderm (Fig. 4D).[116,132] However, as the blastopore closes, the dorsal lip becomes dorsoposterior and the previously dorsal expression of *Xwnt-3a* in the blastopore lip becomes progressively more posterior (Fig. 4E). *Xwnt-3a* appears to play a dominant role in anteroposterior patterning of the central nervous system.[136,142,143] At the early neurula stage, expression of *Xwnt-3a* begins along the neural folds and in a diffuse ring around the blastopore. Subsequently, during neurulation, the expression around the blastopore and in the posterior neural folds becomes very strong (Fig. 4E, F).[136] Several lines of experimental evidence support roles for *Xwnt-3a* in anteroposterior patterning of the neuroectoderm. Overexpression of *Xwnt-3a* or *β-catenin* suppresses the expression of anterior neural genes and promotes the expression of posterior neural genes, suggesting that *Xwnt-3a* functions in establishing and/or maintaining posterior identity in the nerve cord.[132,136,144] The roles of *Xwnt-8* and *Xwnt-3a* may be interchangeable, because in other vertebrates, the roles of *Xwnt-3a* in *Xenopus* appear to be performed by *Wnt-8* genes: for example, the phenotype of transgenic mice expressing chicken *Cwnt-8c* is similar to that of *Xenopus* overexpressing

*Xwnt-3a*.[145] Conversely, expression of a dominant negative *Wnt-8* reduces the expression of posterior neural genes and increases that of anterior neural genes.[136] In addition, zebrafish *Wnt-8* has been shown to be required for the posteriorization of the early neuroectoderm.[146-148]

In vertebrates, Wnt signals are kept low anteriorly by secretion of Wnt and BMP antagonists by the anterior endoderm and the prechordal plate (in *Xenopus* and zebrafish) or by the anterior visceral endoderm (in mice).[107,148,149] Thus, injection of secreted Wnt inhibitors (e.g., *frzb-1, dickkopf-1, cerberus*, or *crescent*) anteriorizes the neuroectoderm.[150-154] These results are in agreement with inhibition of the development of dorsoanterior structures, like the head and the notochord by injection of *Xwnt-8* after the mid-blastula transition.[115,135] Anterior suppression of Wnt signaling is also necessary for proper anteroposterior patterning of the mesoderm and endoderm. Mouse mutants have shown that *Wnt-3a* is important for both anteroposterior patterning of the vertebral column[155] and posterior somite formation.[83,156] Moreover, in chicken as well as mice, the Wnt/β-catenin signaling pathway is implicated in patterning along the anteroposterior axis of the mesoderm and endoderm as well as the nerve cord.[157,158] Thus for heart formation, inhibition of Wnt/β-catenin signaling is required anteriorly in the cardiac mesoderm in the chick[158] and overexpression of the Wnt antagonists *Dickkopf-1* and *Crescent* can induce heart formation in explants of the ventral marginal zone in *Xenopus*.[159]

As in amphioxus and sea urchins, in *Xenopus*, Wnt, Notch, and Brachyury probably act together to induce posterior structures.[160,161] Starting at the gastrula stage, *Xenopus Notch* and *delta-1* are expressed in mesodermal cells posterior to the neural plate. Increasing amounts of Dishevelled, a component of both the Notch and Wnt signaling pathways, posteriorizes neural tissues and activates posterior markers such as *brachyury*.[162] Conversely, in mice, loss of either *Wnt-3a* or *Notch1* function results in posterior defects.[163,164] Expression of *Xwnt-3a* together with active *Notch* (the *Notch* intracellular domain) induces animal cap explants grafted at any anteroposterior level onto the neural plate to form ectopic tails with neural tubes, but without notochords. Active *Notch* alone only induces ectopic tails in animal caps grafted onto the posterior neural plate near the zone where *Xwnt-3a* is expressed.[160,161] These ectopic tails induced by active *Notch* express *brachyury (Xbra), Xwnt-3a, caudal (Xcad), FGF-8, Xdelta-1*, and *lunatic fringe (lfng)*. The *brachyury* and *caudal* genes are downstream targets of Xwnt-3a.[83,155] *Brachyury* is expressed around the blastopore in early *Xenopus* embryos, and expression continues in the tail bud.[165,166]

## Evolutionary Conservation of *Wnt* Gene Involvement in Axial Patterning of Protostomes, Deuterostomes, and Cnidarians

In insects (*Drosophila*, grasshoppers, and beetles), the first expression of *wingless*, the homolog of *Wnt-1*, is in a ring around the posterior end of the early embryo.[39,45,167] *Brachyury* has a similar pattern.[168-170] Similarly, in *Hydra*, genes of the Wnt/β-catenin pathway are expressed together with *brachyury* at the apical tip of the body around the single gut opening.[54,171,172] Co-expression of *Wnt* and *brachyury* at one end of the embryo in protostomes, deuterostomes, and cnidarians suggests that the activation of *brachyury* by *Wnt* genes of the Wnt/β-catenin class is a component of a highly conserved signaling network mediating axial patterning, which evolved very early during metazoan evolution.

There are additional similarities in the roles of *Wnt* genes in anteroposterior patterning between *Drosophila* and deuterostomes. The *teashirt* gene encodes a zinc-finger transcription factor, which associates with and requires Armadillo (i.e., *Drosophila* β-catenin) for its function.[173,174] Expression is limited to the trunk, where it is required to maintain *wingless* expression, presumably in a feedback loop. Manipulation of *teashirt* expression shows that it is required for specification of trunk identity. When *teashirt* expression is lost in the trunk of the embryo, the ventral part of the trunk takes on anterior head identity.[175,176] Similarly, the two orthologs of teashirt in mice, *mtsh1* and *mtsh2*, are expressed in the neural tube and the somites of the trunk and the tail.[177] Thus, it seems likely that in vertebrates as well as *Drosophila*, *teashirt* may be acting to maintain trunk and tail expression of *Wnt* genes of the Wnt/β-catenin class.

## *Wnt* Genes and Body Elongation from the Chordate Tail Bud

During amphioxus and *Xenopus* development, much of the tail bud is derived from the blastoporal lip. From the neurula stage on, the tail bud is responsible for much of the elongation of the embryo and larva.[22,166,178] In contrast, in ascidians, the tail elongates by rearrangement of preexisting cells,[179] and a definitive tail bud is lacking. The vertebrate tail bud is subdivided into a mosaic of different regions that originate during gastrulation from the blastoporal lip and are subsequently incorporated into the tail bud. In vertebrates, cells from particular tail bud regions can be traced to specific differentiated structures in the tail. Cell tracing experiments and gene expression patterns have shown that the arrangement of tissues in the tail bud is similar in *Xenopus* and the chick, even though the chick tail bud has a much higher proportion of mesenchyme than that of *Xenopus*.[166,180,181] At the neurula stage, the blastopore-associated expression of *Xwnt-3a*, *Xwnt-5a*, *Xwnt-8*, and *Xwnt-11* is carried over into the tail bud and maintained during subsequent elongation of the tail (Fig. 4E-G). *Wnt*, *Notch*, and *brachyury* genes are expressed in specific regions of the tail bud. For example, *Xwnt-3a* and *Xwnt-5a* are both expressed in the most dorsoposterior region of the neural tube and in the overlying ectoderm.[160,161] *Brachyury* is expressed in the chordoneural hinge, the posterior part of the notochord, the posterior wall of the neurenteric canal, and the roof plate of the spinal cord, and *Notch* and its ligand *delta-1/2* in the distal tip of the tail bud and the posterior wall of the neurenteric canal.[160,165,166] Posteriorly expressed *Wnt* genes function in elongation of the tail bud as well as in anteroposterior patterning (discussed above).[182] In mice, loss of *Wnt-3a* leads to a complete loss of the tail bud and the caudal somites,[83,163] while *Wnt-5a* deficiency leads to a reduced extension of the primary body axis.[183] In zebrafish, both *Wnt-5* and *Wnt-8* are required for tail outgrowth.[147,148,184] *Wnt* genes signaling through the Wnt/$Ca^{2+}$ pathway and/or the Wnt/JNK pathway (e.g., *Wnt-5* and *Wnt-11*) have been implicated in convergent extension movements in both *Xenopus* and zebrafish (see below).[9,82] In *Xenopus*, at the gastrula and early neurula stages, the expression patterns of *Xwnt-11* and *brachyury* are congruent, with *Xwnt-11* being a target of *brachyury*,[9] and it seems likely that they may function together in elongation of the body axis.

A recent analysis of the amphioxus tail bud has revealed both similarities and differences between the amphioxus and vertebrate tail buds.[22] As in vertebrates, the amphioxus tail bud represents a developmental continuation of the blastoporal lip, provides cells for the caudal outgrowth of other tissues, and is subdivided into specific regions. However, in contrast to vertebrates, the amphioxus tail bud does not have any mesenchymal component. Moreover, the specific *Wnt* genes expressed in the tail bud and the regions of the tail bud where they are expressed are similar in some respect but differ in others between amphioxus and *Xenopus*. Similarities include the expression of both *Wnt-8* and *Wnt-11* in the tail bud (Figs. 3D, E; 4F, G). In addition, the expression of mouse *Wnt-5b* in the posterior hindgut and notochord[92] resembles the tail bud-associated expression of amphioxus *AmphiWnt-5*. However, while *Wnt-3* and *Wnt-5* are expressed in partially overlapping domains in the tail bud of both amphioxus and *Xenopus*,[22,160,161] their expression domains are not identical in the two organisms. Both amphioxus *AmphiWnt-3* and *Xenopus Xwnt-3a* are expressed in the posterior neural tube and in the dorsoposterior ectoderm. However, the expression of *AmphiWnt-3* in the posterior wall of the neurenteric canal has no counterpart in *Xenopus*. Moreover, while *AmphiWnt-5* is expressed in the chordoneural hinge and the posterior wall of the neurenteric canal, *Xenopus Xwnt-5a* is expressed in the posterior ectoderm and posterior neural tube. In addition, in amphioxus, *AmphiWnt-1* is expressed in the blastoporal lip and tail bud (Fig. 3B-G), but *Xwnt-1* is not expressed at all in either region. Similarly, *Wnt-4* and *Wnt-7* are expressed in the amphioxus tail bud (Fig. 3D, E), but not in that of *Xenopus*. Other genes expressed in the tail buds of both amphioxus and *Xenopus* include *brachyury, Notch, BMP-2/4, caudal, Hox3*, and *hedgehog*.[99,100,160,166,185-188] Although the localization of expression of these genes within the amphioxus tail bud has not been studied in as much detail as for *Wnt* genes, in general, the expression seems comparable to that of their vertebrate homologs. Thus, a combinatorial code of developmental gene expression, which patterns the tail bud and generates specific tissues during posterior elongation, seems to be conserved between amphioxus and vertebrates and was therefore probably already present in the last invertebrate ancestor of the vertebrates.

## Conservation of Wnt Signaling in the Paraxial Mesoderm in Amphioxus and Vertebrates

In amphioxus and the vertebrates, the axial and the paraxial mesoderm develops into the notochord and somites, respectively. In vertebrates, the *Wnt* genes with roles in formation and patterning of the axial and paraxial mesoderm are *Wnt-1, Wnt-3a, Wnt-7a, Wnt-8*, and *Wnt-11*. In *Xenopus, Xwnt-8* and *Xwnt-11* are expressed, respectively, in the ventral and lateral mesoderm and the somites (Fig. 4F, G). In *Xenopus* and the zebrafish, loss of *Wnt-8* leads to a loss of posterior and ventral structures.[147,148] The role of *Wnt-11* in paraxial mesoderm patterning is more difficult to assess since in null mutants, gastrulation movements are disrupted and consequently the mesoderm does not form properly.[9,189]

The roles of *Wnt-1, Wnt-3a*, and *Wnt-7a* in mesodermal patterning have been better studied than that of *Wnt-11*. *Wnt-3a* plays an early role in formation of the posterior mesoderm, while *Wnt-1* and *Wnt-7a* act later in patterning of the somites. *Wnt-3a* is expressed in the tail bud in cells that will give rise to paraxial mesoderm[190,191] and is essential for the formation of caudal somites. In a mouse mutant for *Wnt-3a*, neither tail bud, notochord, nor caudal somites form, although the anterior somites are relatively normal.[83,142,156,163,192] Similar results are obtained by injection of *WIF-1* (a secreted inhibitor of Wnt proteins of the Wnt/β-catenin class) into 4-cell *Xenopus* embryos, which antagonizes the formation of posterior presomitic mesoderm.[193] Conversely, co-injection of *Xwnt-3a* or *Xwnt-8* with *noggin* into *Xenopus* animal caps induces dorsal mesodermal markers.[194]

Wnt signaling also acts later in development in the segmentation of the paraxial mesoderm and in the patterning of the somites.[195] Components of the Wnt/β-catenin pathway are expressed in the paraxial mesoderm before the onset of expression of the myogenic factor *MyoD*,[196] and have been shown to function in patterning of the somites. In homozygous *LEF-1/TCF-1* mouse mutants, expression of several regional marker genes in the somites is abnormal, and expression of *Notch1* and *Wnt-5a* in the presomitic mesoderm is eliminated.[197] In addition, up-regulation of the Wnt/β-catenin pathway with lithium results in supernumerary and malformed somites, especially in anterior regions of the paraxial mesoderm.[198] Experimental studies have demonstrated that *Wnt-1* can induce myogenesis in explants of paraxial mesoderm.[199] Additional evidence for a role of *Wnt* genes in somitogenesis is that six *Frizzled (Fz)* genes (*Frizzled 1, 3, 6, 7, 8*, and *9*), which encode Wnt receptors, are expressed in the forming and differentiating somites.[200] Moreover, injection of the Wnt antagonist, *WIF-1*, which is normally expressed in paraxial mesoderm in vertebrates, into *Xenopus* neurulae results in disorganized somites with impaired segmentation.[193] A model has been presented for the chick, whereby Wnt-1, secreted by the neural tube, binds to Fz-1 and, acting via β-catenin, activates transcription of the myogenic factor *Myf5*, while Wnt-7a, secreted by the ectoderm overlying the somites binds to Fz-7 and, acting through a β-catenin independent pathway, leads to the activation of myogenesis.[199-203] In *Xenopus*, expression of *Xwnt-7a* appears to be restricted to the neural tube,[204] but the gene could be playing a similar role in somitogenesis. In addition,

*Wnt-11* may play a role in the regionalization of the somite. In the mouse, *Wnt-11* is expressed at the junction of the dermatome and myotome,[205] while in the zebrafish, expression in the somites is limited to the dermatome.[82]

Although, it seems likely that Wnt signaling plays similar roles in the formation and patterning of the paraxial mesoderm in amphioxus and vertebrates, there are some differences between them in the formation of the mesoderm and in the *Wnt* genes that are mesodermally expressed. The first major difference in mesoderm formation is in the origin of ventral mesoderm. In amphioxus, the ventral mesoderm (also called lateral plate mesoderm) arises relatively late in development (at the mid-neurula stage) as ventral outgrowths from the somites. In contrast, in vertebrates, the ventral mesoderm arises earlier (at the gastrula stage) about the same time the paraxial mesoderm initially forms. Nevertheless, in spite of this offset in developmental timing, *Vent* gene expression marks nascent ventral mesoderm in both amphioxus and vertebrates.[206] The second major difference in mesoderm formation is that in amphioxus, unlike vertebrates, the somites extend to the anterior tip of the animal. The anteriormost eight somites pinch off enterocoelically from grooves of the dorsolateral mesendoderm during the late gastrula, whereas the posterior somites are formed from the tail bud, as are vertebrate somites. Nevertheless, the fate of the anterior and posterior somites is the same: they all form muscles used in undulatory swimming. Thus, while some genes (e.g., *engrailed* and *Wnt-8*) are expressed early in formation of the anteriormost eight somites, but not in the more posterior somites, for the most part, gene expression in the anterior and posterior somites is the same.

Experiments blocking Wnt signaling have not been done in amphioxus. However, the expression patterns of amphioxus *Wnt* genes suggest that, as in vertebrates, Wnt proteins are required for formation of the axial and paraxial mesoderm as well as for somitogenesis. As noted above, during the gastrula stage, all of the amphioxus *Wnt* genes are expressed around the blastopore, and *AmphiWnt-4, AmphiWnt-5, AmphiWnt-7,* and *AmphiWnt-8* are expressed in the dorsal mesoderm and/or ectoderm as well. In addition, several *Wnt* genes continue to be expressed in the tail bud as somites are budded off from it. These include *AmphiWnt-1, AmphiWnt-3, AmphiWnt-4, AmphiWnt-5, AmphiWnt-6,* and *AmphiWnt-11* (Fig. 3D, E).[18-20,22] Of these, *AmphiWnt-5* and *AmphiWnt-6* are expressed in the posterior mesoderm. The signaling pathway of *Wnt-6* genes has not been determined; however, in phylogenetic analyses, *Wnt-6* genes group with *Wnt-8* genes,[22] which signal via β-catenin. Thus, it is likely that *Wnt-6* signals via the Wnt/β-catenin pathway and that blocking this pathway at the tailbud stage would prevent the formation of new somites in amphioxus as it does in vertebrates.

Expression of *Wnt* genes during somitogenesis in amphioxus suggests that, as in vertebrates, *Wnt* genes have roles in patterning both the anterior and posterior somites. However, in amphioxus, *Wnt-1* cannot be the Wnt/β-catenin-class gene that performs this function. Rather, *Wnt-6* and *Wnt-8* may function in somitogenesis in amphioxus as *Wnt-1* does in vertebrates. *AmphiWnt-8* is expressed in the anterior presomitic (early paraxial) mesoderm, which gives rise to somites two through eight that form by enterocoely. It is possible that in the presomitic mesoderm, *AmphiWnt-8* is interacting with *engrailed (AmphiEn)*, which is expressed in the posterior portion of each of the first eight somites before there is overt segmentation.[207] Segmentation and patterning of the more posterior somites may be mediated by AmphiWnt-6 and AmphiWnt-7, secreted by the nerve cord, as is known for *Wnt-1* and *Wnt-7a* in vertebrates. As in vertebrates, *AmphiWnt-7* is expressed in the dorsal mesoderm and ectoderm at the gastrula stage, in the edges of the neural plate at the neurula stage, and in much of the neural tube in the larva. *AmphiWnt-6*, which, as discussed above, may signal via the Wnt/β-catenin pathway, is expressed in the nerve cord from the neurula stage on and could be performing the same function as *Wnt-1* does in vertebrate somitogenesis. *AmphiWnt-4* and *AmphiWnt-11* are expressed in the somites in amphioxus; *AmphiWnt-11* in the myotomal compartment (there is no subdivision into dermatome and myotome in amphioxus) and *AmphiWnt-4* in the dorsal part of the somite.[19,20] Thus, as in vertebrates, *AmphiWnt-11* may be involved in regionalization of the somite; however, the expression of *AmphiWnt-4* in amphioxus somites has no counterpart in vertebrates.

In addition to *Wnt* genes, the expression of *Notch* genes in the tail bud and in forming somites is similar in amphioxus and the vertebrates. In vertebrates, *Notch* expression is required for normal somitogenesis and segmentation of vertebrate embryos.[208-210] Although a direct link between the Notch and Wnt signaling pathways in vertebrate somitogenesis and segmentation has not been shown, the absence of *Notch1* expression in the presomitic mesoderm in homozygous *LEF-1/TCF-1* mouse mutants suggests a possible interaction between the two pathways.[197] Moreover, Notch signaling through *hairy-related* genes appears to specifically regulate expression of *MyoD*, a gene, which has also been shown to be downstream of Wnt/β-catenin signaling.[211] In amphioxus, *Notch* is expressed in the posterior mesoderm in the gastrula and neurula, the tail bud and the dorsoposterior portion of all somites,[99] suggesting that its role in somitogenesis may be similar to that in vertebrates. In both vertebrates and amphioxus, *brachyury* is also expressed around the blastopore and in the tail bud. In vertebrates, *brachyury* is a target of Wnt-3a,[83] but is upstream of *Wnt-11*.[9] Co-expression of these two *Wnt* genes and *brachyury* in amphioxus suggests that a similar relation between those genes may exist in amphioxus and vertebrates.

Formation of the muscle in ascidians differs considerably from that in amphioxus. While ascidian larvae have a notochord, the tail muscle is not organized into somites, and there is no tail bud. Instead, the tail muscle lineage is determined shortly after fertilization, and the muscle cell lineage is partitioned into the tail. Nevertheless, ascidian *Wnt-7*, the only ascidian homolog cloned to date of a *Wnt* gene involved in vertebrate (and possibly amphioxus) somitogenesis and segmentation, is expressed exclusively in cells of the ascidian nervous system,[16] and it could be acting to pattern the paraxial mesoderm in ascidians as in other chordates. Thus, even though there is no tail bud in ascidians and the tail musculature is unsegmented, a role of *Wnt-7* in patterning the paraxial musculature might be conserved among chordates.

## The Roles of *Wnt* Genes in Patterning the Chordate Notochord

Zebrafish *Wnt-11* is the only *Wnt* gene described to date that is expressed in the notochord, and the expression of this gene depends on that of *brachyury*.[82] *Wnt-11* signals via the Wnt/JNK pathway, which has been shown to mediate convergent extension movements.[7,9] In other vertebrates, *Wnt* genes are not expressed in the notochord and a down-regulation of signaling via the Wnt/β-catenin pathway is necessary for specification of the notochord.[212] In agreement with this idea, ectopic expression of *Xwnt-8* inhibits expression of the notochord marker *Xnot*[138] and can transform notochord into somitic muscle.[130] Moreover, the *Wnt* antagonist *frzb* is expressed in the *Xenopus* notochord during the gastrula stage.[212] In amphioxus, the only *Wnt* gene expressed in the notochord is *AmphiWnt-5*, which is in the Wnt/$Ca^{2+}$ class. *AmphiWnt-5* is expressed in the anterior and posterior ends of the notochord.[22] This suggests that, even though the amphioxus notochord is a special type of muscle, as in vertebrates, the Wnt/β-catenin pathway must be repressed for notochord formation.

## Is There an Evolutionary Conservation of Wnt Signaling in Segmentation between Chordates and Protostomes?

As discussed above, in amphioxus, *engrailed* and *Wnt-8* are expressed in the anteriormost eight somites.[21,207] The expression of *engrailed* in the posterior half of each nascent somite is reminiscent of the pattern of engrailed in the *Drosophila* ectoderm. During the segmentation of the *Drosophila* embryo, *engrailed* cooperates with *wingless, cubitus interruptus, hedgehog* and other gap/pair rule and segment polarity genes.[213] The expression of several segmentation genes in similar patterns in other insects, such as grasshoppers and beetles, suggests that the genetic basis of segmentation of the ectoderm is conserved among all insects.[45,167,214] However, it is now clear that amphioxus homologs of many of the genes, which cooperate with *engrailed* in segmentation of the *Drosophila* ectoderm (e.g., *Wnt-1, hh, gli*)[18,188,215] (SM Shimeld pers. com.), although they are expressed in the amphioxus mesoderm, are not expressed together with *AmphiEn* in a pattern that suggests an evolutionary conservation of the mechanisms for segmenting the ectoderm in *Drosophila* and the mesoderm in amphioxus. Indeed, whether, a genetic mechanism involving *engrailed* and *Wnt* genes of the Wnt/β-catenin class acts in segmentation of the ectoderm in any protostomes other than arthropods is controversial. *Engrailed* is expressed in the ectoderm of embryos and regenerates of the annelid *Platynereis dumerilii* in a pattern suggesting a role in segmentation,[216] but it is not similarly expressed in the ectoderm of other annelids[217-220] or of the onychophoran *Peripatus*,[221] which has led to the suggestion that a role for *engrailed* in ectodermal segmentation is limited to arthropods.[222]

However, evolutionary conservation of the roles of *engrailed* and *wingless* in mesodermal segmentation remains a possibility. As in amphioxus, *engrailed* is segmentally expressed in the *Drosophila* mesoderm.[223,224] The genetic networks controlling segmental expression of *engrailed* in the *Drosophila* mesoderm are not entirely understood, but do not appear to be identical to those for segmenting the ectoderm.[225] In the current models, segmental *wingless (wg)* expression in the ectoderm induces striped expression of downstream targets (e.g., the forkhead gene *sloppy paired*) in the mesoderm. Expression of *sloppy paired* is required for specification of somatic muscle. *Engrailed* is downstream of *sloppy* paired.[226] Although the *wingless* signal for mesodermal segmentation appears to come from the ectoderm, *wingless* is also expressed in the invaginating mesoderm up to stage 9 in a rather patchy pair-rule pattern.[227] The roles of this *wingless* expression have not been precisely determined in *Drosophila*. However, because overexpression of *wingless* throughout the mesoderm up-regulates and prolongs the striped mesodermal expression of *engrailed*, it has been suggested that mesodermal wingless may act to maintain mesodermal *engrailed* expression.[228] Since amphioxus *AmphiWnt-8*, which, like *wingless*, signals via β-catenin, is uniformly expressed in the presomitic and somitic mesoderm of somites two through eight,[21] it could be that there is a similar mechanism for segmenting the anterior mesoderm of amphioxus and the mesoderm of *Drosophila*, involving *Wnt* and *engrailed* genes. Whether this mechanism is evolutionarily conserved in arthropods and amphioxus awaits a thorough understanding of the interactions between *Wnt/wingless* in mesodermal segmentation in the anterior somites of amphioxus and the somatic mesoderm of *Drosophila*.

## Evolutionary Conservation of the Roles of *Wnt* Genes in Convergent Extension and Planar Polarity Signaling

Convergent extension movements lead to a narrowing and lengthening of tissues mainly by compaction and intercalation of cells. Among the chordates, the mechanism of convergent extension has been well studied only in vertebrates. In amphioxus, cell tracing and time-lapse analyses have not been done to address the question of convergent extension during elongation of the body axis. In ascidian tunicates, elongation of the tail is solely due to convergent extension. *Wnt-5* may play a role in this process. It is expressed in the notochord and, when overexpressed, the anteroposterior body axis of the embryo is shortened due to the abnormal movement of notochord cells. Thus, ascidian *Wnt-5* may act to limit the extent of convergent extension.[229] The possible roles of other *Wnt* genes in convergent extension have not been studied in ascidians.

In vertebrates, convergent extension is important in gastrulation and in elongation of the body axis.[230] Wnt signaling, both through the two Dishevelled-mediated pathways (JNK/SAPK and β-catenin) and through the Wnt/$Ca^{2+}$ pathway is important in convergent extension.[4]

*Wnt-1*, signaling through the β-catenin pathway, promotes cell adhesion and enhances convergent extension due to regulation of the *nodal* gene *Xnr3*.[4] In addition, *Wnt-11*, which acts via the Wnt/JNK pathway, potentiates convergent extension movements in the neuroectoderm and the mesoderm.[9,189,231-234] In contrast, *Xwnt-5a*, signaling through the Wnt/$Ca^{2+}$ pathway decreases cell adhesion and depresses convergent extension. It has been suggested that signaling via this pathway functions to balance signaling between the two Dishevelled-mediated pathways.[4] In *Xenopus*, defects in convergent extension of the mesoderm can be rescued by *Dishevelled*, but not by molecules that specifically activate the Wnt/β-catenin pathway.[235,236] In addition, time-lapse analyses of cell movements in *Xenopus* have shown that *Dishevelled* controls the polarity of the lamellipodia driving convergent extension, which is consistent with a role for planar polarity signaling in convergent cell extension.[233] Experiments with deletions of *Xenopus Dishevelled* suggest that the *Wnt-11* pathway controlling convergent extension movements in vertebrates may be similar to the *Drosophila* planar cell polarity cascade.[9,233] In *Drosophila*, the planar cell polarity cascade controls the organization of cells within the plane of an epithelium, for example the coordinated orientation of hairs and bristles in the ectoderm. The transduction of the planar polarity signal in *Drosophila* is mediated, among other factors, by Frizzled and Dishevelled.[237] In *Drosophila*, separate functional domains of the Dishevelled protein transduce the signal for the Wnt/β-catenin and the planar polarity cascade, respectively,[238,239] which is comparable to the function of the *Xenopus* Dishevelled protein.[9,233] Thus, although the Wnt/JNK pathway has not been as well characterized as the Wnt/β-catenin pathway, it is possible that the former pathway may have mediated convergent extension movements in the ancestral bilaterian.

## The Involvement of *Wnt* Genes in Patterning the Developing Central Nervous System

Vertebrate *Wnt* genes are involved in patterning several regions of the central nervous system. Of these regions, the midbrain-hindbrain boundary region (MHB), which expresses *Wnt-1*, together with *engrailed* genes, *Pax2* and *Pax5*, has received the most attention.[116,132,137,240-242] Loss of murine *Wnt-1* leads to loss of the midbrain and anterior hindbrain and to a complete ablation of *engrailed* expression in this region of the central nervous system.[243-246] These defects can be rescued by overexpression of *engrailed-2*.[246] In addition, the *Xenopus engrailed-2* promoter contains three binding sites for the transcription factors TCF/LEF in the Wnt/β-catenin signaling pathway, suggesting that *engrailed* is a direct target of Wnt/β-catenin signaling.[247]

In amphioxus, *Wnt* genes of the Wnt/β-catenin class, *engrailed*, and *Pax2/5/8* are not expressed in the nerve cord in patterns consistent with there being a homolog of the MHB.[18,207,248,249] Thus, *AmphiWnt-1* is not expressed at all in the amphioxus nerve cord,[18] while *engrailed* is expressed in two domains in the amphioxus nerve cord, one in the equivalent of rhombomere 5, and the second in the equivalent of the forebrain.[207] The latter domain is within that of *AmphiWnt-3* in the forebrain and close to that of *AmphiWnt-8*.[21,22,207] Thus, the functional relationship between *engrailed* and Wnt/β-catenin *Wnt* genes may be conserved between amphioxus, vertebrates and *Drosophila*, even though the structures being patterned are not homologous.

It is controversial whether the absence of a MHB in amphioxus represents a loss or is an indication that the MHB first evolved within the vertebrate lineage. In ascidians, *Pax-2/5/8* is expressed in two cells just posterior to the *Otx* domain and just anterior to the *Hox1* domain, where a MHB would be expected to be if there were one.[250] The expression of *Pax-2/5/8* has led to the suggestion that ascidians have a homolog of the MHB and that amphioxus must have evolutionarily lost it. However, expression of *Wnt-1* and *engrailed* has not yet been determined in ascidian tunicates, and, therefore, the possibility that the *Pax-2/5/8* domain corresponds to hindbrain cannot be ruled out.

In vertebrates, *Wnt* genes are also involved in the induction of neural crest. In *Xenopus, Xwnt-7b* is expressed in the dorsal marginal zone in the early gastrula and later in the dorsal midline of the neural tube and the MHB.[137] When injected into embryos or when co-injected with the neural inducer *noggin* into ectodermal explants, *Xwnt-7b* induces the neural crest markers *Xslug* and *Xtwist.*[137] Moreover, blocking *Xwnt-8* inhibits the expression of neural crest markers.[251] In addition to the MHB region, *Xenopus Xwnt-1* is transcribed in the dorsal midline of midbrain, hindbrain, and anterior spinal cord,[132,137,241] and loss of both *Wnt-1* and *Wnt-3a* in mice leads to deficiencies in neural crest derivatives.[252] The Wnt receptor *Frizzled 3,* which is expressed in the dorsal part of the neural tube, interacts with Xwnt-1 and can induce neural crest in embryos and explants.[253] Thus, *Wnt* genes of the Wnt/β-catenin class probably function in induction and/or maintenance and/or proliferation of neural crest. In addition, *Wnt* genes play several roles in later neural patterning. For example, *Wnt-3a* is necessary for the development of the murine hippocampus,[143] and *Wnt-7a* is required for axonal remodeling and synaptic differentiation in the cerebellum.[254] Loss of both *Wnt-1* and *Wnt-3a* in mice leads to a reduction of neural precursors within the neural tube as well as defective neural crest.[252]

In *Drosophila, wingless* is expressed in all three regions of the brain, most conspicuously in the protocerebrum. In *wingless* null mutants, approximately half the protocerebrum is missing.[255] It has been proposed that *wingless* functions in the determination of neural cell fate in the brain and in other areas of the central nervous system where it is expressed.[255-258] *Wingless* is also expressed in the eye, where it appears to act in specification of the dorsoventral axis.[259] Ectopic expression of another *Drosophila Wnt* gene, *DWnt-3,* which is expressed along the ventral midline of the nerve cord, results in abnormal commissural axon tracts in the developing central nervous system.[42,43,46] In addition, *DWnt-4* and *Dwnt-10* are expressed in the *Drosophila* central nervous system;[47,50] however, their function in the development of the central nervous system has not been studied.

The expression of *Wnt* genes of the Wnt/β-catenin class in specific subsets of developing neurons in both chordates and *Drosophila* might reflect a conserved evolutionary role in neuroblast specification. However, since these roles have not been studied in detail in either chordates or *Drosophila,* it remains possible that *Wnt* genes have been repeatedly co-opted for roles in neuroblast specification.

## Conclusions

Sequence comparisons and genome analyses suggest that the last common ancestor of protostomes and deuterostomes had at least six *Wnt* genes (*Wnt-1, Wnt-3, Wnt-5, Wnt-6, Wnt-7,* and *Wnt-10*). However, this number could be an underestimate, since phylogenetic analyses do not adequately resolve the relationships between *Wnt* subfamilies. Specific *Wnt* genes signal preferentially through one of three major pathways: (1) Wnt/$Ca^{2+}$, (2) Wnt/JNK, and (3) Wnt/β-catenin.

The Wnt/β-catenin pathway is the best studied of the three Wnt pathways, and its major conserved role in metazoan development appears to be in patterning the longitudinal axis of the body. In *Hydra, Wnt-3* and other components of this pathway are expressed around the common gut opening at the oral end of the animal, which corresponds to the posterior end of the larva. In both deuterostomes and protostomes, one or more of the *Wnt* genes signaling through this pathway (*wingless/Wnt-1, Wnt-3,* and/or *Wnt-8*) are expressed at the posterior/vegetal end of the early embryo. Experimental evidence from deuterostomes indicates that this posterior Wnt signaling center creates an anteroposterior gradient of *Wnt* activity that is down-regulated anteriorly to allow for normal anterior development. *Notch* genes and *brachyury* are typically co-expressed with *Wnt* genes in this posterior signaling center, and there is evidence to demonstrate that they all interact. Another conserved role of the Wnt/β-catenin pathway appears to be in boundary formation. Such boundaries include segment boundaries in *Drosophila* and the midbrain-hindbrain boundary in vertebrates.

The Wnt/$Ca^{2+}$ pathway, through which *Wnt-4* and *Wnt-5* preferentially signal, appears to antagonize the effects of the Wnt/β-catenin pathway. During deuterostome development, *Wnt* genes signaling through the Wnt/β-catenin and the Wnt/$Ca^{2+}$ pathway are often expressed in the same or contiguous cells, as would be expected if they had balancing effects on a given developmental process. The Wnt/JNK pathway appears to play a major role in convergent extension. It has been suggested that the role of *Wnt* genes in convergent extension may be evolutionarily related to the role of *Wnt* genes in the planar cell polarity pathway in *Drosophila*.

*Wnt* genes signaling through all three pathways function in patterning numerous structures in developing embryos. Such functions include specification of particular areas of the brain and neuroblast identity in both protostomes and deuterostomes and roles in somitogenesis and induction of neural crest in chordates. However, a role of the Wnt/β-catenin pathway in establishing the dorsoventral axis of the frog *Xenopus* appears to have no counterpart in invertebrate chordates, and may have arisen only during vertebrate evolution.

## *Acknowledgements*

The authors would like to thank ND Holland for comments and advice on the manuscript. In addition, we are indebted to V Laudet and the members of his laboratory for supporting M Schubert while writing this book chapter at the Ecole Normale Supérieure de Lyon. This work was supported by a fellowship within the PostDoc-Program of the German Academic Exchange Service (DAAD) to M Schubert and by NSF grant number IBN00-78599 to LZ Holland and ND Holland.

## References

1. Christian JL. BMP, Wnt and Hedgehog signals: How far can they go? Curr Opin Cell Biol 2000; 12:244-249.
2. Hamilton FS, Wheeler GN, Hoppler S. Difference in XTCF-3 dependency accounts for change in response to β-catenin-mediated Wnt signaling in Xenopus blastula. Development 2001; 128:2063-2073.
3. Huelsken J, Birchmeier W. New aspects of Wnt signaling pathways in higher vertebrates. Curr Opin Genet Dev 2001; 11:547-553.
4. Kühl M, Geis K, Sheldahl LC et al. Antagonistic regulation of convergent extension movements in Xenopus by Wnt/β-catenin and Wnt/$Ca^{2+}$ signaling. Mech Dev 2001; 106:61-76.
5. McEwen DG, Peifer M. Wnt signaling: the naked truth? Curr Biol 2001; 11:R524-R526.
6. Niehrs C. Solving a sticky problem. Nature 2001; 413:787-788.
7. Winklbauer R, Medina A, Swain RK et al. Frizzled-7 signalling controls tissue separation during Xenopus gastrulation. Nature 2001; 413:856-860.
8. Landesman Y, Sokol SY. Xwnt-2b is a novel axis-inducing Xenopus Wnt, which is expressed in embryonic brain. Mech Dev 1997; 63:199-209.
9. Tada M, Smith JC. Xwnt11 is a target of Xenopus Brachyury: regulation of gastrulation movements via Dishevelled, but not through the canonical Wnt pathway. Development 2000; 127:2227-2238.
10. Kühl M, Sheldahl LC, Malbon CC et al. $Ca^{2+}$/calmodulin-dependent protein kinase II is stimulated by Wnt and Frizzled homologs and promotes ventral cell fates in Xenopus. J Biol Chem 2000; 275:12701-12711.
11. Ferkowicz MJ, Stander MC, Raff RA. Phylogenetic relationships and developmental expression of three sea urchin Wnt genes. Mol Biol Evol 1998; 15:809-819.
12. Angerer LM, Angerer RC. Animal-vegetal axis patterning mechanisms in the early sea urchin embryo. Dev Biol 2000; 218:1-12.

13. Ferkowicz MJ, Raff RA. Wnt gene expression in sea urchin development: heterochronies associated with the evolution of developmental mode. Evol Dev 2001; 3:24-33.
14. Zhu X, Mahairas G, Illies M et al. A large-scale analysis of mRNAs expressed by primary mesenchyme cells of the sea urchin embryo. Development 2001; 128:2615-2627.
15. Sasakura Y, Ogasawara M, Makabe KW. HrWnt-5: a maternally expressed ascidian Wnt gene with posterior localization in the embryos. Int J Dev Biol 1998; 42:573-580.
16. Sasakura Y, Makabe KW. Ascidian Wnt-7 gene is expressed exclusively in the tail neural tube of tailbud embryos. Dev Genes Evol 2000; 210:641-643.
17. Makabe KW, Kawashima T, Kawashima S et al. Large-scale cDNA analysis of the maternal genetic information in the egg of Halocynthia roretzi for a gene expression catalog of ascidian development. Development 2001; 128:2555-2567.
18. Holland LZ, Holland ND, Schubert M. Developmental expression of AmphiWnt1, an amphioxus gene in the Wnt1/wingless subfamily. Dev Genes Evol 2000a; 210:522-524.
19. Schubert M, Holland LZ, Holland ND. Characterization of two amphioxus Wnt genes (AmphiWnt4 and AmphiWnt7b) with early expression in the developing central nervous system. Dev Dyn 2000a; 217:205-217.
20. Schubert M, Holland LZ, Holland ND. Characterization of an amphioxus Wnt gene, AmphiWnt11, with possible roles in myogenesis and tail outgrowth. Genesis 2000b; 27:1-5.
21. Schubert M, Holland LZ, Panopoulou, GD et al. Characterization of amphioxus AmphiWnt8: insights into the evolution of patterning of the embryonic dorsoventral axis. Evol Dev 2000c; 2:85-92.
22. Schubert M, Holland LZ, Stokes MD et al. Three amphioxus Wnt genes (AmphiWnt3, AmphiWnt5, and AmphiWnt6) associated with the tail bud: the evolution of somitogenesis in chordates. Dev Biol 2001; 240:262-273.
23. Yasui K, Saiga H, Zhang PJ et al. Early expressed genes showing a dichotomous developing pattern in the lancelet embryo. Dev Growth Differ 2001; 43:185-194.
24. Llimargas M, Lawrence PA. Seven Wnt homologues in Drosophila: A case study of the developing tracheae. Proc Natl Acad Sci USA 2001; 98:14487-14492.
25. Kamb A, Weir M, Rudy B et al. Identification of genes from pattern formation, tyrosine kinase, and potassium channel families by DNA amplification. Proc Natl Acad Sci USA 1989; 86:4372-4376.
26. Shackleford GM, Shivakumar S, Shiue L et al. Two wnt genes in Caenorhabditis elegans. Oncogene 1993; 8:1857-1864.
27. Herman MA, Horvitz HR. The Caenorhabditis elegans gene lin-44 controls the polarity of asymmetric cell divisions. Development 1994; 120:1035-1047.
28. Herman MA, Vassilieva LL, Horvitz HR et al. The C. elegans gene lin-44, which controls the polarity of certain asymmetric cell divisions, encodes a Wnt protein and acts cell nonautonomously. Cell 1995; 83:101-110.
29. Harris J, Honigberg L, Robinson N et al. Neuronal cell migration in C. elegans: regulation of Hox gene expression and cell position. Development 1996; 122:3117-3131.
30. Cadigan KM, Nusse R. Wnt signaling: a common theme in animal development. Genes Dev 1997; 11:3286-3305.
31. Han M. Gut reaction to Wnt signaling in worms. Cell 1997; 90:581-584.
32. Rocheleau CE, Downs WD, Lin R et al. Wnt signaling and an APC-related gene specify endoderm in early C. elegans embryos. Cell 1997; 90:707-716.
33. Thorpe CJ, Schlesinger A, Carter JC et al. Wnt signaling polarizes an early C. elegans blastomere to distinguish endoderm from mesoderm. Cell 1997; 90:695-705.
34. Hawkins N, Garriga G. Asymmetric cell division: from A to Z. Genes Dev 1998; 12:3625-3638.
35. Wodarz A, Nusse R. Mechanisms of Wnt signaling in development. Annu Rev Cell Dev Biol 1998; 14:59-88.
36. Whangbo J, Kenyon C. A Wnt signaling system that specifies two patterns of cell migration in C. elegans. Mol Cell 1999; 4:851-858.
37. Whangbo J, Harris J, Kenyon C. Multiple levels of regulation specify the polarity of an asymmetric cell division in C. elegans. Development 2000; 127:4587-4598.

38. Baker NE. Molecular cloning of sequences from wingless, a segment-polarity gene in Drosophila; the spatial distribution of a transcript in embryos. EMBO J 1987; 6:1765-1773.
39. Baker NE. Localization of transcripts from the wingless gene in whole Drosophila embryos. Development 1988; 103:289-298.
40. Ingham PW, Baker NE, Arias AM. The products of the ftz and eve genes act as positive and negative regulators of engrailed and wingless expression in the Drosophila blastoderm. Nature 1988; 331:73-75.
41. Doe CQ. Molecular markers for identified neuroblasts and ganglion mother cells in the Drosophila central nervous system. Development 1992; 116:855-863.
42. Eisenberg LM, Ingham PW, Brown AMC. Cloning and characterization of a novel Drosophila Wnt gene, DWnt-5, a putative downstream target of the homeobox gene Distal-less. Dev Biol 1992; 154:73-83.
43. Russell J, Gennissen A, Nusse R. Isolation and expression of two novel Wnt/wingless gene homologues in Drosophila. Development 1992; 115:475-485.
44. Chu-LaGraff Q, Doe CQ. Neuroblast specification and formation regulated by wingless in the Drosophila CNS. Science 1993; 261:1594-1597.
45. Nagy LM, Carroll S. Conservation of wingless patterning functions in the short-germ embryos of Tribolium castaneum. Nature 1994; 367:460-463.
46. Fradkin LG, Noordermeer JN, Nusse R. The Drosophila Wnt protein DWnt-3 is a secreted glycoprotein localized on the axon tracts of the embryonic CNS. Dev Biol 1995; 168:202-213.
47. Graba Y, Gieseler K, Aragnol D et al. Dwnt-4, a novel Drosophila Wnt gene acts downstream of homeotic complex genes in the visceral mesoderm. Development 1995; 121:209-218.
48. Kozopas KM, Samos CH, Nusse R. DWnt-2, a Drosophila Wnt gene required for the development of the male reproductive tract, specifies a sexually dimorphic cell fate. Genes Dev 1998; 12:1155-1165.
49. Buratovich MA, Anderson S, Gieseler K et al. DWnt-4 and wingless have distinct activities in the Drosophila dorsal epidermis. Dev Genes Evol 2000; 210:111-119.
50. Janson K, Cohen ED, Wilder EL. Expression of DWnt6, DWnt10, and DFz-4 during Drosophila development. Mech Dev 2001; 103:117-120.
51. Holland PWH, Williams NA, Lanfear J. Cloning of segment polarity gene homologues from the unsegmented brachiopod Terebratulina retusa (Linnaeus). FEBS Lett 1991; 291:211-213.
52. Kostriken R, Weisblat DA. Expression of a Wnt gene in embryonic epithelium of the leech. Dev Biol 1992; 151:225-241.
53. Huang FZ, Bely AE, Weisblat DA. Stochastic WNT signaling between nonequivalent cells regulates adhesion but not fate in the two-cell leech embryo. Curr Biol 2001; 11:1-7.
54. Hobmayer B, Rentzsch F, Kuhn K et al. WNT signaling molecules act in axis formation in the dipoblastic metazoan Hydra. Nature 2000; 407:186-189.
55. Hoshiyama D, Suga H, Iwabe N et al. Sponge Pax cDNA related to Pax-2/5/8 and ancient gene duplications in the Pax family. J Mol Evol 1998; 47:640-648.
56. Escriva H, Delaunay F, Laudet V. Ligand binding and nuclear receptor evolution. BioEssays 2000; 22:717-727.
57. Gauchat D, Mazet F, Berney C et al. Evolution of Antp-class genes and differential expression of Hydra Hox/paraHox genes in anterior patterning. Proc Natl Acad Sci USA 2000; 97:4493-4498.
58. Kourakis MJ, Martindale MQ. Combined-method phylogenetic analysis of Hox and ParaHox genes of the Metazoa. J Exp Zool 2000; 288:175-191.
59. Blaxter ML, De Ley P, Garey JR et al. A molecular evolutionary framework for the phylum Nematoda. Nature 1998; 392:71-75.
60. Giribet G, Edgecombe GD, Wheeler WC. Arthropod phylogeny based on eight molecular loci and morphology. Nature 2001; 413:157-161.
61. Hwang UW, Friedrich M, Tautz D et al. Mitochondrial protein phylogeny joins myriapods with chelicerates. Nature 2001; 413:154-157.
62. Aguinaldo AMA, Turbeville JM, Linford LS et al. Evidence for a clade of nematodes, arthropods and other moulting animals. Nature 1997; 387:489-493.

63. Schubert M, Holland LZ, Holland ND et al. A phylogenetic tree of the Wnt genes based on all available full-length sequences, including five from the cephalochordate amphioxus. Mol Biol Evol 2000d; 17:1896-1903.
64. Jockusch EL, Ober KA. Phylogenetic analysis of the Wnt gene family and discovery of an arthropod Wnt-10 orthologue. J Exp Zool 2000; 288:105-119.
65. Holland PWH, Garcia-Fernàndez J, Williams NA et al. Gene duplications and the origins of vertebrate development. Development 1994; 120S:125-133.
66. Sidow A. Gen(om)e duplications in the evolution of early vertebrates. Curr Opin Genet Dev 1996; 6:715-722.
67. Sidow A. Diversification of the Wnt gene family on the ancestral lineage of vertebrates. Proc Natl Acad Sci USA 1992; 89:5098-5102.
68. Nusse R. An ancient cluster of Wnt paralogues. TIG 2001; 17:443.
69. Kenny AP, Kozlowski DJ, Oleksyn DW et al. SpSoxB1, a maternally encoded transcription factor asymmetrically distributed among early sea urchin blastomeres. Development 1999; 126:5473-5483.
70. Ghiglione C, Lhomond G, Lepage T et al. Cell-autonomous expression and position-dependent repression by $Li^+$ of two zygotic genes during sea urchin early development. EMBO J 1993; 12:87-96.
71. Emily-Fenouil F, Ghiglione C, Lhomond G et al. GSK3β/shaggy mediates patterning along the animal-vegetal axis of the sea urchin embryo. Development 1998; 125:2489-2498.
72. Wikramanayake AH, Huang L, Klein WH. β-Catenin is essential for patterning the maternally specified animal-vegetal axis in the sea urchin embryo. Proc Natl Acad Sci USA 1998; 95:9343-9348.
73. Logan CY, Miller JR, Ferkowicz MJ et al. Nuclear β-catenin is required to specify vegetal cell fates in the sea urchin embryo. Development 1999; 126:345-357.
74. Huang L, Li X, El-Hodiri HM et al. Involvement of TCF/Lef in establishing cell types along the animal-vegetal axis of sea urchins. Dev Genes Evol 2000; 210:73-81.
75. McClay DR, Peterson RE, Range RC et al. A micromere induction signal is activated by β-catenin and acts through Notch to initiate specification of secondary mesenchyme cells in the sea urchin embryo. Development 2000; 127:5113-5122.
76. Vonica A, Weng W, Gumbiner BM et al. TCF is the nuclear effector of the β-catenin signal that patterns the sea urchin animal-vegetal axis. Dev Biol 2000; 217:230-243.
77. Howard EW, Newman LA, Oleksyn DW et al. SpKrl1: a direct target of β-catenin regulation required for endoderm differentiation in sea urchin embryos. Development 2001; 128:365-375.
78. Sweet HC, Hodor PG, Ettensohn CA. The role of micromere signaling in Notch activation and mesoderm specification during sea urchin embryogenesis. Development 1999; 126:5255-5265.
79. Sherwood DR, McClay DR. Identification and localization of a sea urchin Notch homologue: insights into vegetal plate regionalization and Notch receptor regulation. Development 1997; 124:3363-3374.
80. Sherwood DR, McClay DR. LvNotch signaling mediates secondary mesenchyme specification in the sea urchin embryo. Development 1999; 126:1703-1713.
81. Sherwood DR, McClay DR. LvNotch signaling plays a dual role in regulating the position of the ectoderm-endoderm boundary in the sea urchin embryo. Development 2001; 128:2221-2232.
82. Makita R, Mizuno T, Koshida S et al. Zebrafish wnt11: pattern and regulation of the expression by the yolk cell and No tail activity. Mech Dev 1998; 71:165-176.
83. Yamaguchi TP, Takada S, Yoshikawa Y et al. T (Brachyury) is a direct target of Wnt3a during paraxial mesoderm specification. Genes Dev 1999a; 13:3185-3190.
84. Harada Y, Yasuo H, Satoh N. A sea urchin homologue of the chordate Brachyury (T) gene is expressed in the secondary mesenchyme founder cells. Development 1995; 121:2747-2754.
85. Peterson KJ, Harada Y, Cameron RA et al. Expression pattern of Brachyury and Not in the sea urchin: comparative implications for the origins of mesoderm in basal deuterostomes. Dev Biol 1999; 207:419-431.
86. Gross JM, McClay DR. The role of Brachyury (T) during gastrulation movements in the sea urchin Lytechinus variegatus. Dev Biol 2001; 239:132-147.
87. Tagawa K, Humphreys T, Satoh N. Novel pattern of Brachyury gene expression in hemichordate embryos. Mech Dev 1998; 75:139-143.

88. Shoguchi E, Satoh N, Maruyama YK. Pattern of Brachyury gene expression in starfish embryos resembles that of hemichordate embryos but not of sea urchin embryos. Mech Dev 1999; 82:185-189.
89. Nishida H. Cell fate specification by localized cytoplasmic determinants and cell interactions in ascidian embryos. Int Rev Cytol 1997; 176:245-305.
90. Nishida H. Localization of determinants for formation of the anterior-posterior axis in eggs of the ascidian Halocynthia roretzi. Development 1994; 120:3093-3104.
91. Conklin EG. The organization and cell-lineage of the ascidian egg. J Acad Nat Sci Phila 1905; 13:5-119 + plates I-XI.
92. Gofflot F, Hall M, Morriss-Kay GM. Genetic patterning of the developing mouse tail at the time of posterior neuropore closure. Dev Dyn 1997; 210:431-445.
93. Imai K, Takada N, Satoh N et al. β-catenin mediates the specification of endoderm cells in ascidian embryos. Development 2000; 127:3009-3020.
94. Yoshida S, Marikawa Y, Satoh N. Regulation of the trunk-tail patterning in the ascidian embryo: a possible interaction of cascades between lithium/β-catenin and localized maternal factor pem. Dev Biol 1998; 202:264-279.
95. Corbo JC, Fujiwara S, Levine M et al. Suppressor of hairless activates brachyury expression in the Ciona embryo. Dev Biol 1998; 203:358-368.
96. Yasuo H, Kobayashi M, Shimauchi Y et al. The ascidian genome contains another T-domain gene that is expressed in differentiating muscle and the tip of the tail of the embryo. Dev Biol 1996; 180:773-779.
97. Hori S, Saitoh T, Matsumoto M et al. Notch homologue from Halocynthia roretzi is preferentially expressed in the central nervous system during ascidian embryogenesis. Dev Genes Evol 1997; 207:371-380.
98. Coffman CR, Skoglund P, Harris WA et al. Expression of an extracellular deletion of Xotch diverts cell fate in Xenopus embryos. Cell 1993; 73:659-671.
99. Holland LZ, Abi Rached L, Tamme R et al. Characterization and developmental expression of the amphioxus homolog of Notch (AmphiNotch): evolutionary conservation of multiple expression domains in amphioxus and vertebrates. Dev Biol 2001; 232:493-507.
100. Holland PWH, Koschorz B, Holland LZ et al. Conservation of Brachyury (T) genes in amphioxus and vertebrates: developmental and evolutionary implications. Development 1995; 121:4283-4291.
101. Holland ND, Panganiban G, Henyey EL et al. Sequence and developmental expression of AmphiDll, an amphioxus Distal-less gene transcribed in the ectoderm, epidermis and nervous system: insights into the evolution of craniate forebrain and neural crest. Development 1996: 122:2911-2920.
102. Schubert M, Holland ND, Holland LZ. Amphioxus AmphiDRAL encoding a LIM-domain protein: expression in the epidermis but not in the presumptive neuroectoderm. Mech Dev 1998; 76:203-205.
103. Martin B, Schneider R, Starzinski-Powitz A. FHL2/MDRAL interacts with β-catenin and represses TCF/β-catenin dependent transcription. Biol Cell 2001; 93:203.
104. Tung TC, Wu SC, Tung YYF. Experimental studies on the neural induction in amphioxus. Sci Sin 1962; 11:805-820.
105. Torres MA, Yang-Snyder JA, Purcell SM et al. Activities of the Wnt-1 class of secreted signaling factors are antagonized by the Wnt-5A class and by a dominant negative cadherin in early Xenopus development. J Cell Biol 1996; 133:1123-1137.
106. Holland LZ, Schubert M, Holland ND et al. Evolutionary conservation of the presumptive neural plate markers AmphiSox1/2/3 and AmphiNeurogenin in the invertebrate chordate amphioxus. Dev Biol 2000b; 226:18-33.
107. De Robertis EM, Larraín J, Oelgeschläger M et al. The establishment of Spemann's organizer and patterning of the vertebrate embryo. Nat Rev Genet 2001; 1:171-181.
108. Wylie C, Kofron M, Payne C et al. Maternal β-catenin establishes a 'dorsal signal' in early Xenopus embryos. Development 1996; 122:2987-2996.
109. Schneider S, Steinbeisser H, Warga RM et al. β-catenin translocation into nuclei demarcates the dorsalizing centers in frog and fish embryos. Mech Dev 1996; 57:191-198.

110. Larabell CA, Torres M, Rowning BA et al. Establishment of the dorso-ventral axis in Xenopus embryos is presaged by early asymmetries in β-catenin that are modulated by the Wnt signaling pathway. J Cell Biol 1997; 136:1123-1136.
111. Schohl A, Fagotto F. β-catenin, MAPK, and Smad signaling during early Xenopus development. Development 2002; 129:37-52.
112. Ryu S-L, Fujii R, Yamanaka Y et al. Regulation of dharma/bozozok by the Wnt pathway. Dev Biol 2001; 231:397-409.
113. Kao KR, Elinson RP. The legacy of lithium effects on development. Biol Cell 1998; 90:585-589.
114. Heasman J, Kofron M, Wylie C. β-catenin signaling activity dissected in the early Xenopus embryo: a novel antisense approach. Dev Biol 2000; 222:124-134.
115. Cui Y, Brown JD, Moon RT et al. Xwnt-8b: a maternally expressed Xenopus Wnt gene with a potential role in establishing the dorsoventral axis. Development 1995; 121:2177-2186.
116. Gradl D, Kühl M, Wedlich D. Keeping a close eye on Wnt-1/wg signaling in Xenopus. Mech Dev 1999; 86:3-15.
117. Moon RT, Kimelman D. From cortical rotation to organizer gene expression: toward a molecular explanation of axis specification in Xenopus. BioEssays 1998; 20:536-545.
118. Sokol SY. Wnt signaling and dorso-ventral axis specification in vertebrates. Curr Opin Genet Dev 1999; 9:405-410.
119. Sakai Y, Hiraoka Y, Konishi M et al. Isolation and characterization of Xenopus laevis xSox-B1 cDNA. Arch Biochem Biophys 1997; 346:1-6.
120. Mizuseki K, Kishi M, Matsui M et al. Xenopus Zic-related-1 and Sox-2, two factors induced by chordin, have distinct activities in the initiation of neural induction. Development 1998; 125:579-587.
121. Wessely O, Agius E, Oelgeschläger M et al. Neural induction in the absence of mesoderm: β-catenin-dependent expression of secreted BMP antagonists at the blastula stage in Xenopus. Dev Biol 2001; 234:161-173.
122. Foley AC, Skromne I, Stern CD. Reconciling different models of forebrain induction and patterning: a dual role for the hypoblast. Development 2000; 127:3839-3854.
123. Coffman C, Harris W, Kintner C. Xotch, the Xenopus homolog of Drosophila Notch. Science 1990; 249:1438-1441.
124. Niehrs C. Head in the WNT: The molecular nature of Spemann's head organizer. TIG 1999; 15:314-319.
125. Gamse JT, Sive H. Early anteroposterior division of the presumptive neuroectoderm in Xenopus. Mech Dev 2001; 104:21-36.
126. Kiecker C, Niehrs C. A morphogen gradient of Wnt/β-catenin signalling regulates anteroposterior neural patterning in Xenopus. Development 2001; 128:4189-4201.
127. Villanueva S, Glavic A, Ruiz P et al. Posteriorization by FGF, Wnt, and retinoic acid is required for neural crest induction. Dev Biol 2002; 241:289-301.
128. Christian JL, McMahon JA, McMahon AP et al. Xwnt-8, a Xenopus Wnt-1/int-1-related gene responsive to mesoderm-inducing growth factors, may play a role in ventral mesodermal patterning during embryogenesis. Development 1991; 111:1045-1055.
129. Smith WC, Harland RM. Injected Xwnt-8 RNA acts early in Xenopus embryos to promote formation of a vegetal dorsalizing center. Cell 1991; 67:753-765.
130. Christian JL, Moon RT. Interactions between Xwnt-8 and Spemann organizer signaling pathways generate dorsoventral pattern in the embryonic mesoderm of Xenopus. Genes Dev 1993; 7:13-28.
131. Ku M, Melton DA. Xwnt-11: a maternally expressed Xenopus Wnt gene. Development 1993; 119:1161-1173.
132. Wolda SL, Moody CJ, Moon RT. Overlapping expression of Xwnt-3A and Xwnt-1 in neural tissue of Xenopus laevis embryos. Dev Biol 1993; 155:46-57.
133. Kloc M, Etkin LD. Two distinct pathways for the localization of RNAs at the vegetal cortex in Xenopus oocytes. Development 1995; 121:287-297.
134. Glinka A, Delius H, Blumenstock C et al. Combinatorial signalling by Xwnt-11 and Xnr3 in the organizer epithelium. Mech Dev 1996; 60:221-231.
135. Fredieu JR, Cui Y, Maier D et al. Xwnt-8 and lithium can act upon either dorsal mesodermal or neuroectodermal cells to cause a loss of forebrain in Xenopus embryos. Dev Biol 1997; 186:100-114.

136. McGrew LL, Hoppler S, Moon RT. Wnt and FGF pathways cooperatively pattern anteroposterior neural ectoderm in Xenopus. Mech Dev 1997; 69:105-114.
137. Chang C, Hemmati-Brivanlou A. Neural crest induction by Xwnt7b in Xenopus. Dev Biol 1998; 194:129-134.
138. Hoppler S, Moon RT. BMP-2/-4 and Wnt-8 cooperatively pattern the Xenopus mesoderm. Mech Dev 1998; 71:119-129.
139. Yamaguchi Y, Shinagawa A. Marked alteration at midblastula transition in the effect of lithium on formation of the larval body pattern of Xenopus laevis. Dev Growth Differ 1989; 31:531-541.
140. Kinoshita K, Asashima M. Effect of activin and lithium on isolated Xenopus animal blastomeres and response alteration at the midblastula transition. Development 1995; 121:1581-1589.
141. Kelly GM, Greenstein P, Erezyilmaz DF et al. Zebrafish wnt8 and wnt8b share a common activity but are involved in distinct developmental pathways. Development 1995; 121:1787-99.
142. Liu P, Wakamiya M, Shea MJ et al. Requirement for Wnt3 in vertebrate axis formation. Nature Genet 1999; 22:361-365.
143. Lee SMK, Tole S, Grove E et al. A local Wnt-3a signal is required for development of the mammalian hippocampus. Development 2000; 127:457-467.
144. Gamse J, Sive H. Vertebrate anteroposterior patterning: the Xenopus neuroectoderm as a paradigm. BioEssays 2000; 22:976-986.
145. Pöpperl H, Schmidt C, Wilson V et al. Misexpression of Cwnt8c in the mouse induces an ectopic embryonic axis and causes a truncation of the anterior neuroectoderm. Development 1997; 124:2997-3005.
146. Erter CE, Wilm TP, Basler N et al. Wnt8 is required in lateral mesendodermal precursors for neural posteriorization in vivo. Development 2001; 128:3571-3583.
147. Lekven AC, Thorpe CJ, Waxman JS et al. Zebrafish wnt8 encodes two Wnt8 proteins on a bicistronic transcript and is required for mesoderm and neuroectoderm patterning. Dev Cell 2001; 1:103-114.
148. Schier AF. Axis formation and patterning in zebrafish. Curr Opin Genet Dev 2001; 11:393-404.
149. Yamaguchi TP. Heads or tails: Wnts and anterior-posterior patterning. Curr Biol 2001; 11:R713-R724.
150. Bouwmeester T, Kim SH, Sasai Y et al. Cerberus is a head-inducing secreted factor expressed in the anterior endoderm of Spemann's organizer. Nature 1996; 382:595-601.
151. Leyns L, Bouwmeester T, Kim SH et al. Frzb-1 is a secreted antagonist of Wnt signaling expressed in the Spemann organizer. Cell 1997; 88:747-756.
152. Wang S, Krinks M, Lin K et al. Frzb, a secreted protein expressed in the Spemann organizer, binds and inhibits Wnt-8. Cell 1997; 88:757-766.
153. Glinka A, Wu W, Delius H et al. Dickkopf-1 is a member of a new family of secreted proteins and functions in head induction. Nature 1998; 391:357-362.
154. Pera EM, De Robertis EM. A direct screen for secreted proteins in Xenopus embryos identifies distinct activities for the Wnt antagonists Crescent and Frzb-1. Mech Dev 2000; 96:183-195.
155. Ikeya M, Takada S. Wnt-3a is required for somite specification along the anteroposterior axis of the mouse embryo and for regulation of cdx-1 expression. Mech Dev 2001; 103:27-33.
156. Yoshikawa Y, Fujimori T, McMahon AP et al. Evidence that absence of Wnt-3a signaling promotes neuralization instead of paraxial mesoderm development in the mouse. Dev Biol 1997; 183:234-242.
157. Huelsken J, Vogel R, Brinkmann V et al. Requirement for β-catenin in anterior-posterior axis formation in mice. J Cell Biol 2000; 148:567-578.
158. Marvin, MJ, Di Rocco G, Gardiner A et al. Inhibition of Wnt activity induces heart formation from posterior mesoderm. Genes Dev 2001; 15:316-327.
159. Schneider VA, Mercola M. Wnt antagonism initiates cardiogenesis in Xenopus laevis. Genes Dev 2001; 15:304-315.
160. Beck CW, Slack JMW. Analysis of the developing Xenopus tail bud reveals separate phases of gene expression during determination and outgrowth. Mech Dev 1998; 72:41-52.
161. Beck CW, Slack JMW. A developmental pathway controlling outgrowth of the Xenopus tail bud. Development 1999; 126:1611-1620.

162. Itoh K, Sokol SY. Graded amounts of Xenopus dishevelled specify discrete anterioposterior cell fates in prospective ectoderm. Mech Dev 1997; 61:113-125.
163. Takada S, Stark KL, Shea MJ et al. Wnt-3a regulates somite and tailbud formation in the mouse embryo. Genes Dev 1994; 8:174-189.
164. Conlon RA, Reaume AG, Rossant J. Notch1 is required for the coordinate segmentation of somites. Development 1995; 121:1533-1545.
165. Smith JC, Price BMJ, Green JBA et al. Expression of a Xenopus homolog of Brachyury (T) is an immediate-early response to mesoderm induction. Cell 1991; 67:79-87.
166. Gont LK, Steinbeisser H, Blumberg B et al. Tail formation as a continuation of gastrulation: the multiple cell populations of the Xenopus tailbud derive from the late blastopore lip. Development 1993; 119:991-1004.
167. Dearden PK, Akam M. Early embryo patterning in the grasshopper, Schistocerca gregaria: wingless, decapentaplegic and caudal expression. Development 2001; 128:3435-3444.
168. Kispert A, Herrmann BG, Leptin M et al. Homologs of the mouse Brachyury gene are involved in the specification of posterior terminal structures in Drosophila, Tribolium, and Locusta. Genes Dev 1994; 8:2137-2150.
169. Reuter R. The T-related gene (Trg), a Brachyury homologue in insects. Sem Dev Biol 1995; 6:427-435.
170. Wu LH, Lengyel JA. Role of caudal in hindgut specification and gastrulation suggests homology between Drosophila amnioproctodeal invagination and vertebrate blastopore. Development 1998; 125:2433-2442.
171. Technau U, Bode HR. HyBra1, a Brachyury homologue, acts during head formation in Hydra. Development 1999; 126:999-1010.
172. Technau U. Brachyury, the blastopore and the evolution of the mesoderm. BioEssays 2001; 23:788-794.
173. Gallet A, Erkner A, Charroux B et al. Trunk-specific modulation of Wingless signalling in Drosophila by Teashirt binding to Armadillo. Curr Biol 1998; 8:893-902.
174. Gallet A, Angelats C, Erkner A et al. The C-terminal domain of Armadillo binds to hypophosphorylated Teashirt to modulate Wingless signalling in Drosophila. EMBO J 1999; 18:2208-2217.
175. Fasano L, Röder L, Coré N et al. The gene teashirt is required for the development of Drosophila embryonic trunk segments and encodes a protein with widely spaced zinc finger motifs. Cell 1991; 64:63-79.
176. Röder L, Vola C, Kerridge S. The role of the teashirt gene in trunk segmental identity in Drosophila. Development 1992; 115:1017-1033.
177. Caubit X, Coré N, Boned A et al. Vertebrate orthologues of the Drosophila region-specific patterning gene teashirt. Mech Dev 2000; 91:445-448.
178. Davis RL, Kirschner MW. The fate of cells in the tailbud of Xenopus laevis. Development 2000; 127:255-267.
179. Katsuyama Y, Sato Y, Wada S et al. Ascidian tail formation requires caudal function. Dev Biol 1999; 213:257-268.
180. Catala M, Teillet MA, Le Douarin NM. Organization of the tail bud analyzed with quail-chick chimaera system. Mech Dev 1995; 51:51-65.
181. Catala M, Teillet MA, De Robertis EM et al. A spinal cord fate map in the avian embryo: while regressing, Hensen's node lays down the notochord and floor plate thus joining the spinal cord lateral walls. Development 1996; 122:2599-2610.
182. Prinos P, Joseph S, Oh K et al. Multiple pathways governing Cdx1 expression during murine development. Dev Biol 2001; 239:257-269.
183. Yamaguchi TP, Bradley A, McMahon AP et al. A Wnt5a pathway underlies outgrowth of multiple structures in the vertebrate embryo. Development 1999b; 126:1211-1223.
184. Rauch GJ, Hammerschmidt M, Blader P et al. Wnt5 is required for tail formation in the zebrafish embryo. Cold Spring Harbor Symp Quant Biol 1997; 62:227-234.
185. Holland PWH, Holland LZ, Williams NA et al. An amphioxus homeobox gene: sequence conservation, spatial expression during development and insights into vertebrate evolution. Development 1992; 116:653-661.

186. Brooke NM, Garcia-Fernàndez J, Holland PWH. The ParaHox gene cluster is an evolutionary sister of the Hox gene cluster. Nature 1998; 392:920-922.
187. Panopoulou GD, Clark MD, Holland LZ et al. AmphiBMP2/4, an amphioxus bone morphogenetic protein closely related to Drosophila decapentaplegic and vertebrate BMP2 and BMP4: insights into evolution of dorsoventral axis specification. Dev Dyn 1998; 213:130-139.
188. Shimeld SM. The evolution of the hedgehog gene family in chordates: insights from amphioxus hedgehog. Dev Genes Evol 1999a; 209:40-47.
189. Heisenberg CP, Tada M, Rauch GJ et al. Silberblick/Wnt11 mediates convergent extension movements during zebrafish gastrulation. Nature 2000; 405:76-81.
190. Lawson KA, Meneses JJ, Pedersen RA. Clonal analysis of epiblast fate during germ layer formation in the mouse embryo. Development 1991; 113:891-911.
191. Parameswaran M, Tam PP. Regionalisation of cell fate and morphogenetic movement of the mesoderm during mouse gastrulation. Dev Genet 1995; 17:16-28.
192. Greco TL, Takada S, Newhouse MM et al. Analysis of the vestigial tail mutation demonstrates that Wnt-3a gene dosage regulates mouse axial development. Genes Dev 1996; 10:313-324.
193. Hsieh JC, Kodjabachian L, Rebbert ML et al. A new secreted protein that binds to Wnt proteins and inhibits their activities. Nature 1999; 398:431-436.
194. Domingos PM, Itasaki N, Jones CM et al. The Wnt/β-Catenin pathway posteriorizes neural tissue in Xenopus by an indirect mechanism requiring FGF signalling. Dev Biol 2001; 239:148-160.
195. Spörle R. Epaxial-adaxial-hypaxial regionalisation of the vertebrate somite: evidence for a somitic organiser and a mirror-image duplication. Dev Genes Evol 2001; 211:198-217.
196. Schmidt M, Tanaka M, Münsterberg A. Expression of β-catenin in the developing chick myotome is regulated by myogenic signals. Development 2000; 127:4105-4113.
197. Galceran J, Fari'as I, Depew MJ et al. Wnt3a$^{-/-}$like phenotype and limb deficiency in LEF-1$^{-/}$TCF-1$^{-/-}$ mice. Genes Dev 1999; 13:709-717.
198. Linask KK, Ludwig C, Han MD et al. N-cadherin/catenin-mediated morphoregulation of somite formation. Dev Biol 1998; 202:85-102.
199. Stern HM, Brown AMC, Hauschka SD. Myogenesis in paraxial mesoderm: preferential induction by dorsal neural tube and by cells expressing Wnt-1. Development 1995; 121:3675-3686.
200. Borello U, Buffa V, Sonnino C et al. Differential expression of the Wnt putative receptors Frizzled during mouse somitogenesis. Mech Dev 1999; 89:173-177.
201. Münsterberg AE, Kitajewski J, Bumcrot DA et al. Combinatorial signaling by Sonic hedgehog and Wnt family members induces myogenic bHLH gene expression in the somite. Genes Dev 1995; 9:2911-2922.
202. Tajbakhsh S, Borello U, Vivarelli E et al. Differential activation of Myf5 and MyoD by different Wnts in explants of mouse paraxial mesoderm and the later activation of myogenesis in the absence of Myf5. Development 1998; 125:4155-4162.
203. Cossu G, Borello U. Wnt signaling and the activation of myogenesis in mammals. EMBO J 1999; 18:6867-6872.
204. Wolda SL, Moon RT. Cloning and developmental expression in Xenopus laevis of seven additional members of the Wnt family. Oncogene 1992; 7:1941-1947.
205. Christiansen JH, Dennis CL, Wicking CA et al. Murine Wnt-11 and Wnt-12 have temporally and spatially restricted expression patterns during embryonic development. Mech Dev 1995; 51:341-350.
206. Kozmik Z, Holland LZ, Schubert M et al. Characterization of amphioxus AmphiVent, an evolutionarily conserved marker for chordate ventral mesoderm. Genesis 2001; 29:172-179.
207. Holland LZ, Kene M, Williams NA et al. Sequence and embryonic expression of the amphioxus engrailed gene (AmphiEn): the metameric pattern of transcription resembles that of its segment-polarity homolog in Drosophila. Development 1997; 124:1723-1732.
208. Aulehla A, Johnson RL. Dynamic expression of lunatic fringe suggests a link between notch signaling and an autonomous cellular oscillator driving somite segmentation. Dev Biol 1999; 207:49-61.
209. Takke C, Campos-Ortega JA. her1, a zebrafish pair-rule like gene, acts downstream of notch signalling to control somite development. Development 1999; 126:3005-3014.
210. Jouve C, Palmeirim I, Henrique D et al. Notch signalling is required for cyclic expression of the hairy-like gene HES1 in the presomitic mesoderm. Development 2000; 127:1421-1429.

211. Münsterberg AE, Lassar AB. Combinatorial signals from the neural tube, floor plate and notochord induce myogenic bHLH gene expression in the somite. Development 1995; 121:651-660.
212. Yasuo H, Lemaire P. Role of Goosecoid, Xnot and Wnt antagonists in the maintenance of the notochord genetic programme in Xenopus gastrulae. Development 2001; 128:3783-3793.
213. Klingensmith J, Nusse R. Signaling by wingless in Drosophila. Dev Biol 1994; 166:396-414.
214. Patel NH. The evolution of arthropod segmentation: insights from comparisons of gene expression patterns. Development 1994; 120S:201-207.
215. Shimeld SM. The evolution of dorsoventral pattern formation in the chordate neural tube. Am Zool 1999b; 39:641-649.
216. Prud'homme B, De Rosa R, Julien JF et al. Is segmentation ancestral? Looking for conserved segmentation genes in annelids. Abst Ann Meeting SICB 2002; 2002:351.
217. Wedeen CJ, Weisblat DA. Segmental expression of an engrailed-class gene during early development and neurogenesis in an annelid. Development 1991; 113:805-814.
218. Lans D, Wedeen CJ, Weisblat DA. Cell lineage analysis of the expression of an engrailed homolog in leech embryos. Development 1993; 117:857-871.
219. Bely AE, Wray GA. Evolution of regeneration and fission in annelids: insights from engrailed- and orthodenticle-class gene expression. Development 2001; 128:2781-2791.
220. Seaver EC, Paulson DA, Irvine SQ et al. Variation in segment formation in polychaetes and its relationship to life history characteristics. Abst Ann Meeting SICB 2002; 2002:380.
221. Wedeen CJ, Kostriken RG, Leach D et al. Segmentally iterated expression of an engrailed-class gene in the embryo of an australian onychophoran. Dev Genes Evol 1997; 270:282-286.
222. Budd GE. Why are arthropods segmented? Evol Dev 2001; 3:332-342.
223. Lawrence PA, Johnston P. On the role of the engrailed+ gene in the internal organs of Drosophila. EMBO J 1984; 3:2839-2844.
224. Baylies MK, Martinez Arias A, Bate M. wingless is required for the formation of a subset of muscle founder cells during Drosophila embryogenesis. Development 1995; 121:3829-3837.
225. Frasch M, Nguyen HT. Genetic control of mesoderm patterning and differentiation during Drosophila embryogenesis. Adv Dev Biochem 1999; 5:1-47.
226. Riechmann V, Irion U, Wilson R et al. Control of cell fates and segmentation in the Drosophila mesoderm. Development 1997; 124:2915-2922.
227. Bilder D, Scott MP. Hedgehog and wingless induce metameric pattern in the Drosophila visceral mesoderm. Dev Biol 1998; 201:43-56.
228. Lawrence PA, Johnston P, Vincent JP. Wingless can bring about a mesoderm-to-ectoderm induction in Drosophila embryos. Development 1994; 120:3355-3359.
229. Sasakura Y, Makabe KW. Ascidian Wnt-5 gene is involved in the morphogenetic movement of notochord cells. Dev Growth Differ 2001; 43:573-582.
230. Locascio A, Nieto MA. Cell movements during vertebrate development: integrated tissue behaviour versus individual cell migration. Curr Opin Genet Dev 2001; 11:464-469.
231. Moon RT, Campbell RM, Christian JL et al. Xwnt-5a: A maternal Wnt that affects morphogenetic movements after overexpression in embryos of Xenopus laevis. Development 1993; 119:97-111.
232. Sokol SY. Analysis of Dishevelled signalling pathways during Xenopus development. Curr Biol 1996; 6:1456-1467.
233. Wallingford JB, Rowning BA, Vogeli KM et al. Dishevelled controls cell polarity during Xenopus gastrulation. Nature 2000; 405:81-85.
234. Wallingford JB, Harland RM. Xenopus Dishevelled signaling regulates both neural and mesodermal convergent extension: parallel forces elongating the body axis. Development 2001; 128:2581-2592.
235. Djiane A, Riou J, Umbhauer M et al. Role of frizzled 7 in the regulation of convergent extension movements during gastrulation in Xenopus laevis. Development 2000; 127:3091-3100.
236. Wallingford JB, Vogeli KM, Harland RM. Regulation of convergent extension in Xenopus by Wnt5a and Frizzled-8 is independent of the canonical Wnt pathway. Int J Dev Biol 2001; 45:225-227.
237. Strapps WR, Tomlinson A. Transducing properties of Drosophila Frizzled proteins. Development 2001; 128:4829-4835.

238. Axelrod JD, Miller JR, Shulman JM et al. Differential recruitment of Dishevelled provides signaling specificity in the planar cell polarity and Wingless signaling pathways. Genes Dev 1998; 12:2610-2622.
239. Boutros M, Mihaly J, Bouwmeester T et al. Signaling specificity by Frizzled receptors in Drosophila. Science 2000; 288:1825-1828.
240. Joyner AL, Hanks M. The engrailed genes: Evolution of function. Sem Dev Biol 1991; 2:435-445.
241. McGrew LL, Otte AP, Moon RT. Analysis of Xwnt-4 in embryos of Xenopus laevis: a Wnt family member expressed in the brain and floor plate. Development 1992; 115:463-473.
242. Liu A, Joyner AL. Early anterior/posterior patterning of the midbrain and cerebellum. Annu Rev Neurosci 2001; 24:869-896.
243. McMahon AP, Bradley A. The Wnt-1 (int-1) proto-oncogene is required for development of a large region of the mouse brain. Cell 1990; 62:1073-1085.
244. Thomas KR, Capecchi MR. Targeted disruption of the murine int-1 proto-oncogene resulting in severe abnormalities in midbrain and cerebellar development. Nature 1990; 346:847-850.
245. McMahon AP, Joyner AL, Bradley A et al. The midbrain-hindbrain phenotype of $Wnt\text{-}1^{-}/Wnt\text{-}1^{-}$ mice results from stepwise deletion of engrailed-expressing cells by 9.5 days postcoitum. Cell 1992; 69:581-595.
246. Danielian PS, McMahon AP. Engrailed-1 as a target of the Wnt-1 signalling pathway in vertebrate midbrain development. Nature 1996; 383:332-334.
247. McGrew LL, Takemaru KI, Bates R et al. Direct regulation of the Xenopus engrailed-2 promoter by the Wnt signaling pathway, and a molecular screen for Wnt-responsive genes, confirm a role for Wnt signaling during neural patterning in Xenopus. Mech Dev 1999; 87:21-32.
248. Holland LZ, Holland ND. Chordate origins of the vertebrate central nervous system. Curr Opin Neurobiol 1999; 9:596-602.
249. Kozmik Z, Holland ND, Kalousova A et al. Characterization of an amphioxus paired box gene, AmphiPax2/5/8: developmental expression patterns in optic support cells, nephridium, thyroid-like structures and pharyngeal gill slits, but not in the midbrain-hindbrain boundary region. Development 1999; 126:1295-1304.
250. Wada H, Saiga H, Satoh N et al. Tripartite organization of the ancestral chordate brain and the antiquity of placodes: insights from ascidian Pax-2/5/8, Hox and Otx genes. Development 1998; 125:1113-1122.
251. LaBonne C, Bronner-Fraser M. Neural crest induction in Xenopus: evidence for a two-signal model. Development 1998; 125:2403-2414.
252. Ikeya M, Lee SMK, Johnson JE et al. Wnt signalling required for expansion of neural crest and CNS progenitors. Nature 1997; 389:966-970.
253. Deardorff MA, Tan C, Saint-Jeannet JP et al. A role for frizzled 3 in neural crest development. Development 2001; 128:3655-3663.
254. Hall AC, Lucas FR, Salinas PC. Axonal remodeling and synaptic differentiation in the cerebellum is regulated by WNT-7a signaling. Cell 2000; 100:525-535.
255. Richter S, Hartmann B, Reichert H. The wingless gene is required for embryonic brain development in Drosophila. Dev Genes Evol 1998; 208:37-45.
256. Bhat KM, van Beers EH, Bhat P. Sloppy paired acts as the downstream target of Wingless in the Drosophila CNS and interaction between sloppy paired and gooseberry inhibits sloppy paired during neurogenesis. Development 2000; 127:655-665.
257. Song Y, Chung S, Kunes S. Combgap relays wingless signal reception to the determination of cortical cell fate in the Drosophila visual system. Mol Cell 2000; 6:1143-1154.
258. Deshpande N, Dittrich R, Technau GM et al. Successive specification of Drosophila neuroblasts NB 6-4 and NB 7-3 depends on interaction of the segment polarity genes wingless, gooseberry and naked cuticle. Development 2001; 128:3253-3261.
259. Kaphingst K, Kunes S. Pattern formation in the visual centers of the Drosophila brain: wingless acts via decapentaplegic to specify the dorsoventral axis. Cell 1994; 78:437-448.
260. Holland LZ. Body-plan evolution in the Bilateria: early antero-posterior patterning and the deuterostome-protostome dichotomy. Curr Opin Genet Dev 2000; 10:434-442.
261. Holland LZ. Heads or tails? Amphioxus and the evolution of anterior-posterior patterning in deuterostomes. Dev Biol 2002; 241:209-228.

# CHAPTER 14

# Wnt Signaling and Cell Migration

**Alexandra Schambony and Doris Wedlich**


## Abstract

The significance of Wnt signaling for cell migration came into play most recently, and with it, a burst displaying the diversity of Wnt family members in signal transduction and physiological function. Our most comprehensive knowledge of how Wnt signaling controls cell movement becomes obvious in the gastrulation process of vertebrates. Here, the role of Wnt molecules in driving cell migration can be separated from their role in regulating cell differentiation. This discrimination is difficult to make for other migration processes, e.g., neural crest or myocyte migration. Therefore, the role of the different Wnt signaling cascades in gastrulation movements are covered in the main part of this article while other cell migration processes are touched only briefly.


## Introduction

Cell movements during gastrulation have been extensively studied in the amphibian, *Xenopus laevis*, taking advantage of the size of the embryo and its extracorporal development that allows manipulations and explant observations at very early stages of embryogenesis. Pionieering work in the creation of assays to study the different cell behaviors was contributed by Keller and Winklbauer.[1,2,3] With these assays they were able to distinguish between the epiboly of the ectoderm, the active migration of the head mesoderm, the convergent extension movement of the involuting mesoderm and the rotation of the endoderm. Although gastrulation movements express species-specific characteristics, reflecting the topography of yolk and cytoplasm distribution in the egg, common patterns of cell behaviour are observed in different organisms and they are repeatedly used during morphogenesis of an individual. Convergent extension movements, for example, occur ubiquitously in metazoan development. They drive gastrulation but also elongation of the embryo in vertebrates, archenteron elongation in sea urchins, germ band extension and imaginal disc evagination in *Drosophila*. In all these cases, convergence and extension describe mass tissue movements based on narrowing and lengthening of cell populations. Therefore, it seems most likely that the regulative mechanisms inducing and controlling these cell behaviours are also conserved.

Different Wnt subfamily members seem to be involved in controlling gastrulation movement. At first view, the reports appear confusing as increase in Wnt signaling activity but also its inhibition result in disturbance of the migration process. This contradiction may be solved when the complex pattern of tissue movements is temporally and spatially sorted into the individual cell behaviours of the different tissue compartments.

*Wnt Signaling in Development*, edited by Michael Kühl.

## *A Moderate Wnt/PCP Signaling Drives Convergent Extension*

The best example of contradictory observations is given by Wnt-11 which activates the Wnt/PCP (planar cell polarity) pathway (see chapters 1 and 6 in this issue). In *Xenopus*, either activation or inhibition of this pathway by exogenous expression of wildtype (wt) or mutants of the signaling molecule, its receptor or downstream effectors lead to gastrulation defects. [4-8] The induction of cell polarity is one of the first crucial steps in convergent extension movements as cells from both lateral sides of an embryo intercalate and migrate towards the dorsal midline (Fig. 1). Performing time-lapse studies with *Xenopus* open-face Keller explants Wallingford et al [6] were able to demonstrate the consequence of defect Wnt/PCP signaling on cellular level. Expression of the Xdsh mutant (Xdd1) or overexpression of wt Xdsh led to loss of cell polarity formation. The cells showed randomly distributed protrusive activity of high frequency, a behavior normally observed shortly before convergent extension starts (Fig. 2, time point 0). In controls, lamellopodia formation was mediolaterally biased and more stable reflecting the bipolar status (Fig. 2, time point 15-20). An influence of Wnt/β-catenin signaling could be excluded because dnGSK3β failed in re-establishing cell polarity. Interestingly, a rescue effect was observed when wt Xdsh and the mutant were coexpressed. This indicates that a balanced level of Wnt/PCP signaling activity is required because reduced or increased activity led to the same defect. A similar observation was made by Tada and Smith[7] who blocked Wnt/PCP signaling and convergent extension movements by expressing dn-Xwnt-11. Rescue was only achieved by a moderate concentration of Xwnt-11. The idea of a balanced Wnt-11 signaling activity explains previous findings that overexpression of Xwnt-11 blocks convergent extension. [4]

In zebrafish Wnt-11 is encoded by *silberblick (sbl)*. Cell tracking in the *silberblick* mutant revealed that convergent extension movement is affected. [9] This phenotype is rescued by the dsh mutant (dsh-ΔN) which is not able to mediate Wnt/β-catenin signaling but which still functions in the Wnt/PCP pathway. Transplantation of cells between wt and slb-/- embryos additionally revealed that the required Wnt-11 signal acts in a cell non-autonomous manner. The glypican knypek potentiates the Wnt-11 effect on convergent extension movements,[10] most likely by enhancing the ligand receptor interaction. BMP, on the other hand, restricts convergent extension movements by downregulating Wnt-11 and Wnt-5a.[11]

Posttranslational modifications modulating cell-adhesion or cytoskeleton rearrangements but also JNK dependent gene regulation are discussed as downstream effects of Wnt/PCP signaling (see chapter 6 this issue, see also ref. 12). The Xwnt-11-dependent cell polarity formation in the vegetal alignment zone[6] and also the rescue effects by dominant negative cdc42 in case of Xwnt-11 or XFz-7 overexpression[5] argues for a short-term posttranslational regulation. This does not rule out that long-term effects drive convergent extension at the neurula stage. Strabismus (stbm), an interacting partner of dsh in Wnt/PCP signaling, mediates c-Jun and AP-1 dependent transcription of neural marker genes. Expression of a wt-stbm or a dn-stbm inhibited elongation of activin-treated animal caps.[8] Since primary *stbm* expression starts in early neurula and defects of morphogenetic movements in gain or loss of function experiments are deduced from shortened body axis and reduced inter-somite spacing, it seems most likely that long-term effects of Wnt/PCP signaling were monitored here.

In most reports on *Xenopus* gastrulation the effects of Wnt signaling on convergent extension movements were measured in open-face explants after several hours at the end of the elongation process. In these reports, neither distinction between different types of cell polarity formation (bipolar/mesoderm vs. monopolar/neural-notoplate) was made nor contribution of different tissue layers considered (Fig. 2). The analysis of convergent extension movement in activin-treated animal caps[4,5,8] is also handicapped, as this assay poorly reflects the tissue architecture and the combination of the endogenous signaling cascades in the dorsal marginal zone. Depending on the experimental conditions, Wnt signaling cannot be allocated to specific cell

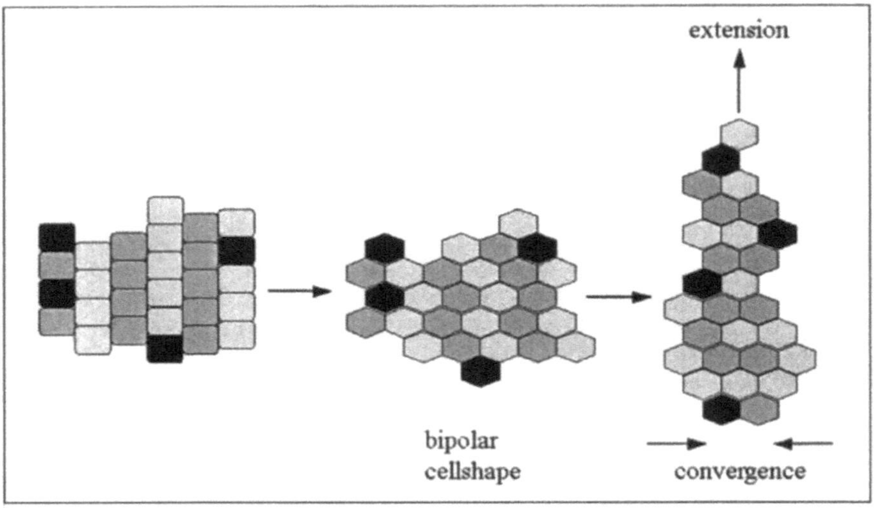


Figure 1. Mediolateral intercalation of the involuting mesoderm. Cells from both lateral sides polarise and migrate towards the dorsal midline and intermingle at the midline which results in elongation. Four cells are labeled black to facilitate orientation.

behaviours or phases within the time course of convergent extension movements summarised in Figure 2. There are two exceptions: Wallingford et al.[6] showed that Wnt/PCP signaling induces the formation of the bipolar cell shape in the deep cells of the dorsal marginal zone when the vegetal alignment zone (VAZ) is formed (Fig. 2). Moreover, Wallingford and Harland[13] used targeted expression of Xdsh mutants and Keller sandwich explants to demonstrate that the Wnt/PCP cascade also controls the extension of the neural plate. However, it remains to be clarified whether Wnt/PCP induces monopolar polarity in lateral neural plate cells. In the future, more effort should be made to distinguish between these short- and long-term effects by analysing cell behaviour at different time points and in different tissue layers of explants.

## *Canonical Wnt/β-Catenin Signaling Is Required for Proper Convergent Extension*

Overexpression of components of the canonical signaling pathway in *Xenopus* never led to disturbed gastrulation movement or inhibition of elongation of the embryo. However, blocking the Wnt/β-catenin pathway by overexpression of either secreted forms of Xfz-8[14], naturally occuring Wnt antagonists[15,16] or dominant negative forms of the transcription factor LEF-1[16] resulted in inhibition of convergent extension movements. Most strikingly, a branching of the canonical Wnt signaling cascade downstream of β-catenin was observed in *Xenopus*. LEF-1 or the target gene Xnr3 restored convergent extension movements in embryos with blocked canonical Wnt pathway while TCF-3 or the target gene siamois did not. These findings implicate that gene expression, at least *Xnr3* expression, is necessary.[16]

Wnt/β-catenin sigaling also regulates cell movements in zebrafish gastrulation.[17] Interestingly, accumulation of β-catenin leads to tyrosine phosphorylation and nuclear localization of Stat3. This activation of Stat3 is TCF-3 dependent, but does not involve the induction of Dharma/Bozozok, Nodal or Wnt-11, the target genes of maternal canonical wnt signaling in zebrafish. Following the migration behavior of prechordal and lateral mesodermal cells with defect Stat3 function in wt embryos, or of wt mesodermal cells in Stat3 deficient host embryos,

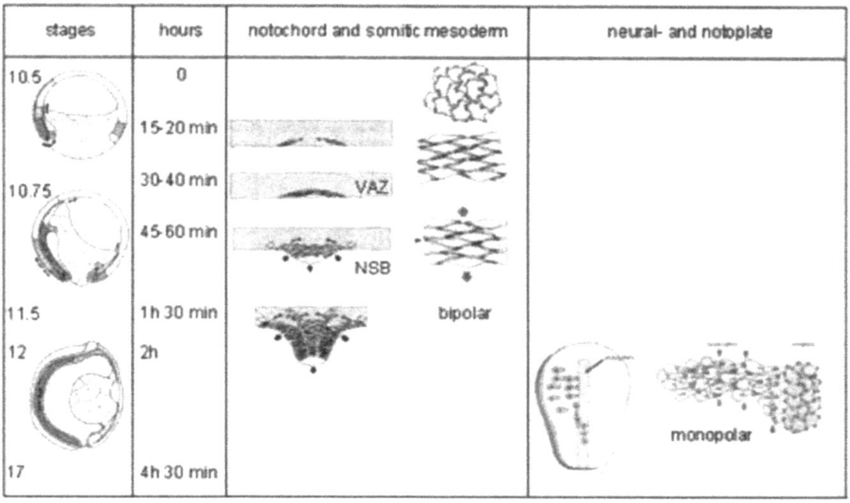


Figure 2. Time-course of convergent extension movements in *Xenopus laevis* according to references 1 and 52. The first half of gastrulation is dominated by the mediolateral intercalation of the involuting presumptive notochord and somitic mesoderm cells that adopt a bipolar shape with intensive protusive activity at both ends. At the end of gastrulation and during neurulation, elongation of the embryo is promoted by medially orientated monopolar protrusive activity of the lateral neural plate cells and ubiquitous lamellopodia formation of the notoplate cells. VAZ: Vegetal Alignment Zone, NSB: Notochord Somite Boundary.

Yamahita et al.[17] demonstrated that Stat3 regulates migration of the prechordal mesoderm cell autonomously while it affects convergent extension movements of the lateral mesoderm in cell non-autonomous manner.

Canoncial wnt signaling seems to affect cell behavior in *Xenopus* and zebrafish very differently. A cell autonomous influence of Wnt/β-catenin on active migration of the anterior mesoendoderm in *Xenopus* has not been shown yet. Instead, FGF and activin promote this substrate dependent cell migration.[18] A cell non-autonomous Wnt/β-catenin influence on convergent extension movements could be assumed as the required target gene Xnr3, a TGFß family member, is a secreted molecule. In zebrafish, however, the *Xnr3* homolog *Nodal* is not necessary for cell movement. The discrepancies between both organisms in regulation of cell motility might reflect that the corresponding target genes are still unknown and/or that different tissues are compared, e.g., the lateral mesoderm in zebrafish with the axial and paraxial mesoderm or the neural plate in *Xenopus*.

## Wnt/$Ca^{2+}$-Signaling Blocks Convergent Extension Movement

Overexpression of Xwnt-5a was found to block convergent extension movements in Xenopus.[19] PKC and a constitutively active form of CamKII mimick the Xwnt-5a phenotype confirming the idea of a separate Wnt/$Ca^{2+}$ pathway.[16,20] PKC and CamKII antagonize the canonical Wnt pathway at different levels. PKC blocks the pathway upstream of β-catenin, whereas CamKII blocks it downstream of β-catenin.[16] Since CamKII is more active at the ventral side in the Xenopus embryo,[20] convergent extension movements will be suppressed in the ventral compartment. Activation of CamKII was also observed by overexpression of Xwnt-11 raising the question whether the Wnt/$Ca^{2+}$ and the Wnt/PCP signaling cascades are separate pathways. In this context it is interesting to note that strabismus, a component of the PCP pathway, blocks Wnt/β-catenin dependent gene expression.[8] The Rho GTPase Cdc42 acts downstream of Xwnt-5a and PKC-α, because the inhibition of convergent extension movements by overexpression of Xwnt-5a or PKC-α was rescued by co-expression of dnXcdc42.[21] Dn Xcdc42 was also able to rescue XFz-7 or Xwnt-11 overexpression, both molecules were assigned to the Wnt/PCP-signaling.[5] This implicates that cdc42 might function in both pathways or a separation between the Wnt/$Ca^{2+}$ and the Wnt/PCP pathway appears to be artificial.

Intracellular calcium waves have been shown to occur during gastrulation in zebrafish[22,23] and *Xenopus*.[24] Blocking the release of calcium from intracellular stores resulted in inhibition of convergent extension movements in *Xenopus* open-face Keller explants. The mechanism of calcium wave initiation and propagation, however, remained unclear. A role of Wnt signaling was proven by expression of an N-terminal fragment of Frizzled-8 (NXFz-8) and yielded little influence, because only the frequency of the waves were found slightly reduced.[24] Further studies by modulating Wnt/$Ca^{2+}$-signaling components will answer this question.

## *Formation of Brachet's Cleft Depends on Wnt Signaling*

In *Xenopus* gastrulation, the anterior involuting mesoderm, also termed head mesoderm, exerts a separate type of migration behaviour. These cells migrate as cohort on the blastocoel roof and use the extracellular matrix, predominantly the fibronectin fibrils, for orientation towards the animal pole.[2] During the involution process they alter their surface properties and lose the ability to integrate into the blastocoel tissue after they have moved inside. This results in the formation of the Brachet's cleft which separates the head mesoderm from the blastocoel roof. Downregulation of XFz-7 by anti-sense morpholino injection resulted in loss of Brachet's cleft formation and abolished the separation behaviour of the head mesoderm cells while convergent extension movements were not blocked.[25] The latter result is unexpected as overexpression of full length or an extracellular fragment of XFz-7 inhibited elongation of activin-treated animal caps.[5] The disturbed separation behaviour upon XFz-7 downregulation could neither be restored by activation of the canonical pathway nor of the PCP signaling pathway which led Winklbauer et al.[25] to speculate about a further branching of Wnt signaling pathways. However, the separation behavior was pertussis toxin sensitive and both, disturbance by the toxin or by XFz-7-receptor decrease was rescued by PKC-α expression. This makes a contribution of the Wnt/$Ca^{2+}$-pathway most likely, particularly, because is has been shown that XFz-7 can signal in the canonical but also in the noncanonical, PKC-α dependent manner.

## *Wnt Signals Controlling Migration of Neuroblasts and Neural Crest Cells*

Wnt signals are known to play a role in the specification and patterning of the nervous system. In *C. elegans*, Wnt signals play a crucial role in differentiation and migration of neuroblasts. During the first larval stage, the P11/P12 pair of ventral epidermal blast cells migrates to align along the antero-posterior axis. This rotation is biased for the P1/P2 and the P11/P12 pairs. In the latter case, 85% of the right cells adopt the fate of the posterior P12 cell. Although lin-44/Wnt signaling does not change this ratio, disturbed Wnt signaling results in a loss of positonal information. In these mutants, both cells are driven into the P11 lineage[26,27] and lack P12.a-derived neurons and P12.p descendants.

Work by Maloof et al.[28] shows that Wnt *signaling* determines cell fate of Q-cell descendants. The QL cell and its descendants migrate posteriorly while the QR cell migrates anteriorly. These opposite migrations are due to canonical Wnt signals. The *egl-20*/Wnt signaling pathway involves *bar-1*, a β-catenin homologue[29] and POP-1/TCF.[30] In the QL cell egl20/Wnt signaling is inhibited by *pry-1*, leading to posterior migration. In the QR cell, Wnt signaling induces the expression of the Hox gene *mab-5*, which has been shown to be sufficient to

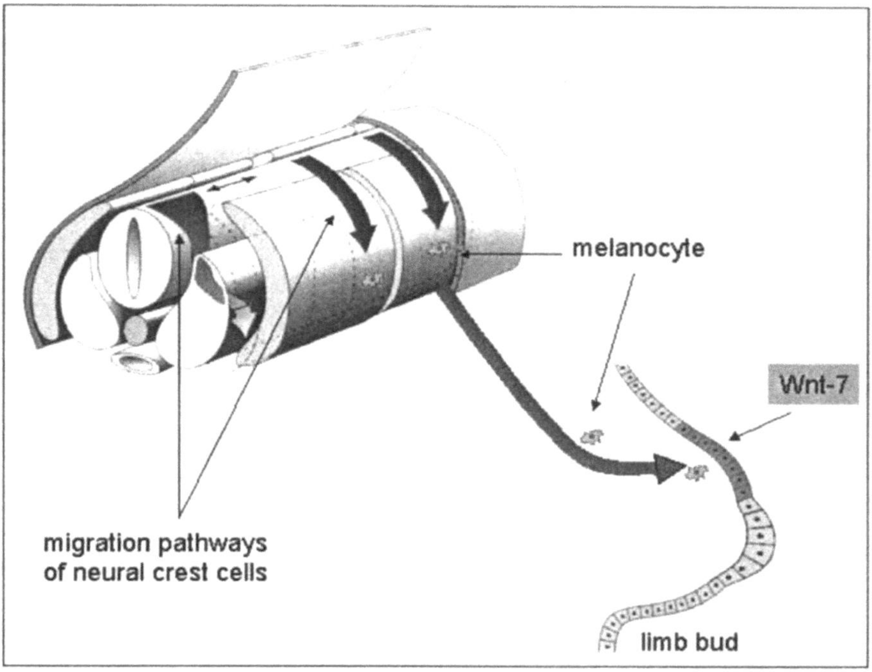


Figure 3. Migrational pathways of neural crest cells. Melanocytes migrate between the myotome and the epidermis, while cells that adopt a neural fate migrate beneath the myotome. When melanocytes reach the developing limb bud, a Wnt-7a signal directs their migration towards the dorsal side of the extremities.

reverse the direction of cell migration. In the absence of *pry-1*, both Q-descendants behave like the wild-type QR cell.[28]

In vertebrates, Wnt signals contribute to the patterning of the brain and the peripheral nervous system. Wnt-1 and Wnt-3A are expressed along the dorsal midline in the mouse.[31,32] In this region neural crest cells derive from neural precursors and start to emigrate from the neural tube towards their destined position in the embryo (see Fig. 3). Neural crest cells contribute to a variety of structures, e.g., they form part of the peripheral nervous system, ganglia and Schwann cells. Another subset of neural crest cells differentiates into melanocytes and migrates all over the body to form a specific pigmentation pattern.

Wnt signals have been shown to be involved in neural crest specification. Wnt-1 and Wnt-3a promote melanocyte formation at the expense of neuronal cell fates in the zebrafish.[33] Later studies indicate a role of canonical Wnt signaling not only in the specification but also in migration of melanocytes. Wnt expression is maintained from the onset of neural crest emigration until the completion of melanocyte migration in the avian embryo.[34] Thus, Wnts are expressed in the right place and at the right time to influence neural crest cell specification and migration. In vitro experiments showing a correlation between Wnt signaling and neural crest migration[35] support this assumption.

The in vivo situation seems to be more complex. Double knockout mice, which lack both Wnt-1 and Wnt-3A, show significant reduction or even absence of neural crest cells. This seems to be due to defects in the induction of neural crest cell fates, as marker gene expression

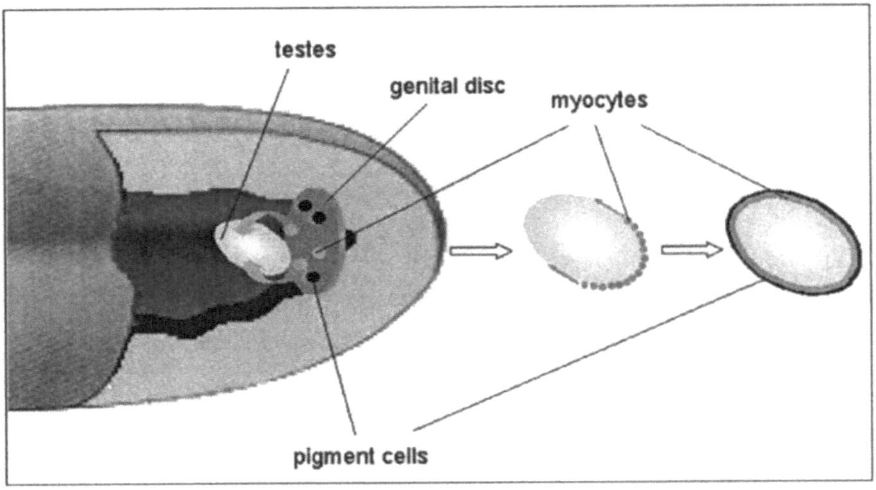


Figure 4. Formation of the muscle and pigment cell sheath of the Drosophila testes. Myocytes and pigment cells emigrate from the genital disc upon contact with the gonad. Myocytes form an enveloping layer over the testes, which in turn is coated by a sex specific layer of pigment cells.

is also altered. A decrease in CRABP-1 as well as AP-2 positive migrating cells was observed, while AP-2 expression persisted in the neural tube. Additionally, a dramatic reduction in Pax-3 expression was noted[36]. Taking into account that Splotch-mice which lack functional Pax-3 show severe defects in neural crest formation probably due to disturbed cell-cell interactions[37], these findings point to an indirect influence of Wnt signals in neural crest migration.

The importance of proper cell-cell interactions is shown by the effect of overexpressing cadherin-11 or deletion mutants of cadherin-11 in *Xenopus*. Overexpression of cadherin-11 constructs containing the extracellular domain, thus enhancing cell adhesion, completely suppresses cranial neural crest cell migration and promotes a neuronal cell fate. Disturbance of cell adhesion by expression of a extracellular deleted mutant leads to premature migration and suppression of twist-expression. The latter is due to a depletion of the endogenous ß-catenin pool and is rescued by co-expression of β-catenin. This strongly indicates the importance of canonical Wnt signaling for the maintenance of cranial neural crest cell fate during migration.[38] Recent studies in chicken embryos support a differential mechanism of specification and migration for neural crest cell subtypes. For melanocytes few TRP-2 (tyrosinase-related protein-2)[39] positive precursor cells form, but are unable to emigrate from the neural tube.[36] Emigration itself follows a certain time course depending on the neural crest cell subtype. Early neural crest cells which differentiate into neuronal and glial precursors are formed under the control of BMP-4 although Wnts are expressed in the neural tube from the very beginning of neural crest differentiation. These neural and glial precursors secrete the Wnt antagonist cFrzB-1. In later stages of neural crest formation, the expression of BMP-4 and cFrzB-1 ceases with the onset of melanocyte migration. The persisting Wnt-signal drives the later neural crest cells into the melanocyte lineage and also seems to be involved in the emigration of melanocyte precursors from the dorsal midline. Furthermore, Wnt-7A directs melanocyte localisation to the dorsal side of developing limbs (Fig. 3). Wnt-7A is expressed on the dorsal side of growing limb buds, while its expression on the ventral side is repressed by engrailed-1. Ectopic expression of Wnt-7A on the ventral side induces dorsal patterning including pigmentation, thus implicating that Wnt-7A signals are involved in the direction of migrating melanocytes to the dorsal side of extremities.[34]

## Wnt Signals Influence the Migrational Behavior of Myocytes

Wnt-1 and Wnt-3 in the chicken[40,41] as well as Wg in *Drosophila*[42,43] have been shown to induce myogenesis. Besides this determining role, Wnts play an additional role in the control of myocyte motility.

In the mouse Wnt-5A and Wnt-3A induce expression of N-cadherin in cardiac myocytes and of E- and M-cadherin in fibroblasts. The enhanced cadherin-mediated cell adhesion seems to be a prerequisite for the formation of myocyte aggregates in the presence of fibroblasts. Therefore, Wnt-5A or Wnt-3A signals decrease the mobility of cardiac myocytes in the mouse and favor their aggregation.[44]

In *Drosophila* DWnt-2 was shown to control myocyte migration in the male genital tract. DWnt-2 expression patterns are sexually dimorphic. Expression in the genital disc and gonadal mesodermal cells is observed only in males. During gonad development, the gonad and the genital disc contact and fuse with each other. In the male, the genital disc forms the external organs of the genital tract and a pigmented sheath that surrounds the testes and the seminal vesicle. In females the genital sheath is not pigmented. The myocytes and the pigment cell precursors in males are formed in the genital disc, emigrate from the disc as it is contacted by the gonad and subsequently migrate all over the gonad and finally envelop it (Fig. 4).

In DWnt-2 null mutants, the formation of the muscle sheath enclosing the reproductive tract is selectively disturbed in males. DWnt-2 induces the differentiation of the male specific pigment cells and the migration of myocytes from the genital disc to and over the testis. Although the molecular mode of this effect has not yet been clarified, it is presumed that DWnt-2 might be involved in the induction of myocyte migration from the genital disc, their attachment to the testis or both events.[45] An even more general role of DWnt-2 in the control of myocyte migration is indicated by a recent study. It has been shown that in DWnt-2 mutants the direct flight muscles are either missing or fail to attach to the correct sites. DWnt-2, which is expressed in the adjacent epithelial cells, might act as a directional signal for myocyte migration.[46]

### *Canonical Wnt Signaling Controls Tracheal Cell Migration*

The *Drosophila* tracheal system originates from ten pairs of ectodermal placodes. The tracheal cells first invaginate from the placodes to form tracheal sacs. The five primary tracheal branches (dorsal branch, anterior and posterior dorsal trunk branch, visceral branch, lateral trunk branch and ganglionic branch) grow out from these sacs. Further branching leads to the formation of a complex network. The invagination of the cells from placodes and the formation of the five primary branches are influenced by different signals: FGF, decapentaplegic (Dpp, a TGF$\beta$ homolog), EGFR signaling, activation of the MAPK pathway and Wnt/Wg. While FGF signaling is required for the migration of all tracheal cells from the cellular sacs, Dpp and Wnt/Wg each regulate the migration of cells to specific primary branches. Failure of Dpp signaling causes the loss of the dorsal branch, lateral trunk branch and ganglionic branch.[47,48] Wnt/Wg signaling directs the cells forming the anterior and posterior dorsal trunk branches.[4950] In *Drosophila* mutants with defects in Wnt/Wg or downstream components of the canonical Wnt signaling pathway, the cells of the dorsal trunk branches fail to leave the transverse connective region. As this phenotype is stronger if downstream components like $\beta$-catenin/armadillo or TCF are affected, Wg might not be the only Wnt molecule involved in the regulation of tracheal cell migration.[50] Additionally, it has recently been shown that ribbon, a BTB/POZ protein, acts in parallel to or downstream of FGF and Wnt/Wg signaling to promote directed cell migration. If one assumes a crosstalk between ribbon and the Wnt/Wg pathway downstream of Wg itself, mutations of downstream components should show stronger phenotypes because of the failure to respond to ribbon.[51]

## Perspectives

In summary, the state of the art on Wnts function in cell migration yields more open questions than answers. The reports provide evidence that different Wnt signaling pathways control cell motility in various morphogenetic movements. While Wnt/PCP seems to act by confering polarity to cells, the roles of the Wnt/β-catenin or Wnt/$Ca^{2+}$ pathways on cellular level are completely unknown. In addition, the studies reveal that different pathways can influence each other, and further branching points in the pathways seem to exist. The current aim is to decipher the network of downstream signal transducers and effectors, predominantly in the Wnt/PCP and Wnt/$Ca^{2+}$ cascade, and to link them with cellular events driving motility. The future challenge will be to depict the spatiotemporal activity of regulators in mass tissue movements that regulates adhesive strength, cytoskeleton re-arrangements, orientation and polarity of cells in different tissue compartments.

### *Acknowledgements*

We thank Dr. K. Astrahantseff for helpful discussions and critical reading of the manuscript.

## References

1. Keller R, Davidson L, Edlund A et al. Mechanisms of convergence and extension by cell intercalation. Phil Trans R Soc Lond B 2000; 355:897-922.
2. Winklbauer R, Nagel M, Selchow A, Wacker S. Mesoderm migration in the Xenopus gastrula. Int J Dev Biol 1996; 40:305-311.
3. Winklbauer R, Schurfeld M. Vegetal rotation, a new gastrulation movement involved in the internalization of the mesoderm and endoderm in Xenopus. Development 1999; 126:3703-3713.
4. Du JS, Purcell SM, Christian JL et al. Identification of Distinct Classes and Functional Domains of Wnts through Expression of Wild-Type and Chimeric Proteins in Xenopus Embryos. Mol Cell Biol 1995; 15:2625-2634.
5. Djiane A, Riou JF, Umbhauer M et al. Role of frizzled 7 in the regulation of convergent extension movements during gastrulation in Xenopus laevis. Developmenz 2000; 127:3091-3100.
6. Wallingford JB, Rowning BA, Vogeli KM et al. Dishevelled controls cell polarity during Xenopus gastrulation. Nature 2000; 405:81-85.
7. Tada M, Smith JC. Xwnt11 is a target of Xenopus Brachyury: regulation of gastrulation movements via Dishevelled, but not through the canonical Wnt pathway. Development 2000; 127,2227-2238.
8. Park M, Moon, RT. The planar cell-polarity gene stbm regulates cell behaviour and cell fate in vertebrate embryos. Nat Cell Biol 2002; 4:20-25.
9. Heisenberg CP, Tada M, Rauch GJ et al. Silberblick/Wnt11 mediates convergent extension movements during zebrafish gastrulation. Nature 2000; 405:76-81.
10. Topczewski J, Sepich DS, Myers DC et al. The zebrafish glypican knypek controls cell polarity during gastrulation movements of convergent extension. Dev Cell 2001; 1:251-264.
11. Myers DC, Sepich DS, Solnica-Krezel L. Bmp activity gradient regulates convergent extension during zebrafish gastrulation. Dev Biol 2002; 243:81-98.
12. Yamanaka H, Moriguchi T, Masuyama N et al. JNK functions in the non-canonical Wnt pathway to regulate convergent extension movements in vertebrates. EMBO Rep 2002; 3:69-75.
13. Wallingford JB, Harland, RM. Xenopus Dishevelled signaling regulates both neural and mesodermal convergent extension: parallel forces elongating the body axis. Development 2001; 128:2581-2592.

14 Deardorff MA, Tan C, Conrad LJ, Klein PS. Frizzled-8 is expressed in the Spemann organizer and plays a role in early morphogenesis. Development 1998; 125:2687-2700.
15 Xu Q, D'Amore PAD, Sokol SY. Functional and biochemical interactions of Wnts with FrzA, a secreted Wnt antagonist. Development 1998; 125:4767-4776.
16 Kühl M, Geis K, Sheldahl LC, Pukrop T, Moon RT, Wedlich D. Antagonistic regulation of convergent extension movements in Xenopus by Wnt/beta-catenin and Wnt/$Ca^{2+}$ signaling. Mech Dev 2001; 106:61-76.
17 Yamashita S, Miyagi C, Carmany-Rampey A et al. Stat3 controls cell movements during zebrafish gastrulation. Dev Cell 2002; 2:363-375.
18 Wacker S, Brodbeck A, Lemaire P et al. Patterns and control of cell motility in the Xenopus gastrula. Development 1998; 125:1931-1942.
19 Torres MA, Yang-Snyder JA, Purcell SM, et al. Activities of the Wnt-1 Class of Secreted Signaling Factors Are Antagonized by the Wnt-5 Class and by a Dominant Negative Cadherin in Early Xenopus Development. J Cell Biol 1996; 133:1123-1137.
20 Kühl M, Sheldahl LC, Malbon CC, Moon RT. Ca(2+)/calmodulin-dependent protein kinase II is stimulated by Wnt and Frizzled homologs and promotes ventral cell fates in Xenopus. J Biol Chem 2000; 275:12701-12711.
21 Choi SC, Han JK. Xenopus Cdc42 regulates convergent extension movements during gastrulation through Wnt/$Ca^{2+}$ signaling pathway. Dev Biol 2002.
22. Gilland E, Miller AL, Karplus E et al. Imaging of multicellular large-scale rhythmic calcium waves during zebrafish gastrulation. Proc Natl Acad Sci USA 1999; 96:157-161.
23. Creton R, Speksnijder JE, Jaffe LF. Patterns of free calcium in zebrafish embryos. J Cell Sci 1998; 111:1613-16, 22.
24. Wallingford JB, Ewald AJ, Harland RM, Fraser SE. Calcium signaling during convergent extension in Xenopus. Curr Biol 2001; 11:652-661.
25. Winklbauer R, Medina A, Swain RK, Steinbeisser H. Frizzled-7 signalling controls tissue separation during Xenopus gastrulation. Nature 2001; 413:856-860.
26. Jiang LI, Sternberg PW . Interactions of EGF, Wnt and HOM-C genes specify the P12 neuroectoblast fate in C. elegans. Development 1998; 125:2337-2347.
27. Delattre M, Félix MA. Development and Evolution of a variable left-right asymmetry in Nematodes: the handedness of P11/P12 migration. Dev Biol 2001; 232:362-271.
28. Maloof JN, Whangbo J, Harris JM et al. A Wnt signaling pathway controls Hox gene expression and neuroblast migration in C. elegans. Development 1999; 126:37-49.
29. Eisenmann DM, Maloof JN, Simske JS et al.The beta-catenin homolog BAR-1 and LET-60 Ras coordinately regulate the Hox gene lin-39 during Caenorhabditis elegans vulval development. Development 1998; 125:3667-3680.
30. Herman MA. C. elegans POP-1/TCF functions in a canonical Wnt pathway that controls cell migration and in a noncanonical Wnt pathway that controls cell polarity. Development 2001; 128:581-590.
31. Wilkinson DG, Bailes JA, McMahon AP. Expression of the proto-oncogene int-1 is restricted to specific neural cells in the developing mouse embryo. Cell 1987; 50:79-88.
32. Parr BA, Shea MJ, Vassileva G, McMahon AP. Mouse Wnt genes exhibit discrete domains of expression in the early embryonic CNS and limb buds. Development 1993; 119:247-261.
33. Dorsky RI, Moon RT, Raible DW. Control of neural crest cell fate by the Wnt signalling pathway. Nature 1998; 396:370-373.
34. Jin EJ, Erickson CA, Takada S, Burrus LW. Wnt and BMP signaling govern lineage segregation of melanocytes in the avian embryo. Dev Biol 2001; 233,22-37.
35. Dickinson ME, Selleck MA, McMahon AP, Bronner-Fraser M. Dorsalization of the neural tube by the non-neural ectoderm. Development 1995; 121:2099-2106.
36. Ikeya M, Lee SMK, Johnson JE, McMahon AP, Takada S. Wnt signalling required for expansion of neural crest and CNS progenitors. Nature 1997; 389:966-970.
37. Serbedzija GN, McMahon AP. Analysis of neural crest cell migration in Splotch mice using a neural crest-specific LacZ reporter. Dev Biol 1997; 185:139-147.
38. Borchers A, David R, Wedlich D. Xenopus cadherin-11 restrains cranial neural crest migration and influences neural crest specification. Development 2001; 128:3049-3060.

39. Wehrle-Haller B, Weston, JA. Soluble and cell-bound forms of steel factor activity play distinct roles in melanocyte precursor dispersal and survival on the lateral neural crest migration pathway. Development 1995; 121:731-742.
40. Munsterberg AE, Kitajewski J, Bumcrot DA et al. Combinatorial signaling by Sonic hedgehog and Wnt family members induces myogenic bHLH gene expression in the somite. Genes Dev 1995; 9:2911-2922.
41. Stern HM, Brown AM, Hauschka, SD. Myogenesis in paraxial mesoderm: preferential induction by dorsal neural tube and by cells expressing Wnt-1. Development 1995; 121:3675-86.
42. Baylies MK, Martinez Arias A, Bate M. wingless is required for the formation of a subset of muscle founder cells during Drosophila embryogenesis. Development 1995; 121:3829-3837.
43. Ranganayakulu G, Schulz RA, Olson EN. Wingless signaling induces nautilus expression in the ventral mesoderm of the Drosophila embryo. Dev Biol 1996; 176:143-148.
44. Toyofuku T, Hong Z, Kuzuya T et al. Wnt/frizzled-2 signalling induces aggregation and adhesion among cardiac myocytes by increased cadherin-β-catenin complex. J Cell Biol 2000; 150:225-241.
45. Kozopas KM, Samos CH, Nusse R. DWnt-2, a Drosophila Wnt gene required for the development of the male reproductive tract, specifies a sexually dimorphic cell fate. Genes Dev 1998; 12:1155-1165.
46. Kozopas KM, Nusse R. Direct flight muscles in Drosophila develop from cells with characteristics of founders and depend on DWnt-2 for their correct patterning. Dev Biol 2002.
47. Llimargas M, Casanova J. ventral veinless, a POU domain transcription factor, regulates different transduction pathways required for tracheal branching in Drosophila. Development 1997; 124:3273-3281.
48. Vincent S, Ruberte E, Grieder NC et al. DPP controls tracheal cell migration along the dorsoventral body axis of the Drosophila embryo. Development 1997; 124:2741-2750.
49. Chihara T, Hayashi S. Control of tracheal tubulogenesis by Wingless signaling. Development 2000; 127:4433-4442.
50. Llimargas M. Wingless and its signalling pathway have common and separable functions during tracheal development. Development 2000; 127 4407-4417.
51. Bradley PL, Andrew DJ. ribbon encodes a novel BTB/POZ protein required for directed cell migration in Drosophila melanogaster. Development 2001; 128:3001-3015.
52. Shih J, Keller R. Patterns of cell motility in the organizer and dorsal mesoderm of Xenopus laevis. Development 1992; 116:915-930.

# CHAPTER 15

# GSK3-Signal Regulation of Pattern Formation in *Dictyostelium*: Wnt-Like Pathways during Non-Canonical Multicellular Development

**Leung Kim and Alan R. Kimmel**

## Abstract

Dictyostelium establishes an asymmetric body axis during development that is dependent upon signaling through distinct 7-TM cAMP morphogen receptors (CARs) that regulate GSK3-mediated downstream effectors. These pathways reflect a similar dependency for metazoan cell fate decisions on Wnt/Frizzled-GSK3 signaling. In Dictyostelium, GSK3 promotes posterior (prespore) cell patterning but antagonizes anterior (prestalk) differentiation. We have shown that during pattern formation in Dictyostelium, distinct CARs alternatively activate a tyrosine kinase, ZAK1 that mediates GSK3 activation or a tyrosine phosphatase that inhibits GSK3 activity. We have identified the essential phospho-tyrosines in Dictyostelium and in mammalian GSK3 and shown that they are required for complete enzymatic activation. Thus, developmentally regulated GSK3 tyrosine phosphorylation/de-phosphorylation determines the activated/de-activated state of GSK3. We also suggest a molecular mechanism for GSK3 activation through tyrosine phosphorylation in the context of its three dimensional structure. Finally, we demonstrate that CARs and Frizzleds are phylogenetically related, providing a comparative understanding of cAMP/CAR-GSK3 and Wnt/Frizzled-GSK3 pathways.

## Introduction

Protein kinase GSK3 is a central element that regulates fate patterns during multicellular development.[1-9] In the metazoa, the morphogen Wnt (Wg) regulates developmental cascades by activation of the family of 7-transmembrane (TM) Frizzled receptors and effective inhibition of GSK3 activity. GSK3 also serves as a core switch for *Dictyostelium* cell fate determination. Although *Dictyostelium* proliferates as unicellular amoeba, it develops as a multicellular organism with an asymmetric body axis and distinctly differentiated compartments.[5,10-11] Regulation of GSK3 activity is central to processing developmental signals to direct these asymmetric cell fate decisions.

The precise molecular mechanism for GSK3 down-regulation by Wnt remains elusive, but it is argued to be more related to control of substrate accessibility than directly associated

*Wnt Signaling in Development*, edited by Michael Kühl.

with GSK3 enzymatic activity.[1-9] However, we have now shown that in *Dictyostelium* the enzymatic activity of GSK3 is specifically and antagonistically regulated by separate 7-TM receptors, and that activated or de-activated states of GSK3 direct different developmental cell fate decisions.[12-15] This developmental orchestration is dependent upon the tyrosine phosphorylation of GSK3.[14,17,18] Antagonistic receptor signaling alternatively activates a specific tyrosine kinase, ZAK1, or a specific protein tyrosine phosphatase.[14,17] We have identified essential phospho-tyrosines in GSK3, confirmed their role in activated/de-activated regulation, and related them to the predicted 3D-structure of GSK3.[17,19,20] These tyrosines are conserved in all GSK3 proteins, suggesting the possibility for similar regulation in more elaborated metazoa.[17,18]

The cell surface 7-TM receptors of *Dictyostelium* do not recognize Wnt, rather they bind a distinct extracellular ligand, cAMP. However, both receptor families nonetheless share topologically organized amino acid sequence similarity through their TM and loop domains, suggesting an ancient phylogenetic relationship for regulation of GSK3. Further, downstream of GSK3, they also target structurally related β-catenin proteins.[21] Thus, the capacity of *Dictyostelium* to differentiate into a multicellular organism may define an ancient developmental paradigm for GSK3-regulated cell fate decisions. Our molecular findings in *Dictyostelium* may provide novel insight into this program for all phyla. In this chapter, we will discuss novel GSK3 regulated pathways in *Dictyostelium* and the potential for their interface with Wnt signaling.

## Antagonistic Regulation of Cell Fate Determination in *Dictyostelium* by the 7-Transmembrane cAMP Receptors CAR3 and CAR4

Extracellular cAMP elicits multiple, even antagonistic, responses during *Dictyostelium* development.[5,10,11,17,22] cAMP initially serves as a chemoattractant to recruit individual, undifferentiated cells to form multicellular structures, aggregation mounds (see Fig. 1). This chemotactic response is directed by CAR1, a high affinity, G protein-coupled, 7-transmembrane (TM) cell surface receptor for cAMP. CAR1 signaling also mediates induction of the cAMP-dependent gene activation pathways required for early development, thus defining the complexity of pathways regulated downstream of receptor activation. Three functionally distinct cAMP receptors, CARs 2, 3, and 4, become sequentially expressed as development proceeds.[23] Their temporally and spatially restricted expressions are required for the differentiation of the various cell types and proper patterning. Following aggregate formation, the prestalk and prespore progenitor cells begin to differentiate. They eventually become spatially separated from each other, with the prestalk cells localized primarily at the tip of the aggregation mound, while the prespore cells form the base of the aggregate. The multicellular mound undergoes morphogenesis, elongating into a pseudoplasmodium, the migrating slug, that is organized along a prestalk/anterior and prespore/posterior horizontal axis (see Fig. 1). This pattern formation precedes developmental culmination and terminal differentiation of the progenitor prestalk and prespore cells into the stalk and spores of the mature fruiting body.[5,10,11]

There are essential and distinct roles for each of the four cAMP receptors during *Dictyostelium* development (see Fig. 1). The primary defects observed with car1- and car2-nulls involve developmental arrest. CAR1 is required for multi-cell formation, progression through the earliest stages of development, and even induction of CARs 2, 3, and 4. car2-null cells arrest at the next major stage, establishment of the patterned aggregate prior to slug morphogenesis (Fig. 1). Both CAR1 and CAR2 regulate patterns of cell movement, but also, gene expression during development.[24]

CAR3 is the major cAMP receptor of the posterior prespore cells, while CAR4 is enriched in the anterior prestalk population.[15,25] car3⁻ cells have decreased prespore gene expression, but up-regulated prestalk gene expression (Fig. 1). Conversely, car4⁻ cells have enhanced prespore gene expression patterns, with a concomitant inhibition of prespore differentiation (Fig. 1).[5,12,17] Thus, a single morphogen, cAMP, can elicit antagonistic responses for cell fate determination through distinct expression patterns of CAR3 and CAR4 and the differential action on their downstream signaling components.

## GSK3, a Developmental Switch Regulating Anterior/Posterior Axis Formation

GSK3, a serine/threonine protein kinase, was originally identified as a regulator of glycogen metabolism, but it has recently witnessed an experimental renaissance with the discovery of its role as an effective switch regulating diverse developmental fate choices.[1-9] GSK3 has basal kinase activity, and inactive or activated GSK3 will antagonistically regulate dorsal/ventral patterns in vertebrates, posterior/anterior segment polarity in *Drosophila*, mesoderm/endoderm cell fates in *C. elegans*, and anterior/posterior cell fates in *Dictyostelium*, among others.[1,4-8,12-17,26-29]

Genetic studies on canonical Wnt signaling suggested an inhibition of GSK3 activity through Frizzled receptor-dependent pathways, while in *Dictyostelium*, GSK3 is regulated by both inhibiting, as well as, activating pathways.[5,12-17] During development the specific kinase activity of GSK3 undergoes cAMP-dependent up-regulation, followed by an inhibitory, de-activating response (Fig. 2). The activation process is mediated by CAR3 and the major cell fate defects observed in car3-null cells are phenocopied in cells that lack GSK3. Indeed, GSK3 activation is blocked in $car3^-$ cells (Fig. 2). Thus, CAR3 activation of GSK3 is required to establish prespore/spore fate patterns, but is inhibitory to prestalk/stalk development. Conversely, GSK3 specific activity is antagonized by CAR4. The car4-null phenotype can be rescued by expression of a kinase-inactive form of GSK3. The dominant-negative nature of this inactive enzyme, places GSK3 as a negative target, downstream of CAR4.[14,17] Furthermore, repression of GSK3 specific activity is compromised in cells that lack CAR4 (Fig. 2). Given that, activated GSK3 will inhibit prestalk patterns, it is not surprising that car4-nulls exhibit significant repression of prestalk-specific gene expression.

cAMP will both stimulate and repress GSK3 through the antagonistic actions of the CAR3 and CAR4 receptors during *Dictyostelium* development. A CAR3-mediated GSK3 stimulating signal would prevail in the prespore cells, whereas the CAR4-dependent inhibitory pathway would dominate in the prestalk cells.

Although the CAR and Frizzled (Fz) 7-TM receptors are activated by distinct ligands, cAMP and Wnt, respectively, and the Fz members have been considered a strictly metazoan receptor, the resemblance of the CAR and Fz signaling pathways converging at GSK3 invited a more rigorous phylogenetic analysis. Fz family members are defined by the Fz structural domain (pfam01392) within transmembrane domains 2-6. This is especially highlighted by sequence comparison the Fz domain with the Hedgehog co-receptor Smoothened (Smo), a divergent member of the Fz family.[30] Two recent reports indicate that Hedgehog signaling from Smoothened requires GSK3 and emphasize the molecular connection between Fz domain-containing receptors and GSK3.[31,32] This relationship is further underscored by sequence comparison of CAR3 and CAR4 with the consensus Fz transmembrane domain (Fig. 3). Significant alignments ($9e^{-04}$) of the CARs and Fz receptors using ClustalW and PROTPARS (in PHILYP 3.5) confirm topological, spatial, and sequence preservation throughout the entire Fz domain, and spanning both transmembrane and loop domains 2-6 of the CAR and Fz families (Fig. 3). The phylogenetic kinship among the Fz and CAR receptors, in concert with their common functional linkage to GSK3 and β-catenin, underlies the conservation of a Wnt-like pathway for regulating cell fate decisions and patterning among these diverse developmental systems.[21] This suggests an ancient origin for this essential pathway.

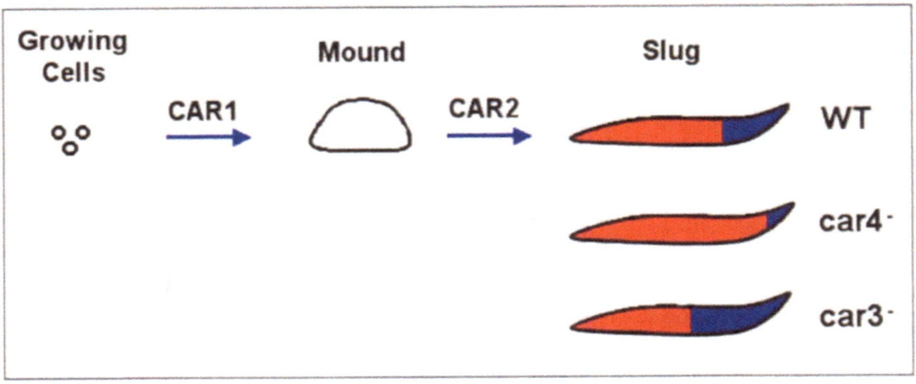


Figure 1. Regulation of *Dictyostelium* pattern formation by CAR3 and CAR4. *Dictyostelium* grow as single cells and initiate multicellular development in response to nutrient deprivation. Developing cells secrete cAMP which acts as both a chemoattractant and a morphogen. CAR1 is required for aggregation and multicellular mound formation, while CAR2 is required for morphogenesis to the slug stage. CAR4 and CAR3 serve as morphogen receptors to establish the anterior/prestalk (blue) and posterior/prespore (red) axis. In the absence of CAR4 the posterior/prespore zone is expanded and in the absence of CAR3 the anterior/prestalk zone is expanded.

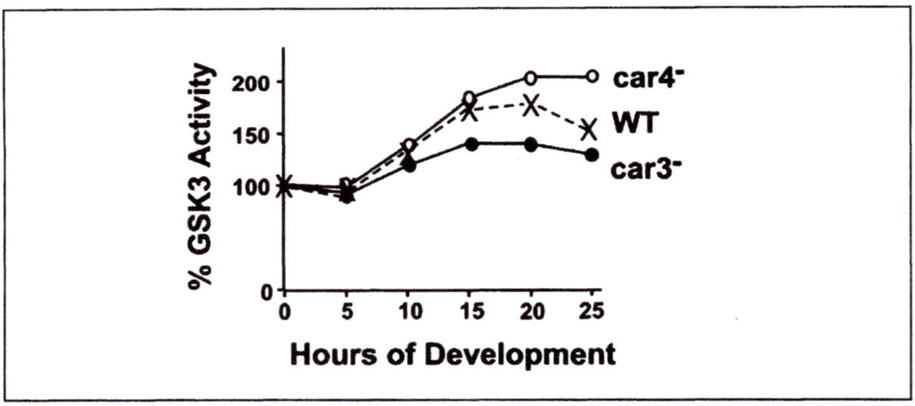


Figure 2. CAR3 and CAR4 antagonistically regulate GSK3 activity. GSK3 specific activity increases during multicellular mound formation (see Figure 1) and subsequently declines as the patterned slug becomes committed to terminal differentiation (see Fig. 1). This developmental up-regulation is compromised in car3-nulls, while the repressive phase of GSK3 regulation is lost in car4-nulls.

## ZAK1 and PTPase Regulate Tyrosine Phosphorylation and Activity of GSK3

We had previously identified the novel tyrosine kinase ZAK1 in *Dictyostelium* that is essential for normal prespore/spore differentiation.[14] The developmental defects of zak1-nulls are very similar to those observed for car3- and gsk3-null cells. We have additionally shown that cAMP activates ZAK1 tyrosine kinase activity in wild-type cells, but not in cells lacking CAR3. These data indicate that ZAK1 is a downstream effector of CAR3. Furthermore, ZAK1 is required for cAMP/CAR3-mediated activation of GSK3. We, therfore, suggested that GSK3 may be a direct substrate of ZAK1 and that GSK3 activation during *Dictyostelium* development may be regulated by tyrosine phosphorylation.[14,17]

We have, in fact, demonstrated that GSK3 is tyrosine phosphorylated in response to cAMP and that phosphorylation anticipates GSK3 activation (Fig. 4). These responses are not permanent; with time, GSK3 is both de-phosphorylated and de-activated.[17] We further showed that phosphorylation and activation of GSK3 is absent in car3-nulls, while both are persistent in car4-nulls. We have also demonstrated that ZAK1 can both phosphorylate and activate *Dictyostelium*, as well as, mammalian GSK3β (see below) in vitro.[14] Finally, we showed that

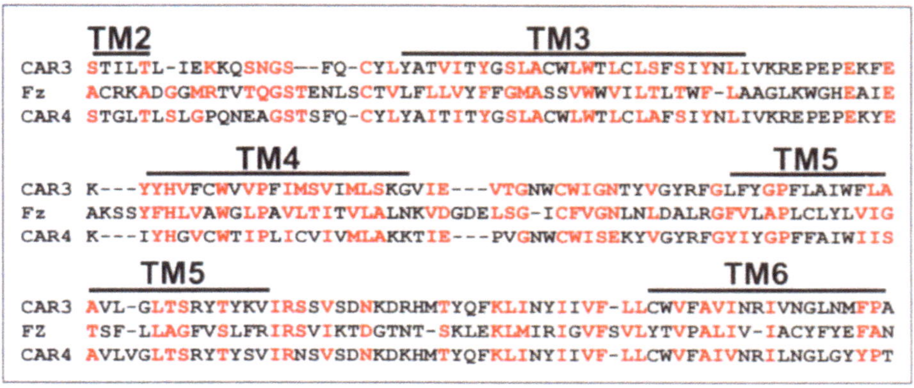


Figure 3. CARs possess a Frizzled domain within their 7-TM regions. The consensus Frizzled transmembrane domain was identified by the NCBI Conserved Domain search program. CAR3 and CAR4 sequences were aligned to the consesus Fz sequence by Clustal W and manually modified. Amino acid identities and similarities between the consensus Fz domain and each CAR are in red. The TM bars above the sequences indicate the predicted transmembrane domains of the aligned sequences.

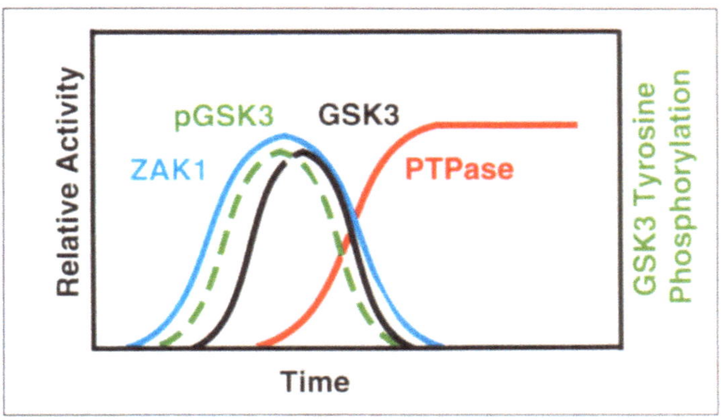


Figure 4. cAMP regulation of ZAK1, GSK3, the CAR4-PTPase, and pGSK3. Following a cAMP stimulus, ZAK1 specific activity exhibits a rapid but transient increase, which is paralleled closely by a transient increase in GSK3 tyrosine phosphorylation. GSK3 specific activity is also induced, but with kinetics that are slightly delayed relative to the increases of ZAK1 and the tyrosine phosphorylation of GSK3 (pGSK3). ZAK1 and GSK3 activation are regulated by CAR3. CAR4 regulates the activation of PTPase that de-phosphorylates and de-activates GSK3. CAR4 does not regulate ZAK1 activation or de-activation.

activation of GSK3 during *Dictyostelium* development is regulated in vivo by the cAMP/CAR3/ZAK1-mediated pathway (Fig. 4), while inhibition of GSK3 is mediated by a de-phosphorylation mechanism regulated by CAR4.[14,17]

Potentially, CAR4 could inhibit the CAR3-mediated activation of ZAK1 and in the absence of an activated ZAK1, GSK3 would be de-phosphoryated by a constitutively active PTPase and, hence, repressed. Alternatively, CAR4 may regulate directly the PTPase activity responsible for de-phosphorylating and de-activating GSK3. We measured the cAMP-mediated activation of ZAK1 in both wild-type and car4⁻ cells and found identical results.[17] In both stains, ZAK1 is transiently activated, with temporal kinetics that parallel that of GSK3 tyrosine phosphorylation in vivo (Fig. 4). Furthermore, using a tyrosine phosphorylated GSK3 substrate, we were able to detect cAMP-mediated activation of PTPase in wild-type cells, whose activation was undetectable in car4⁻ cells (Fig. 4). These data strongly suggest that CAR4 regulates a PTPase that specifically de-activates GSK3.[17]

The amino acid sequences of mammalian and *Dictyostelium* GSK3 are highly conserved within their catalytic domains; 14 tyrosine residues are identically conserved (Fig. 5). The ability to phosphorylate GSK3 in vitro, has permitted us to map the phosphorylated tyrosines and to show that they are essential for activation by ZAK1. Figure 5 indicates the two tyrosine residues within the activation loop of GSK3 that are substrates for phosphorylation by ZAK1. Mutation of either tyrosine to phenylalanine reduces basal activity of GSK3, and expression of either Y/F mutant form in vivo phenocopies expression of kinase-inactive, dominant-negative GSK3 and mimics that of car3-, zak1-, or gsk3-nulls.[17]

Recent resolution of the crystal structure of unphosphorylated GSK3 provides valuable insights on the role of the tyrosine phosphorylation on GSK3.[19,20] The activation loops of inactive kinases, such as unphosphorylated ERK, adopt a closed conformation that limits substrate access. In contrast, the activation loop of phospho-ERK has an open conformation. Here, the phosphate group in the activation loop forms stable ionic interactions with neighboring arginine residues, which re-orients the position of the phopho-tyrosine, thus stabilizing the open structure and increasing substrate access. It is interesting that the activation loop of unphosphorylated, basally active GSK3 is already structurally open. However, the un-phosphorylated tyrosine residues of the activation loop of GSK3 are unable to interact with the neighboring arginines but instead are free to sterically restrict substrate access to the active site. Upon phosphorylation by ZAK1, these tyrosine residues may form stable ionic interactions with the positively charged arginine neighbors (see Fig. 5) which would permit full substrate access and hyper-activation.

Inhibitory homodimers of GSK3 have also been reported, with the activation loop as a major interaction surface. We speculate that the phosphorylation of tyrosine residues within the activation loop may disrupt the dimer formation, potentially providing a secondary mechanism for GSK3 activation through tyrosine phosphorylation.

## Downstream Targets of GSK3 Aar (β-Catenin)

β-catenin is a multifunctional protein with transcriptional co-factor properties and is a principal target immediately downstream of GSK3. In the canonical Wnt pathway, GSK3 phosphorylated β-catenin is subject to rapid degradation by a ubiquitin/proteosome pathway involving APC, βTrCP, among others;[1-9] Wnt-stimulated inhibition of GSK3, thus, stabilizes β-catenin and activates transcriptional pathways further downstream. In *Dictyostelium*, the armadillo-repeat protein Aardvark, Aar, which is structurally related to β-catenin, also appears to function downstream of GSK3. As with β-catenin, Aar has multiple, consensus GSK3 phosphorylation sites in its amino-terminal region (Fig. 6), although Aar lacks the C-terminal transactivating domain found in β-catenin.

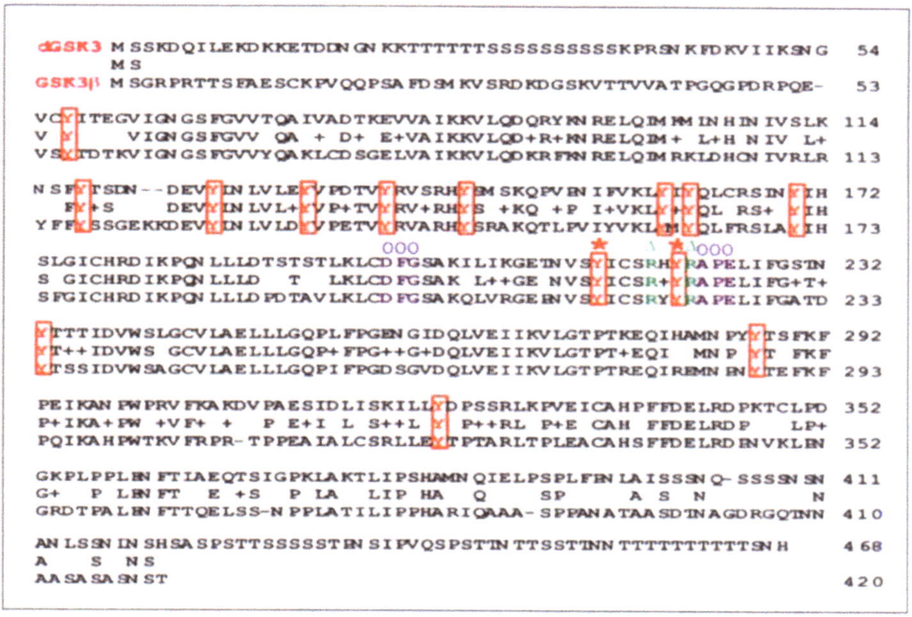


Figure 5. Conservation of essential tyrosines in *Dictyostelium* and mammalian GSK3. *Dictyostelium* (dGSK3) and mouse (GSK3β) GSK3 sequences were aligned and conserved tyrosines indicated by red boxes. The essential DFG and APE amino acids that bracket the activation loop are indicated by a purple circle. The conserved activating tyrosines that are phosphorylated by ZAK1 in vivo and in vitro within the activation loop are indicated by a red star. The neighboring argenines which may interact with phosphorylated tyrosines to stabilize the activated conformation of GSK3 are indicated by a green triangle.

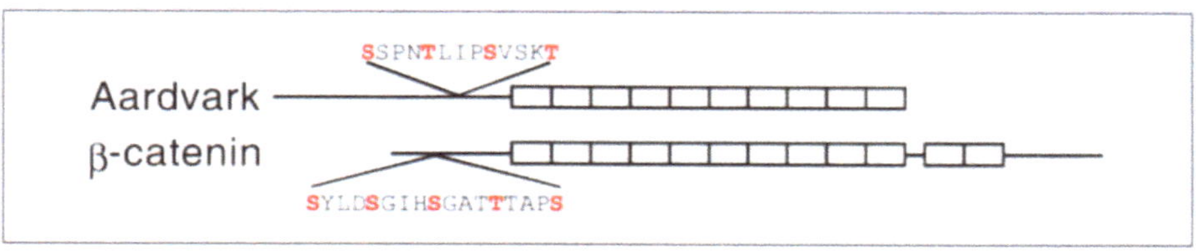


Figure 6. Structural conservation between *Dictyostelium* Aar and β-catenin. *Dictyostelium* Aar and β-catenin were aligned by their armadillo repeats. Aar has 10 repeats while β-catenin has 12. The C-terminus of β-catenin has a transactivating domain that is absent in Aar. Both proteins have a consensus sequence for GSK3 phosphorylation that is just N-terminal to the armadillo repeat region.

While β-catenin serves as a negative target of GSK3 in the canonical pathway, cAMP-induced prespore gene expression appears dependent upon Aar, placing it in a pathway positively regulated by GSK3 signaling. This has parallel with the positive regulation of the β-catenin-like Wrm-1 by GSK3 during *C. elegans* endoderm differentiation (see Fig. 7).[29] In addition, while GSK3 acts as a developmental switch regulating both prestalk and prespore pathways, disruption of Aar only alters prespore differentiation without inducing the concomitant increase in prestalk specific gene expression that is observed in car3-, zak1- and gsk3-nulls.[33] These data indicate that *Dictyostelium* requires a prestalk-specific, GSK3-sensitive transcriptional factor that is distinct from Aar. This may suggest the presence of an additional, novel β-catenin-like protein in *Dictyostelium*.

## Perspectives—Integrating Tyrosine Phosphorylation with Other Signaling Components

Multicellular organisms share a common dependency on a GSK3/β-catenin signaling cascade to regulate cell fate patterns (Fig. 7). In vertebrates, β-catenin acts as a transcriptional co-factor to positively regulate a series of developmental events, including, dorsal/ventral or tumorigenic pathways, among many others.[1,4-9,34-36] GSK3 phosphorylation of β-catenin promotes an APC/βTrCP-dependent path for degradation; Axin scaffolds GSK3, APC and β-catenin, while CKIα serves as a priming kinase for GSK3 phosphorylation.[37,38] PP2A/B56 may also facilitate degradative events (Fig. 7). The morphogen Wnt initiates an inhibitory signal cascade that requires dishevelled (dsh) to functionally down-regulate GSK3, potentially by decreasing GSK3 accessibility to β-catenin. Secreted Wnt may be specifically antagonized by a variety of extracellular molecules, including sFRP. Caesin kinases I and II have been implicated in the inhibitory pathway, but their actions have not been explicitly defined; additional components, including the LRP co-receptors are not shown. In the presence of Wnt, unphosphorylated β-catenin can accumulate; in complex with LEF/TCF, it can potentiate transcriptional activation.

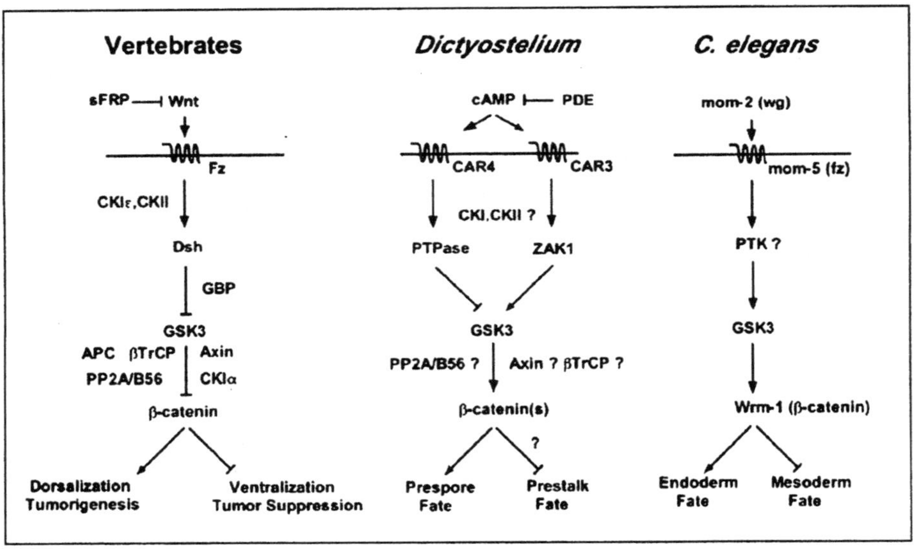


Figure 7. A partial comparison of some CAR/GSK3 and Fz/GSK3 pathways in vertebrates, *Dictyostelium*, and *C. elegans*. In all systems, 7-TM receptor signaling converges at GSK3 to regulate cell fate decisions; see text for details. In the vertebrates (and in *Drosophila*), Wnt/Fz signaling is antagonistic to GSK3 function. Dorsal/ventral patterning and tumorigenic fate represent only 2 examples of opposing choices determined by the presence or absence of an activated Wnt pathway. In contrast, endoderm/mesoderm formation in *C. elegans* is regulated by an activating Wnt/Fz signal through GSK3. In other pathways (e.g., vulval development) of *C. elegans*, Wnt signaling reflects the more canonical inhibitory state. *Dictyostelium* incorporates both activating and de-activating receptor signals through GSK3. The activating pathway requires tyrosine phosphorylation of GSK3 by ZAK1, while these tyrosines are de-phosphorylated in the CAR4-dependent inhibitory pathway. Potentially, tyrosine phosphorylation of GSK3 in *C. elegans* may mediate GSK3 activation via Wnt signaling. In contrast, an activating tyrosine kinase that is an effective equivalent to ZAK1, may antagonize canonical Wnt signaling and serve as a functional tumor suppressor.

Thus, Wnt signaling in the vertebrates promotes dorsal fate patterning, for example, in vertebrates, but antagonizes ventral pattern formation (Fig. 7). In *Drosophila*, Wg signaling promotes naked cuticle formation and posterior segment polarity, but antagonizes denticle formation and anterior segment polarity.

In the endoderm/mesoderm fate pathway of *C. elegans*, Wnt-signaling does not inhibit GSK3 (Fig. 7). The Wnt homologue mom-2 initiates a positive signal cascade from the Fz homologue mom-5 to GSK3. Active GSK3 is required for endoderm formation, while loss of GSK3 function will lead to a mesodermal fate.[29] Additional molecular intermediates in this pathway have not been identified.

By contrast, *Dictyostelium* exhibits both positive and negative signaling cascades downstream of their morphogen cAMP receptors to regulate the common target GSK3. Stimulation of CAR3 activates GSK3 and promotes prespore fate patterns, whereas CAR4 stimulation will inhibit GSK3 and promote prestalk differentiation. CAR3 activation of GSK3 is mediated by the tyrosine kinase ZAK1, while CAR4 inhibits GSK3 by a tyrosine phosphatase (PTPase)-dependent pathway. Thus, tyrosine phosphorylation/de-phosphorylation of GSK3 mediates antagonistic receptor regulation for GSK3 activation/de-activation and cell fate determination. Both pathways are downregulated by secreted cAMP phosphodiesterase (PDE), which will remove the activating extracellular cAMP signal. We have recently turned our focus to the potential roles of CKI, CKII, and PP2A/B56 in *Dictyostelium*.[39,40] Their genes have been isolated and their developmental functions are being assessed. Further, since expression of kinase-inactive GSK3 in *Dictyostelium* functions as a dominant-negative mutant to phenocopy the gsk3-null, we argue that enzymatic activity alone can not explain the entire regulatory pathway.[41] We, therefore, suggest that a scaffolding protein functionally equivalent to Axin, must also be present in *Dictyostelium*. In addition, pathways dependent on the function of βTrCP-related F-box proteins are also essential for *Dictyostelium* development.[42,43]

It remains an open question if tyrosine phosphorylation of GSK3 can regulate cell fate decisions in other systems. Potentially, tyrosine phosphorylation and, hence, hyper-activation of GSK3 could directly antagonize Wnt signaling to provide additional regulation for differentiation and tumor suppression; furthermore, an activated PTPase may also be functionally equivalent to Wnt signaling to promote tumorigenesis.

## *Acknowledgements*

We are indebted to Drs. J. Brzostowski, T. Khurana, L. Kreppel, X. Liu, and D. Rosel for their enthusiasm and discussion. We also appreciate the efforts of Jingchun Liu and Kate Tifft during the course of these studies and continued collaboration with our colleague Dr. Adrian Harwood.

## References

1. Kuhl M, ed. Wnt Signalling in development. Georgetown: Landes Biosciences, 2003.
2. Ali A, Hoeflich KP, Woodgett JR. Glycogen synthase kinase-3: properties, functions, and regulation. Chem. Rev. 2001; 101:2527-2540.
3. Cohen P, Frame S. The renaissance of GSK3. Nat Rev Mol Cell Biol 2001; 2:769-776.
4. Ferkey DM, Kimelman D. GSK-3: new thoughts on an old enzyme. Dev Biol 2000; 225:471-479.
5. Kim L, Kimmel AR. GSK3, a Master switch regulating cell fate specification and tumorigenesis. Curr Opin Genet Dev 2000; 10:508-514.
6. McEwen DG, Peifer M. Wnt signaling: Moving in a new direction. Curr Biol 2000; 10:R562-564.
7. Moon RT, Bowerman B, Boutros M et al. The promise and perils of Wnt signaling through beta-catenin. Science 2002:1644-1646.
8. Wodarz A, Nusse R. Mechanism of Wnt signaling in Development. Ann Rev Cell Dev Biol 1998; 14:59-88; See also www.stanford.edu/~rnusse/wntwindow.html.

9. Woodgett JR. Judging a protein by more than its name: GSK-3. Sci STKE 2001; 100:RE12.
10. Aubry L, Firtel R. Integration of signaling networks that regulate Dictyostelium differentiation. Ann Rev Cell Dev Biol 1999; 15:469-517.
11. Kay RR. Development at the edge of multi-cellularity: Dictyostelium discoideum. Int J Dev Biol 2000; 44:35-38.
12. Ginsburg GT, Kimmel AR. Autonomous and non-autonomous regulation of axis formation by antagonistic signalling via 7-span cAMP receptors and GSK3 of Dictyostelium. Genes Dev 1997; 11:2112-2123.
13. Harwood AJ, Plyte SE, Woodgett J et al. Glycogen synthase kinase 3 regulates cell fate in Dictyostelium. Cell 1995; 80:139-148.
14. Kim L, Liu J, Kimmel AR. The novel tyrosine kinase ZAK1 activates GSK3 to direct cell fate specification. Cell 1999; 99:399-408.
15. Louis JM, Ginsburg GT, Kimmel AR. The cAMP receptor CAR4 regulates axial patterning and cellular differentiation during late development of Dictyostelium. Genes Dev 1994; 8:2086-2096.
16. Plyte SE, O'Donovan E, Woodgett JR et al. Glycogen synthase kinase-3 (GSK-3) is regulated during Dictyostelium development via the serpentine receptor cAR3. Development 1999; 126:325-333.
17. Kim L, Harwood A, Kimmel AR. Receptor-dependent and tyrosine phosphatase-mediated inhibition of GSK3 regulates cell fate choice. Developmental Cell 2002; 3:523-532.
18. Hughes K, Nikolakaki E, Plyte SE, et al. Modulation of the glycogen synthase kinase-3 family by tyrosine phosphorylation. EMBO J 1993; 12:803-808.
19. Dajani R, Fraser E, Mark SM et al. Crystal structure of glycogen synthase kinase 3β: structural basis for phosphate-primed substrate specificity and autoinhibition. Cell 2001; 105:721-732.
20. ter Haar E, Coll JT, Austen DA et al. Structure of GSK3β reveals a primed phosphorylation mechanism. Nat Struct Biol 2001; 8:593-596.
21. Grimson MJ, Coates JC, Reynolds, JP et al. Adherens junctions and beta-catenin-mediated cell signalling in a non-metazoan organism. Nature 2000; 408:27-731.
22. Berks M, Kay RR. Combinatorial control of cell differentiation by cAMP and DIF-1 during development of Dictyostelium discoideum. Development 1990; 110:977-984.
23. Saxe CL, Johnson RL, Devreotes PN et al. Expression of a cAMP receptor gene of Dictyostelium and evidence for a multigene family. Genes Develop 1991; 5:1-8.
24. Saxe CL, Ginsburg GT, Louis JL et al. CAR2, a prestalk cAMP receptor required for tip formation and late development in Dictyostelium. Genes Develop 1993; 7:262-272.
25. Gollop R, Kimmel AR. Control of cell-type specific gene expression in Dictyostelium by the general transcription factor GBF. Development 1997; 124:3395-3405.
26. Dominguez I, Itoh K, Sokol SY. Role of glycogen synthase kinase 3 beta as a negative regulator of dorsoventral axis formation in Xenopus embryos. Proc Natl Acad Sci 1995; 92:8498-8502.
27. He X, Saint-Jeannet JP, Woodgett JR et al. Glycogen synthase kinase-3 and dorsoventral patterning in Xenopus embryos. Nature 1995; 374:617-22.
28. Yost C, Torres M. Miller JR et al. The axis-inducing activity, stability, and subcellular distribution of beta-catenin is regulated in Xenopus embryos by glycogen synthase kinase 3. Genes Dev 1996; 10:1443-1454.
29. Schlesinger A, Shelton, CA, Maloof JN et al. Wnt pathway components orient a mitotic spindle in the early Caenorhabditis elegans embryo without requiring gene transcription in the responding cell. Genes Dev 1999; 13:2028-2038.
30. Alcedo J, Ayzenzon M, Von Ohlen T et al. The Drosophila smoothened gene encodes a seven-pass membrane protein, a putative receptor for the hedgehog signal. Cell 1996; 86(2):221-32.
31. Jia J, Amanai K, Wang G et al. Shaggy/GSK3 antagonizes Hedgehog signalling by regulating Cubitus interruptus. Nature 2002; 416:548-552.
32. Price MA, Kalderon D. Proteolysis of the Hedgehog signaling effector Cubitus interruptus requires phosphorylation by Glycogen Synthase Kinase 3 and Casein Kinase 1. Cell 2002; 108:823-35.
33. Coates JC, Grimson MJ, Williams RS et al. Loss of the beta-catenin homologue aardvark causes ectopic stalk formation in Dictyostelium. Mech Dev 2002; 116:117-27.
34. Bienz M, Clevers H. Linking colorectal cancer to Wnt signaling. Cell 2000; 103:311-320.

35. Manoukian AS, Woodgett JR. ole of glycogen synthase kinase-3 in cancer: regulation by wnts and other signaling pathways. Adv Cancer Res 2002; 84:203-229.
36. Polakis P. Wnt Signaling and Cancer. Genes Dev 2000; 14:1837-1851.
37. Amit S, Hatzubai A, Birman Y et al Axin-mediated CKI phosphorylation of beta-catenin at Ser 45: a molecular switch for the Wnt pathway. Genes Dev 2002; 16:1066-1076.
38. Liu C, Li Y, Semenov M et al. Control of beta-catenin phosphorylation/degradation by a dual-kinase mechanism. Cell 2002; 108:837-847.
39. Kikkawa U, Mann SK, Firtel RA et al. Molecular cloning of casein kinase II alpha subunit from Dictyostelium discoideum and its expression in the life cycle. Mol Cell Biol 1992; 12:5711-23.
40. Moreno-Bueno G, Cales C, Behrens MM et al. Isolation and characterization of casein kinase I from Dictyostelium discoideum. Biochem J 2000; 349:527-37.
41. Fraser E, Young N, Dajani R et al. Identification of the Axin and FRAT/GBP binding region of Glycogen Synthase Kinase-3. J Biol Chem 2002; 277:2176-2185.
42. Ennis HL, Dao DN, Pukatzki SU et al. Dictyostelium amoebae lacking an F-box protein form spores rather than stalk in chimeras with wild type. Proc Natl Acad Sci 2000; 97:3292-3297.
43. Nelson MK, Clark A, Abe T et al. An F-Box/WD40 repeat-containing protein important for Dictyostelium cell-type proportioning, slug behaviour, and culmination. Dev Biol 2000; 224:42-59.

# Index

## T



## U



## V



## W



## X



## Z

MIX
Papier aus verantwortungsvollen Quellen
Paper from responsible sources
FSC® C105338

If you have any concerns about our products,
you can contact us on
**ProductSafety@springernature.com**

In case Publisher is established outside the EU,
the EU authorized representative is:
**Springer Nature Customer Service Center GmbH**
**Europaplatz 3, 69115 Heidelberg, Germany**

Printed by Libri Plureos GmbH
in Hamburg, Germany